Lecture Notes in Geoinformation and Cartography

Monika Sester · Lars Bernard ·
Volker Paelke (Eds.)

Advances in GIScience

Proceedings of the 12th AGILE Conference

Springer

Editors
Prof. Monika Sester
Leibniz Universität Hannover
Inst. Kartographie und
Geoinformatik
Appelstraße 9a
30167 Hannover
Germany
monika.sester@ikg.uni-hannover.de

Prof. Lars Bernard
TU Dresden
Fak. Forst-, Geo- und
Hydrowissenschaften
Helmholtzstr. 10
01069 Dresden
Germany
lars.bernard@tu-dresden.de

Prof. Volker Paelke
Leibniz Universität Hannover
Inst. Kartographie und
Geoinformatik
Appelstraße 9a
30167 Hannover
Germany
volker.paelke@ikg.uni-hannover.de

ISSN 1863-2246 e-ISSN 1863-2351
ISBN 978-3-642-00317-2 e-ISBN 978-3-642-00318-9
DOI 10.1007/978-3-642-00318-9
Springer Dordrecht Heidelberg London New York

Library of Congress Control Number: 2009926690

Cover design: deblik, Berlin

Printed on acid-free paper

Springer is part of Springer Science+Business Media (www.springer.com)

Preface

The Association of Geographic Information Laboratories for Europe (AGILE) was established in early 1998 to promote academic teaching and research on GIS at the European level. Since then, the annual AGILE conference has gradually become the leading GIScience conference in Europe and provides a multidisciplinary forum for scientific knowledge production and dissemination.

GIScience addresses the understanding and automatic processing of geospatial information in its full breadth. While geo-objects can be represented either as vector data or in raster formats these representations have also guided the research in different disciplines, with GIS researchers concentrating on vector data while research in photogrammetry and computer vision focused on (geospatial) raster data. Although there have always been small but fine sessions addressing photogrammetry and image analysis at past AGILE conferences, these topics typically played only a minor role. Thus to broaden the domain of topics the AGILE 2009 conference it is jointly organized with a Workshop of the International Society of Photogrammetry and Remote Sensing (ISPRS), dedicated to High Resolution Satellite Imagery, organized by Prof. Christian Heipke of the Leibniz Universität Hannover.

This collocation provides opportunities to explore commonalities between research communities and to ease exchange between participants to develop or deepen mutual understanding. We hope that this approach enables researchers from the different communities to identify common interests and research methods and thus provides a basis for possible future cooperations.

The topics of the 2009 AGILE conference span the wide spectrum of GIScience ranging from data acquisition, data fusion and integration over spatio-temporal modelling and analysis to generalization, visualization and perception. The Programme Committee conducted a very selective double blind review process to choose the best quality papers for inclusion in these proceedings. A total of 71 full papers were submitted, of which 22 were selected for inclusion in this book (acceptance rate 32%).

Organizing the programme of an international conference and editing a volume of scientific papers requires significant time and effort. This is only possible with the commitment and help of many people. We would like to thank all those helping hands who assisted with various tasks in preparing AGILE 2009. Above all we thank Birgit Kieler who managed the conference tool together with Tobias Dahinden, designed the AGILE 2009 logo and the front cover of this book and integrated all submissions into a coherently formatted document.

We gratefully acknowledge the efforts of the AGILE Programme Committee for their excellent support and for adhering to our strict timetable. Finally, we would like to thank all authors for handing in their best work to AGILE 2009.

Monika Sester, Lars Bernard, Volker Paelke

Hannover and Dresden
March 2009

Programme Committee

Programme Chair Monika Sester,
Leibniz Universität Hannover (Germany)

Programme Co-Chair Lars Bernard,
TU Dresden (Germany)

Programme Co-Chair Volker Paelke,
Leibniz Universität Hannover (Germany)

Local Committee

Tobias Dahinden, Birgit Kieler
Leibniz Universität Hannover (Germany)

Scientific Committee

Pragya Agarwal, University College London (United Kingdom)
Peggy Agouris, George Mason University (USA)
Itzhak Benenson, Tel Aviv University (Israel)
Michela Bertolotto, University College Dublin (Ireland)
Ralf Bill, Universität Rostock (Germany)
Lars Bodum, Aalborg University (Denmark)
Arnold Bregt, Wageningen University (The Netherlands)
Claus Brenner, Leibniz Universität Hannover (Germany)
Christoph Brox, University of Muenster (Germany)
Gilberto Camara, National Institute for Space Research (Brazil)
Christophe Claramunt, Naval Academy Research Institute (France)
Arzu Cöltekin, University of Zurich (Switzerland)
Max Craglia, Joint Research Centre (Italy)
Arie Croitoru, The University of Alberta (Canada)
Isabel Cruz, University of Illinois - Chicago (USA)
Tobias Dahinden, Leibniz Universität Hannover (Germany)
Leila De Floriani, University of Genova (Italy)
Jürgen Döllner, HPI-Institut an der Universität Potsdam (Germany)
Matt Duckham, University of Melbourne (Australia)
Sara Irina Fabrikant, University of Zurich (Switzerland)
Peter Fisher, University of Leicester (United Kingdom)

Anders Friis-Christensen, National Survey and Cadastre (Denmark)
Lars Harrie, Lund University (Sweden)
Francis Harvey, University of Minnesota (USA)
Christian Heipke, Leibniz Universität Hannover (Germany)
Gerard Heuvelink, Wageningen University (The Netherlands)
Stephen Hirtle, University of Pittsburgh (USA)
Hartwig Hochmair, University of Florida (USA)
Bin Jiang, University of Gävle (Sweden)
Chris Jones, Cardiff University (United Kingdom)
Didier Josselin, Université d'Avignon et des Pays du Vaucluse (France)
Marinos Kavouras, National Technical University of Athens (Greece)
Eva Klien, Frauenhofer Institute for Computer Graphics - IGD (Germany)
Thomas Kolbe, Technical University Berlin (Germany)
Menno-Jan Kraak, ITC (The Netherlands)
Antonio Krüger, University of Muenster (Germany)
Lars Kulik, University of Melbourne (Australia)
Roger Longhorn, Geo:connexion (United Kingdom)
Michael Lutz, Joint Research Centre (Italy)
Hans-Gerd Maas, Dresden University of Technology (Germany)
Bela Markus, University of West Hungary (Hungary)
Gábor Mezosi, University of Szeged (Hungary)
Pedro Rafael Muro Medrano, University of Zaragoza (Spain)
Javier Nogueras, Universidad de Zaragoza (Spain)
Atsuyuki Okabe, University of Tokyo (Japan)
Dieter Pfoser, Research Academic Computer Technology Institue (Greece)
Lutz Plümer, Universität Bonn (Germany)
Poulicos Prastacos, Foundation for Research and Technology (Greece)
Florian Probst, SAP Research CEC Darmstadt (Germany)
Hardy Pundt, University of Applied Sciences Harz (Germany)
Ross Purves, University of Zürich (Switzerland)
Martin Raubal, University of California - Santa Barbara (USA)
Tumasch Reichenbacher, University of Zurich (Switzerland)
Wolfgang Reinhardt, Universität der Bundeswehr München (Germany)
Jochen Renz, The Australian National University (Australia)
Claus Rinner, Ryerson University (Canada)
Jorge Rocha, University of Minho (Portugal)
Andrea Rodríguez, Universidad de Concepción (Chile)
Michael Rohs, Deutsche Telekom Laboratories - TU Berlin (Germany)
Tiina Sarjakoski, Finnish Geodetic Institute (Finland)
Tapani Sarjakoski, Finnish Geodetic Institute (Finland)
Johannes Schöning, University of Muenster (Germany)
Takeshi Shirabe, Technical University Vienna (Austria)

Spiros Skiadopoulos, University of Peloponnese (Greece)
Uwe Sörgel, Leibniz Universität Hannover (Germany)
Bettina Speckmann, TU Eindhoven (The Netherlands)
Thérèse Steenberghen, Katholieke Universiteit Leuven (Belgium)
Antony Stefanidis, George Mason University (USA)
John Stell, University of Leeds (United Kingdom)
Kathleen Stewart Hornsby, The University of Iowa (USA)
Juan C. Suárez, Forestry Commission (United Kingdom)
Fred Toppen, University of Utrecht (The Netherlands)
Danny Vandenbroucke, Katholieke Universiteit Leuven (Belgium)
Monica Wachowicz, Universidad Politécnica de Madrid (Spain)
Stephan Winter, The University of Melbourne (Australia)
Alexander Wolff, TU Eindhoven (The Netherlands)
Mike Worboys, University of Maine (USA)
Bisheng Yang, Wuhan University (China)
May Yuan, University of Oklahoma (USA)
Francisco Javier Zarazaga-Soria, University of Zaragoza (Spain)
Marc van Kreveld, University of Utrecht (The Netherlands)
Peter van Oosterom, Delft University of Technology (The Netherlands)

Contributing Authors

Hartmut Asche
University of Potsdam, Germany

Miriam Baglioni
University of Pisa, Italy

Alberto Belussi
Università di Verona, Italy

Karin Berkhoff
Leibniz Universität Hannover, Germany

Stefania Bertazzon
University of Calgary, Canada

Jannes Bolling
Technische Universität Berlin, Germany

Claus Brenner
Leibniz Universität Hannover, Germany

Kevin Buchin
Utrecht University, The Netherlands

Sergio Cabello
Inst. for Math., Physics and Mechanics, Ljubljana, Slovenia

Cláudio Carneiro
LASIG - Ecole Polytechnique Fédérale de Lausanne, Switzerland

Nicolas Champion
MATIS – Institut Géographique National, France

Gilles Desthieux
HEPIA – University of Applied Sciences Western Switzerland

Jürgen Döllner
Hasso-Plattner-Institute, University of Potsdam, Germany

Hongchao Fan
Technische Universität München, Germany

Tassilo Glander
Hasso-Plattner-Institute, University of Potsdam, Germany

Joachim Gudmundsson
NICTA Sydney, Australia

Eric Guilbert
Hong Kong Polytechnic University, Hong Kong

Jan-Henrik Haunert
Leibniz Universität Hannover, Germany

Sylvia Herrmann
Leibniz Universität Hannover, Germany

Markus Holopainen
University of Helsinki, Finland

Wei Huang
National Geomatics
Center of China, Beijing, China

Hannu Hyyppä
Helsinki University of Technology, Finland

Juha Hyyppä
Finnish Geodetic Institute, Finland

Mathias Jahnke
Technische Universität München, Germany

Jie Jiang
National Geomatics
Center of China, Beijing, China

Mika Karjalainen
Finnish Geodetic Institute, Finland

Birgit Kieler
Leibniz Universität Hannover, Germany

Birgit Kleinschmit
Technische Universität Berlin, Germany

Arjan Kuijper
Fraunhofer Institute for Computer Graphics Research, Germany

Yohei Kurata
Universität Bremen, Germany

Javier Lacasta
University of Zaragoza, Spain

Miguel Ángel Latre
University of Zaragoza, Spain

Federica Liguori
Politecnico di Milano, Italy

Maarten Löffler
Institute for Information and Computing Sciences, Utrecht University, The Netherlands

Jun Luo
Chinese Academy of Sciences, China

José Antônio Fernandes de Macêdo
École Polytechnique Fédéral de Lausanne, Switzerland

Jody Marca
Politecnico di Milano, Italy

Himanshu Mathur
University of Calgary, Canada

Martijn Meijers
Delft University of Technology, The Netherlands

Liqiu Meng
Technische Universität München, Germany

Sara Migliorini
Università di Verona, Italy

Eddy Mojica
Universidad de Zaragoza, Spain

Eugenio Morello
SENSEable City Laboratory, MIT, USA

Mauro Negri
Politecnico di Milano, Italy

Javier Nogueras-Iso
University of Zaragoza, Spain

Toshihiro Osaragi
Graduate School of Information Science and Engineering,Tokyo Institute of Technology, Japan

Peter van Oosterom
Delft University of Technology, The Netherlands

Giuseppe Pelagatti
Politecnico di Milano, Italy

Denise Peters
Universität Bremen, Germany

Marc Pierrot-Deseilligny
MATIS – Institut Géographique National, France

Pavel Propastin
Georg-August-University Göttingen, Germany

Ross S. Purves
GIS Division, University of Zurich-Irchel, Switzerland

Wilko Quak
Delft University of Technology, The Netherlands

Thorsten Reitz
Fraunhofer Institute for Computer Graphics Research, Germany

Chiara Renso
KDDLAB, ISTI-CNR, Pisa, Italy

Lutz Ross
Technische Universität Berlin, Germany

Günter Rote
Freie Universität Berlin, Germany

Rodrigo I. Silveira
Institute for Information and Computing Sciences, Utrecht University, The Netherlands

Bettina Speckmann
TU Eindhoven, The Netherlands

Georges Stamon
Université Paris Descartes, France

Martin Tomko
GIS Division, University of Zurich-Irchel, Switzerland

Matthias Trapp
Hasso-Plattner-Institute, University of Potsdam, Germany

Roberto Trasarti
KDDLAB, ISTI-CNR, Pisa, Italy

Friedjoff Trautwein
GIS Division, University of Zurich-Irchel, Switzerland

Sakari Tuominen
Finnish Forest Research Institute, Finland

Mikko Vastaranta
University of Helsinki, Finland

Paolo Visentini
Politecnico di Milano, Italy

Monica Wachowicz
Technical University of Madrid, Spain

Markus Wolff
University of Potsdam, Germany

Thomas Wolle
NICTA Sydney, Australia

Francisco Javier Zarazaga-Soria
University of Zaragoza, Spain

Table of Contents

Data Acquisition, Classification and Interpretation

Data Fusion and Integration

Spatio Temporal Modelling and Analysis

Generalization, Visualization and Perception

Identification of Practically Visible Spatial Objects in Natural Environments

Martin Tomko, Friedjoff Trautwein, Ross S. Purves

GIS Division, Department of Geography, University of Zurich-Irchel, Winterthurerstr. 190, CH-8057 Zurich, Switzerland,
{martin.tomko, trautwein, ross.purves}@geo.uzh.ch

Abstract. Image retrieval of landscape photographs requires accurate annotation using multi-faceted descriptions relating to the subject and content of the photograph. The subject of such photographs is dominantly the terrain and spatial objects visible from the photographer's viewpoint. While some spatial objects in the background may be obscured by foreground vegetation, other visible spatial objects beyond a certain distance may not present noteworthy elements of the captured scene (such as distant houses). Our aim is to assess approaches to improve the identification of practically visible spatial objects for image annotation. These approaches include the consideration of the apparent spatial object size and land-cover information about occluding vegetation. These inputs are used to enhance viewshed analysis to accurately identify only spatial objects practically visible and therefore likely to be notable subjects of a photograph. The two approaches are evaluated in an experiment in a semi-rural area of Switzerland, whose results indicate that visual magnitude is key in accurate identification of visible spatial objects.

1 Introduction

Landscape photographs are records of the visible portion of the terrain and the objects and vegetation positioned on top of it. Current efforts in spatial image annotation, such as project TRIPOD (http://tripod.shef.ac.uk/) aim at accurate annotation and captioning of landscape photographs for image search and retrieval. Photographs can be annotated using multi-faceted descriptions relating to, among others, the subject of the photograph (Shat-

M. Sester et al. (eds.), *Advances in GIScience*, Lecture Notes in Geoinformation and Cartography,
DOI 10.1007/978-3-642-00318-9_1,

ford, 1986). Therefore, the objects visible from a viewpoint contained within a photograph's viewport need to be reliably identified.

Consider a photograph of a rural landscape. Typically, objects in the middle distance or background are partially obscured by vegetation and other proximal objects. Furthermore, distant objects may be barely identifiable due to their small apparent size and reduced contrast from background as a consequence of atmospheric conditions. Hence, while visible, objects beyond a certain distance may not present noteworthy elements of the captured scene. Finally, photographs are printed or viewed on screen and the resolution of this visualization further reduces the number of noteworthy elements of the scene.

The aim of this paper is to assess approaches to improve the identification of *practically visible* objects for image annotation. Apparent object size and enhancement of the digital elevation model with information about vegetation occlusion need to be considered during the calculation of the viewshed in order to accurately identify the objects practically visible from the origin of the photograph and therefore likely to be the subject of the photograph. We test improvements brought about by limiting the computation to a distance beyond which the visual impact of objects is negligible and compare it to the improvements from DEM data enhanced by landcover information from global multispectral remote sensing imagery to infer the presence or absence of occluding vegetation.

As we wish to develop techniques which do not require detailed spatial data, since we wish to process photographs from a large area (such as Europe), only general purpose datasets with large-area coverage are practically usable for image annotation. Furthermore, parameters of the camera sensor and display system further impact on the visibility of an object in the photograph and its relevance to the captured scene.

This paper is structured as follows: in the next section, we review past research pertinent to visual impact analysis of landscapes from the perspective of image information retrieval. In Section 3 we present two methods that may improve the inference of practically visible spatial objects. We put these methods to a test in Section 4 and we present the results of the individual methods. In Section 5 the results are discussed and conclusions are drawn in Section 6, along with suggestions for further work.

2 Background

2.1 Information Retrieval

Accurately annotated documents improve the relevance of results during information search (Salton & Buckley, 1988; van Rijsbergen, 1979) and thus improve user experience. In recent years, the importance of the geographical scope of digital documents was widely recognized (Larson, 1996; Purves et al., 2007). Geographic Information Retrieval (GIR) emerged as a specific area of interest, where methods to infer and use the geographic scope of the documents – their footprint – are researched. Once a footprint is assigned to a document, spatial objects found within it can be used as source of highly contextual information for the annotation of the documents (Naaman et al., 2006; Purves et al., 2008). Such topical and accurate annotation is then used in retrieval to identify documents matching the query by geographic and thematic scope.

Digital photography is an emerging field of interest for GIR. Urban and rural landscape photographs have a clear geographic context provided by the photograph's origin (the location, focus and orientation of the camera) and the subject of the photograph. Photographers' annotations frequently reflect this geographic scope – consider Fig. 1, with an example caption: "*A country house seen across an orchard, near Zurich, Switzerland*". A photographer might annotate this photograph with keywords such as *house*, *orchard*, *Zurich*, and *Switzerland*. One could also refer to the individual trees, grassy lawn, footpath in the foreground and forest in the background. These are, however, not prominent elements of the scene and inclusion in the annotation would reduce the precision of the search results by including this picture in the result sets for photographs of forests or footpaths.

To improve annotation of photographs, we focus on the determination of the *practically* visible portion of a rural landscape, to identify spatial objects of substantial visual impact contained in a photograph. This should lead to annotation accuracy superior to that resulting from the use of simple circular buffer regions around a photograph's origin, or viewsheds computed purely based on the terrain. Parallel research focusing on urban environments is being undertaken by De Boer et al. (2008), and related work identifying other qualities of the scene captured through multifaceted image descriptions is presented in (Edwardes & Purves, 2007).

Fig. 1. A country house seen across an orchard, near Zurich, Switzerland (Photo and caption Martin Tomko)

2.2 Viewshed

The computation of a viewshed – the visible portion of a terrain and objects on top of it (De Floriani & Magillo, 2003), is a geographic analysis task applied to problems from urban planning to archaeology. Viewshed computation typically assumes that an object is visible to an observer if an unobstructed line of sight can be constructed between the observer's eye and the object. The computation is usually performed on an interpolated digital elevation model devoid of surface objects or vegetation (Fisher, 1996; Kaučič & Zalik, 2002; Maloy & Dean, 2001). Viewshed calculation can then be used to identify objects situated in the visible portions of the surface. The calculation of a viewshed can be limited to a specific direction and distance (by specifying, for instance, the maximum length of the line of sight).

As noted by Ervin and Steinitz (Ervin & Steinitz, 2003), simple computation of viewsheds is not sufficient to assess the visual quality of a landscape. The way visual quality of a landscape impacts on a human observer is determined by a wide variety of factors, intrinsic to the landscape but also dependent on the observer's context (Litton, 1968).

2.3 Landscape Perception and Visual Impact

Typically, people are able to summarize the visual quality of a landscape in a few words. While some aspects of the visual quality are highly subjective and reflected in adjectives such as romantic, peaceful, serene, others are more tangible and relate to visible objects and landcover. These different facets of the landscape are similar to the facets of image descriptions, as studied in Shatford (1986).

The material aspects of landscape quality and its change (such as introduction of anthropogenic objects or landuse change) has been the focus of multiple studies (Bishop, 2003; Daniel, 2001; Gret-Regamey et al., 2007; Magill, 1990). These studies relied on the assessment of the visual impact of the introduced objects based on computer visualizations and digital photographs altered by computer animations (Bishop, 2002; Hadrian et al., 1988; Shang & Bishop, 2000) and are restricted to parameters that can be objectively determined, for example by measurement of physical qualities (Groß, 1991).

For an object to be notable in a scene, its apparent size must exceed a certain visual magnitude, also known as visual threshold (Iverson, 1985; Magill, 1990). Three different visual magnitudes derived from the parameters of human visual acuity (approximately 1') determine the thresholds for object detection, recognition (or identification) and visual impact (Shang & Bishop, 2000).

An object with a visual magnitude of 1' can just be detected by the retina (as a single dot, or pixel), but not recognized or have visual impact. Depending on the type of object and viewing conditions, a simple, well known object has to exceed a visual magnitude of approximately 5.5' in order to be recognized (Luebke et al., 2003). At this visual magnitude the most salient elements of the object's structure can be differentiated. This is reflected in common cartographic guidelines (for example (Spiess et al., 2005)) where map symbols are rendered as 5x5 pixels at least.

In natural landscapes, few objects have well defined familiar shapes. Furthermore, the viewer does not know *a priori* which objects will be visible (uninformed recognition). Studies performed on digital images of faces, outdoor and indoor objects and complex scenes showed that a natural object had to be rendered with a higher resolution to be recognized (Cai, 2004).

Visual thresholds based on visual magnitude can be used to limit the length of the line of sight during viewshed calculation. However, recognition of objects in natural settings is a much more complex task than the simple recognition of letters or symbols in controlled laboratory conditions. While it can be limited to the determination of visual magnitude for practical reasons, experience, personal objectives and atmospheric conditions play a strong role in recognition of objects (Pitchford & Malm, 1994). Furthermore, when the objects are to be detected or recognized in photographs as opposed to viewed in natural settings as such, the resolutions of the sensor, lens (optical) and display systems affect the visual thresholds as detailed in Section 3.2.

2.4 Visibility and Occlusion by Vegetation

Little research has directly addressed the influence of vegetation on the visibility of the surrounding space. Dean (1997) proposed a method to improve the prediction of object visibility in forests based on estimates of the vegetation's opacity, characterized by a visual permeability value. The study combined DEM data with extruded vegetation from detailed forest inventory data, including accurate tree heights. All evaluation was limited to lines of sight of 50 to 500m, with an orange air balloon as an artificial target.

Another method was proposed for object visibility prediction in paleoarcheology by Llobera (2007). It is based on principles derived from light attenuation by particles and relies on highly accurate data about spatial distribution of individual plants in the area studied. While plausible, the model has only been tested on a synthetic DEM using simulated vegetation coverage and relies on data of too high an accuracy for practical image annotation.

An attempt to use widely available, global coverage vegetation information of relatively high resolution for realistic visualization of terrain was proposed by Roettger (2007). Based on a classification of the well-known Normalized Difference Vegetation Index (NDVI) values, they infer the presence of vegetation at a particular location. Furthermore, they map NDVI values to vegetation height based on a linear interpolation between user defined maximum and minimum values. While not tested in a field experiment, the method could provide a simple and efficient way of estimating the distribution of vegetation over large areas at acceptable resolution and thus provide a viable basis for the consideration of vegetation occlusion in object visibility analysis.

3 Method

We propose two methods to improve the results of viewshed calculations. First, we determine a visual impact threshold for landscape images viewed on LCD displays. Second, we enhance the DEM used to calculate these viewsheds by adding extruded vegetation information.

3.1 Visual Impact Determination for Photographs

For the annotation of photographs, the impact of the sensor and display parameters to the determination of the visual impact threshold have to be considered. The acuity of human vision, as well as the resolution of consumer grade digital camera sensors is beyond the resolution of typical LCD displays. Photographs are displayed on displays at a fraction of their actual resolution. The display thus represents the effective limit to the identification of objects in photographs. The resampling *r* is equivalent to the ratio between the sizes of the sensor (*sensordim*) and the screen (*screendim*, in pixels)(Figure 2):

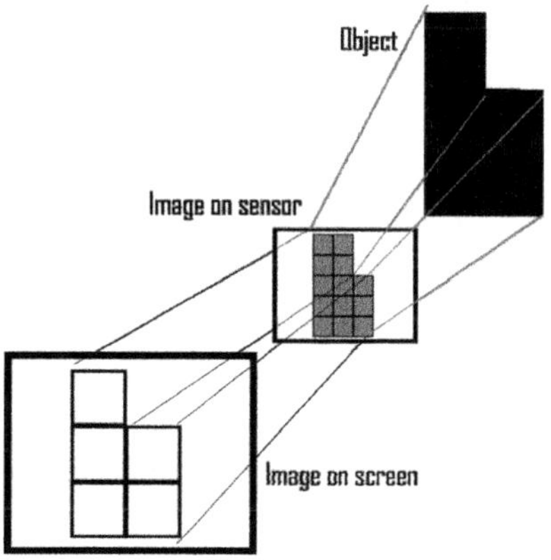

Fig. 2. Resampling occurring in the object-sensor-display system.

$$r = \frac{screendim_{pix}}{sensordim_{pix}} \tag{1}$$

The angular field of view *afov* captured by a camera is characterized by the focal length *f* of the lens used and the physical size of the sensor, in *mm*:

$$afov = 2atan\left(\frac{sensordim_{mm}}{2f}\right) \tag{2}$$

Images of recognizable natural objects consist of at least 1024 pixels (32 x 32 pixels), compared to only 289 pixels(17x17 pixels) for familiar faces (Cai, 2004). If the object is to be recognized on screen, this is the size of the object's rendered image and not that image captured by the sensor. As the resolution of the screen is the limiting factor of the sensor-display system the image of the object has to be captured as a square of side $is = wr$ (is – image size on sensor, w – image size on screen, in pixels).

The density of pixels on the sensor determines the angular resolution of the sensor. The angular resolution *ares* of the sensor – lens combination is the fraction of the angular field of view that is captured by one pixel of the sensor. The higher the sensor pixel density (or, the smaller the pixel size), the more pixels will capture the same extent of *afov*.

From the image size *is* and the angular resolution of the sensor – lens combination it is possible to determine the minimal angular field of view α occupied by an object of known size to exceed the visual impact threshold. The maximal distance d at which this magnitude is exceeded by the object of size o for a given sensor-lens-display combination is:

$$d = \frac{\frac{o}{2}}{tan(\frac{\alpha}{2})} \tag{3}$$

In Section 4.3, we use the approach outlined to compute the distance d for the combination of sensor, lens and display used in a set of field experiments. The value of d is then used to limit the computation of the viewsheds for observation points, in order to identify only practically visible objects for photographs of the given landscape scenes.

3.2 Occlusion by Vegetation

The second method explored aims at accurate inference of vegetation occlusion. This requires reliable information about the spatial distribution of vegetation and its height. In order to be practical for image annotation, the method should use general-purpose datasets of large-area coverage. Furthermore, accurate information about vegetation height is, usually, not available.

We build on the approach of Roettger (2007) using NDVI extracted from remote sensing imagery. NDVI values are computed from sampling the Earth's surface in the near infra-red (NIR) and visible red (VIS) bandwidth of the Landsat ETM+ sensor. The index is calculated as follows:

$$NDVI = (NIR - VIS)/(NIR + VIS) \tag{4}$$

The index gives an estimate of healthy vegetation land cover. While values beyond a given threshold are likely to relate to dense foliage and allow inference of the presence of forests or shrubs, it is impossible to directly relate the value of the index to the height of vegetation. We therefore chose a single threshold value to indicate the presence of dense vegetation, without relating the index values of the vegetated areas to vegetation height. The index value of 0.2 of Roettger (2007) was taken as a starting point and tested in 0.01 increments up to 0.3. Best matches between the vegetation layer derived from NDVI and thematic landcover datasets of the Swiss national mapping agency Swisstopo were achieved for values of 0.27 (Vector200 dataset) and 0.28 (Vector25 dataset) and confirmed by visual comparison with photogrammetric records of the area. The value of 0.28 was chosen for the extrusion of vegetation in the experiment due to its best match in the direct vicinity of the experiments' observation points.

As no detailed datasets of vegetation heights is associated with the vegetation layer derived from NDVI, and our motivation does not allow for specialized spatial datasets, we built on the knowledge of the forest types in the area of interest (mostly mixed beech and spruce forests), three tree heights were used to extrude the vegetation layer - 10, 20 and 30m (for more information on forest types, see `http://www.gis.zh.ch` and (BAFU, 2005)). The extruded vegetation was then added to the DEM of the area studied and viewshed were calculated. Results of the visibility analysis are reported in Section 4.4.

4 Experiment and Results

4.1 Overview

In two experiments we evaluated the possibility to identify visible objects for image annotation. Two approaches are tested - viewshed analysis enriched with heuristics about object's visual magnitude and viewshed analysis including consideration of occlusion by vegetation using an extruded layer of landcover information. The workflow of the two methods and their evaluation is outlined in Fig. **3**.

In the right strand, the workflow for experiment 1 is shown in parallel to experiment 2 (left strand). Joint data or analytical procedures overlap both strands.

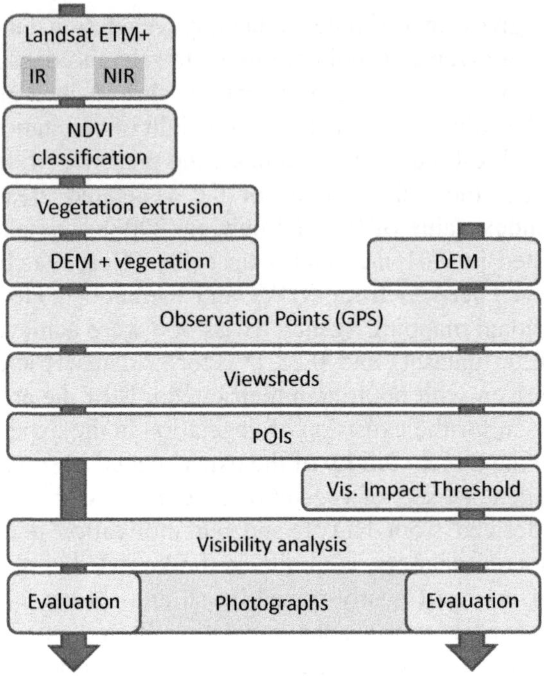

Fig. 3. Workflow schema.

4.2 Data

We limit our analysis to datasets that are available at low costs and provide large area or global. For our experiments, the following datasets covering the region around Zurich, Switzerland, were used (all Swisstopo datasets in the Swiss *CH1903* national grid coordinate system):

- Orthorectified Landsat 7 ETM+ band 3 and 4 dataset (image p194r027_7), acquired on August 24th, 2001, referenced in WGS84 (transformed into CH1903), spatial resolution of 28.5m;
- A raster DEM raster dataset Swisstopo DHM25 with a spatial resolution of 25m. The height accuracy varies from 1.5m in flat lands to 3m in Alpine regions (Swisstopo, 2005);
- A dataset containing centroids of all named objects present on the 1:25000 Swisstopo maps (Swissnames);

While the Swissnames dataset is not an ideal source of point of interest (POI) data due to its explicit focus on cartographic content (it contains the centroids and labels of all toponyms on Swistopo maps), it is the best available dataset with comprehensive coverage in rural areas. The dataset was filtered to include only 29 categories of objects that can be considered point-like for the purpose of our assessment (excluding names of forests, meadows, hills etc.), with the exception of settlements and ponds, included due to their easy visual identification in photographs. Note that no information is available about the objects' size and height, and therefore their projective size cannot be computed.

Furthermore, the following data were collected:

- Coordinates of 12 points from which photographs of the surroundings were taken. These points served as centroids for the generation of viewshed and POI visibility analysis;
- 83 georeferenced photographs with directional information, taken from the 12 observation points, taken with an 8.13 Mpix Ricoh Caplio 500G digital camera (sensor size 3264 x 2448 pixels, physical sensor size 7.18 x 5.32 mm) with direct Bluetooth link to a GPS receiver. Image azimuths were measured with a handheld digital compass. All photographs were taken with a focal length of 5mm (wide angle) reported in EXIF data, equivalent to a field of view of 71°. The 360° panoramas for each of the observation points are shown in Figure 4. The photographs were viewed on an LCD display with resolution of 1280*1024 pixels (Philips Brilliance 200W) with a physical pixel size of approximately 0.294mm.

(a) Point 1

(b) Point 2

(c) Point 3

(d) Point 4

(e) Point 5

(f) Point 6

(g) Point 7

(h) Point 8

(i) Point 9

(j) Point 10

(k) Point 11

(l) Point 12

Fig. 4. Views from the 12 test sites as panoramic collages of the photographs taken.

4.3 Experiment 1: Objects Exceeding the Visual Impact Threshold

The visibility of POI objects was analyzed by calculating a 360° binary viewshed on the DEM and counting the POIs found in visible cells. For comparison of the results with Experiment 2, the location of each POI was rasterized to match the cells of the vegetation layer (spatial resolution of 28.5m). As no information about the real size of the spatial objects was available, this value was taken as input for the calculation of the visual impact threshold. We assert that 28.5m represent a reasonable size estimate for man-made spatial objects such as farm houses. The counts of POIs evaluated as visible in the viewshed analysis without distance limitation are shown in Table 1 (DEM).

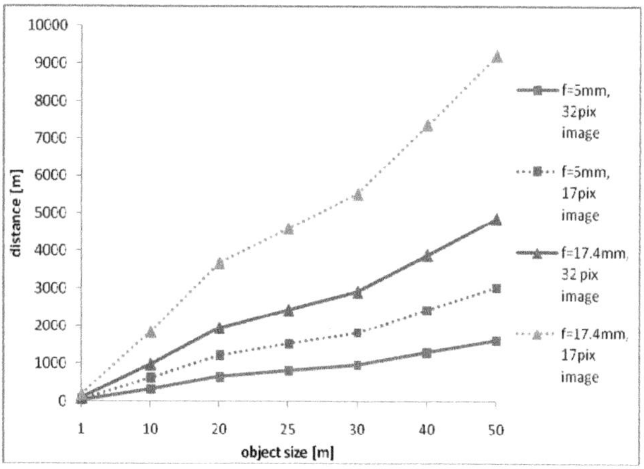

Fig. 5. Dependence of minimum distance to object from object size, visual impact threshold and parameters of the sensor-lens system. For an object to be above visual impact threshold, it must be closer than the distance related to its size.

For comparison, the objects exceeding the visual impact threshold were identified. First, the distance at which the visual impact threshold for the

POIs is exceeded was determined. An object of 28.5m occupies a screen space of 17x17pix to 32x32pix (approximately 0.5cm to 0.94cm on the screen used and 43x43 to 82x82 sensor pixels) when closer than 914m - 1730m, if photographed with *f*=5 mm lenses (wide angle lens). This is equivalent to an apparent visual magnitude of 0.94° to 1.78° for an object observed by naked eye. For the plot of dependencies between the focal length, object size and object distance to exceed the visual impact threshold see Figure 5. As shown, the visual impact threshold distance for the same object, but captured using a *f*=17.5mm lens is between 4 to 10km. A single value of 1km has been taken as a conservative substitute of the interval identified for *f*=5mm lens, allowing for degradation of visual impact due to, for example, contrast reduced by haze and unfamiliar object shapes. The counts of the objects exceeding the visual impact threshold are reported in Table 1.

Each object that was evaluated as visible in either of the two viewshed analyses was searched for in the corresponding photograph and marked as visible or invisible. Only objects considered large enough to be of visual impact to the subject of the image were identified as visible (executed as an image labeling exercise similar to that from Russell et al. (2008), Figure 6). The counts of the visible objects are reported in Table 1 (Image).

The results reported can be interpreted using the standard measures to assess the quality of remote sensing classifications through contingency tables. As none of the points visible in the photograph were reported as invisible in the DEM or not present in the 1km buffer region, the full contingency table can be reconstructed by the interested reader. As shown, the results of viewshed analysis neglecting vegetation information greatly exaggerate the number of visible POIs in all cases. The limitation of the visibility analysis to the distance at which the objects exceed the visual impact threshold achieves significantly higher precision of detection. Only in two out of 12 cases, an extra POI has been reported for a given image.

The apparent size of the smallest object considered of significant visual impact found in the labeled photographs is approximately 170 sensor pixels. The object has an apparent height of approximately 5.8mm on the screen, or approximately 20 screen pixels. This size is slightly inferior to the theoretical visual impact threshold used in this study. The corresponding object is a barely visible radio tower and hence it has a particular, familiar elongated shape and it is positioned on a prominent hill on the horizon. Radio antennas are prominent spatial objects frequently used as landmarks due to their good visibility and their high figure-ground contrast.

Table 1. Counts of visible POIs based on viewshed analysis without distance limitation and with a distance limitation of 1km based on visual impact threshold (DEM without vegetation). Image – POI visible in the photograph. DEM – POI evaluated visible using the DEM. 1km buffer– POI within 1km of the observation point. 1km buffer + DEM – POI predicted to be visible using the DEM within 1km of the ob-servation point.

Observation Point	Image	DEM	1km buffer	1km buffer + DEM
p1	1	33	3	1
p2	0	36	2	0
p3	0	44	0	0
p4	1	91	4	1
p5	0	92	3	0
p6	1	91	5	1
p7	4	261	5	4
p8	0	82	3	1
p9	1	46	5	1
p10	0	57	6	0
p11	1	70	9	1
p12	0	34	7	1

4.4 Experiment 2: Visibility Analysis Simulating Occlusion by Vegetation

The dataset based on NDVI classification provides information about presence or absence of vegetation. A threshold NDVI value of 0.28 was selected for vegetated areas and the extruded vegetation was added to the DEM and used for viewshed computation. The pixel incident with the observation point used for the calculation of a viewshed was kept at the original altitude of the DEM (the observation points were all on the ground or man–made structures).

Fig. 6. Example of detection of visible objects in a labeled photograph. The train station building (image right) is contained in the POI database.

The results indicate that the consideration of vegetation does not perform as well as the simple combination of DEM with a visual magnitude threshold consideration (Table 2). Only for seven out of 12 observation points the counts of visible POIs are accurate, in all cases where there were no visible objects in the photograph and hence the effect of over-filtering cannot be detected. No significant dependence was found for the different values of extruded vegetation height, beyond the minimal value of 10m, lower than the mean height of the typical vegetation in the area.

The results indicate that the method is prone to over-filtering – the elimination of objects that are actually visible and can be identified in photographs (see values for Image[V]/model[NV] in Table 2). This is mostly due to the binary classification of the terrain surface as vegetated and not vegetated. As a result, sparse vegetation is extruded as an opaque cell (Figure 7). Thus, while the vegetation classification may be spatially correct, a simple extrusion of the vegetation layer may not present the most appropriate method for vegetation modeling. It also appears that positional accuracy of the vegetation dataset has higher impact on the results than accurate information about vegetation height.

While the results are often over-filtered, they also contain frequent false matches. POIs are reported as visible while they are not visible. This is likely due to occlusion by objects in the foreground, close to the observer. Hence, we conclude that the method is extremely sensitive and highly dependent on accurate vegetation information, as well as requiring complex data processing. As such, it is not suited for automated annotation of images for GIR.

Table 2. Visibility analysis of POIs with vegetation occlusion. Horizontal reading: counts of spatial objects identified as visible (V) or not visible (NV) on a DEM with extruded vegetation of 10m, 20m and 30m, without distance limitation are shown for each point and vegetation combination. Reading by column: corresponding counts of the same spatial objects visible or not visible in photographs.

p1		image		p1		image		p1		image	
		V	NV			V	NV			V	NV
10m	V	0	3	20m	V	0	0	30m	V	0	0
veg	NV	1	NA	veg	NV	1	NA	veg	NV	1	NA
p2		image		p2		image		p2		image	
		V	NV			V	NV			V	NV
10m	V	0	0	20m	V	0	0	30m	V	0	0
veg	NV	0	NA	veg	NV	0	NA	veg	NV	0	NA
p3		image		p3		image		p3		image	
		V	NV			V	NV			V	NV
10m	V	0	8	20m	V	0	0	30m	V	0	0
veg	NV	0	NA	veg	NV	0	NA	veg	NV	0	NA
p4		image		p4		image		p4		image	
		V	NV			V	NV			V	NV
10m	V	1	17	20m	V	1	5	30m	V	0	1
veg	NV	0	NA	veg	NV	0	NA	veg	NV	1	NA
p5		image		p5		image		p5		image	
		V	NV			V	NV			V	NV
10m	V	0	12	20m	V	0	3	30m	V	0	0
veg	NV	0	NA	veg	NV	0	NA	veg	NV	0	NA
p6		image		p6		image		p6		image	
		V	NV			V	NV			V	NV
10m	V	0	6	20m	V	0	0	30m	V	0	0
veg	NV	0	NA	veg	NV	0	NA	veg	NV	0	NA
p7		image		p7		image		p7		image	
		V	NV			V	NV			V	NV
10m	V	3	103	20m	V	0	43	30m	V	0	7
veg	NV	1	NA	veg	NV	4	NA	veg	NV	4	NA
p8		image		p8		image		p8		image	
		V	NV			V	NV			V	NV
10m	V	0	21	20m	V	0	13	30m	V	0	9
veg	NV	0	NA	veg	NV	0	NA	veg	NV	0	NA
p9		image		p9		image		p9		image	
		V	NV			V	NV			V	NV
10m	V	0	45	20m	V	0	0	30m	V	0	0
veg	NV	1	NA	veg	NV	1	NA	veg	NV	1	NA
p10		image		p10		image		p10		image	
		V	NV			V	NV			V	NV
10m	V	0	0	20m	V	0	0	30m	V	0	0
veg	NV	0	NA	veg	NV	0	NA	veg	NV	0	NA
p11		image		p11		image		p11		image	
		V	NV			V	NV			V	NV
10m	V	0	4	20m	V	0	0	30m	V	0	0
veg	NV	0	NA	veg	NV	0	NA	veg	NV	0	NA
p12		image		p12		image		p12		image	
		V	NV			V	NV			V	NV
10m	V	0	2	20m	V	0	0	30m	V	0	0
veg	NV	0	NA	veg	NV	0	NA	veg	NV	0	NA

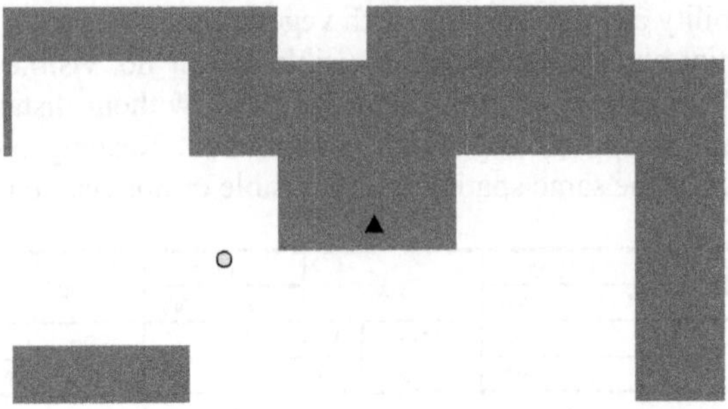

Fig. 7. Visibility of an object (circle) obstructed by vegetation (adjacent pixel). In reality, this is an orchard and the vegetation is visually permeable (Dean, 1997). The observation point P11 is shown as triangle. The photograph of the scene is shown in Figure 1.

5 Case Study and Discussion

5.1 Case Study

In order to verify our findings indicating that visual magnitude thresholds and viewshed analysis based on DEM data (without vegetation) provide sufficient inputs for the inference of practically visible objects, we tested 4 arbitrarily selected georeferenced landscape photographs from different authors, similar to those available from photo-sharing sites such as Flickr. The photographs were selected from the area covered by identical datasets to those used earlier. All photographs were acquired within the last 2 years for the project TRIPOD. The photographs did not contain directional information and this information was therefore computed by relating the edges of the photographs with available spatial data and consequent computation of azimuths.

For each photograph, the viewshed, 1km buffer and its directional field of view were calculated (Figure 8). The results, shown in Table 3 confirm that the combination of visibility calculation based on DEM (without vegetation), combined with a visual impact threshold value (expressed as 1km buffer) and field of view information provide together a reliable means to identify objects captured in a photograph. Note that alone, neither the viewshed analysis, nor the distance limitation within the available field of view yield optimal results. Their combination, however, allows for reliable identification of visible objects. The result for image C (containing one visible object but resulting in a prediction of no objects) points to method's dependence on accurate estimate of the objects size – image C contains a

Table 3. Number of POIs evaluated as visible using five combinations of viewsheds (calculated on DEM), distance thresholds and the actual photograph's field of view (FOV). The values of POI Image indicate the number of POIs actually visible in the photographs.

Image	POI Viewshed	POI buffer 1km	POI Viewshed 1km	POI within 1km in FOV	POI FOV viewshed	POI Image
A	484	11	3	1	1	1
B	130	6	2	4	2	2
C	90	9	2	2	0	1
D	96	1	0	0	0	0

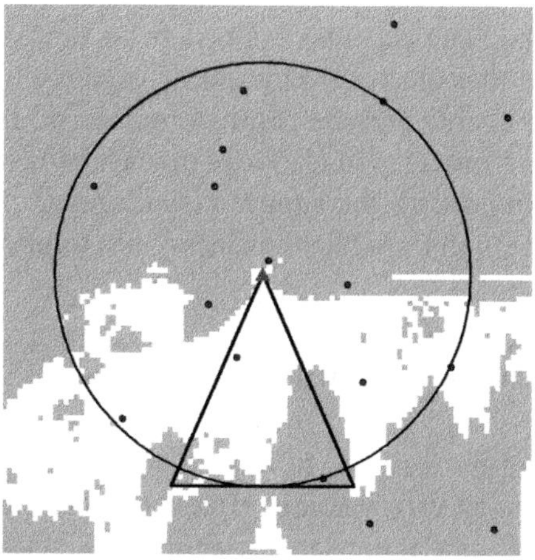

Fig. 8. Viewshed of point A overlaid with the 1km buffer and the visual field of view of the image. POIs are represented as points. Visible cells of the DEM are white, invisible cells are grey.

distant airport, with a building exceeding the size of 25m. As such, the airport is a significant element in the photograph.

5.2 Discussion

Two experiments were performed in a semi-rural environment with abundant vegetation and sparse man-made objects – POIs. In the experiments, a substantial reduction in the counts of spatial objects incorrectly classified as visible was achieved by limiting the visibility calculation to a distance at which an object has a visual magnitude above a visual impact threshold.

Such a limitation based on a simple heuristic determination of the visibility impact threshold allows the elimination of objects that do not present a significant element of the observed and photographed scene if rendered on a computer screen.

Similarly, the visibility analysis including landcover information shows a reduction in the number of spatial objects visible compared to viewsheds calculated on pure DEM. The consideration of vegetation should allow the elimination of objects occluded by foreground vegetation and leads to more realistic results of the visibility analysis. The vegetation in the foreground has high impact on the results compared to background vegetation, as objects in the foreground occlude a larger proportion of the visual field. It seems therefore that the accuracy of the data about the presence or absence of vegetation is more important than the exact knowledge of the vegetation's height. The variation of the vegetation height has had little impact on the results. The results obtained from the experiment performed, however, indicate that the consideration of vegetation is much more sensitive to the data available and the results obtained do not justify the computationally intensive process. In a follow-up case-study, we have shown that DEM data, combined with the simple visual impact threshold allows us to infer the objects actually visible in arbitrary photographs.

6 Conclusions and Future Work

Limiting the visibility analysis to objects appearing larger than the visual impact threshold is an efficient and effective method to reduce the computation of viewsheds and at the same time identify spatial objects relevant to image annotation. The visual magnitude of photographed objects is significantly influenced by the display on which the photographs are viewed, and the consideration of the resampling between the sensor and the display influence the estimate of the visual magnitude of the photographed object. It is important to note that the object's shape and the observer's position in relation to the object alter the visual impact of the observed object.

We further presented a simple method to enhance the estimate of the visual impact of an object with information about occlusion by foreground vegetation. The consideration of vegetation information may, in some cases, further improve the veracity of the visibility analysis, but care has to be taken not to over-filter the visible objects. Further research on vegetation visual permeability could lead to improved results, as suggested by Dean (1997). Note, however, that such approaches seem to be less reliable and more data expensive than a simple heuristic about the visual magnitude of the photographed objects.

The visual impact of an object can be further deteriorated by external factors altering the contrast of the object from the background, such as atmospheric conditions and the surface properties of the object. The consideration of atmospheric influences on visual threshold may be more practical than that of vegetation and could further improve the results. Meteorological services broadcast weather information including visibility range and haze information (for instance, METAR (OFCM, 2005)) that could be included in the threshold determination similar to (Pitchford & Malm, 1994). Heuristics allowing for accurate inference of the objects' size will, however, provide the greatest improvement. Such heuristics could be based, for instance, on the analysis of the category of spatial objects and the use of a mean size value per category.

Image annotation is an important step for the organization and management of searchable image libraries. Images annotated only with keywords related to the image content of practical visual impact allow for better image search relevance. Previously, Tomko and Purves (2008) focused on the analysis of the spatial distribution of POI in a given region as a means to infer an object's relevance for the annotation of the region. The identification of only practically visible spatial objects is a necessary requirement for such a classification method, providing inputs for multifaceted image descriptions (Edwardes & Purves, 2007).

Acknowledgments

The research reported in this paper is part of the project TRIPOD supported by the European commission under contract No. 045335.

References

BAFU. (2005). *Waldtypen der Schweiz* Bern, Switzerland: Bundesamt für Umwelt BAFU.

Bishop, I. (2002). Determination of Thresholds of Visual Impact: the Case of Wind Turbines. *Environment and Planning B: Planning and Design, 29*, 707-718.

Bishop, I. (2003). Assessment of Visual Qualities, Impacts, and Behaviours, in the Landscape, by Using Measures of Visibility. *Environment and Planning B: Planning and Design, 30*(5), 677-688.

Cai, Y. (2004). Minimalism Context-Aware Displays. *CyberPsychology and Behavior, 7*(6), 635-644.

Daniel, T. C. (2001). Whither Scenic Beauty? Visual Landscape Quality Assessment in the 21st Century. *Landscape and Urban Planning, 54*, 267-281.

De Boer, A., Dias, E., & Verbree, E. (2008). Processing 3D Geo-Information for Augmenting Georeferenced and Oriented Photographs with Text Labels. In A. Ruas & C. Gold (Eds.), *Headway in Spatial Data Handling* (pp. 351-365). Berlin, Heidelberg: Springer-Verlag.

De Floriani, L., & Magillo, P. (2003). Algorithms for Visibility Computation on Terrains: a Survey. *Environment and Planning B: Planning and Design, 30*(5), 709-728.

Dean, D. J. (1997). Improving the Accuracy of Forest Viewsheds Using Triangulated Networks and the Visual Permeability Method. *Canadian Journal of Forest Research, 27*, 969-977.

Edwardes, A. J., & Purves, R. S. (2007). *Eliciting Concepts of Place for Text-based Image Retrieval.* Paper presented at the 4th ACM Workshop On Geographic Information Retrieval, GIR 2007, Lisbon, Portugal.

Ervin, S., & Steinitz, C. (2003). Landscape Visibility Computation: Necessary, but not Sufficient. *Environment and Planning B: Planning and Design, 30*(5), 757-766.

Fisher, P. F. (1996). Extending the Applicability of Viewsheds in Landscape Planning. *Photogrammetric Engineering & Remote Sensing, 62*(11), 1297-1302.

Gret-Regamey, A., Bishop, I. D., & Bebi, P. (2007). Predicting the Scenic Beauty Value of Mapped Landscape Changes in a Mountainous Region Through the use of GIS. *Environment and Planning B: Planning and Design, 34*(1), 50-67.

Groß, M. (1991). The Analysis of Visibility—Environmental Interactions Between Computer Graphics, Physics, and Physiology *Computers & Graphics, 15*(3), 407-415.

Hadrian, D. R., Bishop, I. D., & Mitcheltree, R. (1988). Automated Mapping of Visual Impacts in Utility Corridors. *Landscape and Urban Planning*(3), 261-282.

Iverson, W. D. (1985). And that's about the Size of it: Visual Magnitude as a Measurement of the Physical Landscape. *Landscape Journal 4*(1), 14-22.

Kaučič, B., & Zalik, B. (2002). *Comparison of Viewshed Algorithms on Regular Spaced Points.* Paper presented at the 18th Spring Conference on Computer Graphics, Budmerice, Slovakia

Larson, R. R. (1996). *Geographic Information Retrieval and Spatial Browsing.* Paper presented at the Geographic information systems and libraries: patrons, maps, and spatial information. 1995 Clinic on Library Applications of Data Processing, April 10-12, 1995.

Litton, R. B., Jr. (1968). *Forest Landscape Description and Inventories - a Basis for Landplanning and Design.* Berkeley, CA: Pacific Southwest Forest and Range Experiment Station, Forest Service, U.S. Department of Agriculture.

Llobera, M. (2007). Modeling Visibility Through Vegetation. *International Journal for Geographical Information Science, 21*(7), 799-810.

Luebke, D., Reddy, M., Cohen, J. D., Varshney, A., Watson, B., & Huebner, R. (2003). *Level of Detail for 3D Graphics.* Amsterdam, NL: Morgan Kaufmann Publishers.

Magill, A. W. (1990). *Assessing Public Concern for Landscape Quality: A Potential Model to Identify Visual Thresholds.* Berkeley, CA: Pacific Southwest Research Station, Forest Service, U.S. Depratment of Agriculture.

Maloy, M. A., & Dean, D. J. (2001). An Accuracy Assessment of Various GIS_Based Viewshed Delineation Techniques. *Photogrammetric Engineering & Remote Sensing, 67*(11), 1293-1298.

Naaman, M., Songa, Y. J., Paepckea, A., & Garcia-Molina, H. (2006). Assigning Textual Names to Sets of Geographic Coordinates. *Computers, Environment and Urban Systems, 30*(4), 418-435

OFCM. (2005). *Federal Meteorological Handbook No. 1: Surface Weather Observations and Reports*. Washington, D.C., USA: Federal Coordinator for Meteorological Services and Supporting Research, National Oceanic and Atmospheric Administration, U.S. Department of Commerce.

Pitchford, M. L., & Malm, W. C. (1994). Development and Applications of a Standard Visual Index. *Atmospheric Environment, 28*(5), 1049-1054.

Purves, R., Clough, P., Jones, C. B., Arampatzis, A., Bucher, B., Finch, D., et al. (2007). The Design and Implementation of SPIRIT: a Spatially Aware Search Engine for Information Retrieval on the Internet. *International Journal for Geographical Information Science, 21*(7), 717-745.

Purves, R., Edwardes, A. J., & Sanderson, M. (2008). Describing the Where – Improving Image Annotation and Search Through Geography. In G. Csurka (Ed.), *Proceedings of the 1st Intl. Workshop on Metadata Mining for Image Understanding (MMIU 2008)* (pp. 105-113). Funchal, Madeira – Portugal.

Roettger, S. (2007). *NDVI-based Vegetation Rendering*. Paper presented at the Computer Graphics and Imaging CGIM '07, Innsbruck, Austria.

Russell, B. C., Torralba, A., Murphy, K. P., & Freeman, W. T. (2008). LabelMe: a Database and Web-based Tool for Image Annotation *International Journal of Computer Vision, 77*(1-3), 157-173.

Salton, G., & Buckley, C. (1988). Term-Weighting Approaches in Automatic Text Retrieval. *Information Processing & Management, 24*(5), 513-523.

Shang, H., & Bishop, I. D. (2000). Visual Thresholds for Detection , Recognition and Visual Impact in Landscape Settings. *Journal of Environmental Psychology, 20*, 125-140.

Shatford, S. (1986). Analyzing the Subject of a Picture: A Theoretical Approach. *Cataloging and Classification Quarterly, 6*(3), 39-62.

Spiess, E., Baumgartner, U., Arn, S., & Vez, C. (2005). *Topographic Maps. Map Graphics and Generalization*. Wabern, Switzerland: Swiss Society of Cartography.

Swisstopo. (2005). *DHM25. Das digitale Hohenmodell der Schweiz. Produktinformation.* Bern, Switzerland: Swisstopo, Bundesamt fur Landestopografie.

Tomko, M., & Purves, R. S. (2008). *Categorical Prominence and the Characteristic Description of Regions* Paper presented at the Semantic Web meets Geospatial Applications, held in conjunction with AGILE 2008, Girona, Spain.

van Rijsbergen, C. J. (1979). *Information Retrieval*: Butterworth.

Extraction of Features from Mobile Laser Scanning Data for Future Driver Assistance Systems

Claus Brenner

Institute of Cartography and Geoinformatics,
Leibniz Universität Hannover, Appelstr. 9A, 30167 Hannover,
claus.brenner@ikg.uni-hannover.de

Abstract. Research and development of driver assistance systems is currently a very active field. One building block of future systems will be an accurate and reliable positioning, which can be realized by relative measurement, using on-board sensors and maps of the environment. However, a prerequisite will be that such maps can be produced fully automatically.

This paper explores the use of dense laser scans from mobile laser scanning systems for the production of such maps. After presenting the problem and the matching approach, we introduce our test field which consists of a 22 km scan of roads, both inner city streets as well as highways. It is shown how suitable features can be extracted fully automatically. Finally, for a given trajectory, we evaluate how positioning will perform and draw conclusions regarding applicability and future work.

1 Introduction

There is currently a busy research and development in car driver information and assistance systems. Besides the general goal to provide useful information to drivers in order to make well informed decisions, a major motivation is the wish to reduce the number of injuries and fatalities in traffic. During the last years, especially mobile (or personal) navigation systems have become very popular, and a doubling or tripling of the number of sold units from year to year has not been uncommon.

M. Sester et al. (eds.), *Advances in GIScience*, Lecture Notes in Geoinformation and Cartography,
DOI 10.1007/978-3-642-00318-9_2, © Springer-Verlag Berlin Heidelberg 2009

While such devices may be perceived as 'yet another (in-car) electronics', there is a significant difference to all other in-car systems, practically all of which are self-contained. In contrast, navigation systems are completely useless without external information in the form of digital maps. Even if one could argue that any old-style car radio is also dependent on external infrastructure, it is clear that digital maps contain highly interpreted and structured information, which is not required by any other system so far. Thus, car navigation systems are unique in the sense that they are dependent on the probably most complicated data structures that are disseminated and used by the wide public. Also, car navigation maps are the first widely used geographic data structures which are designed to perform computations – like searching for the shortest / fastest route, route guidance, and map matching – instead of just providing a means to store a visual representation in terms of vector graphics.

As for driver assistance systems, such as anti-locking brakes or electronic stability programs, they have up to now been completely dependent on information measured in-car, e.g. obtained by gyroscopes or wheel odometers. However, since the possibilities for passive safety systems (which mitigate the consequences of accidents or driving errors) are basically exploited, active systems (which prevent such situations altogether) are required to improve safety beyond the current level (Stiller, 2005). In order to prevent possibly dangerous situations, such systems must be able to judge the current situation, depending on the vehicle and its environment. Essentially, looking into the distance (the environment) provides the ability to look into the (near) future. For example, adaptive cruise control systems (ACC), mostly based on radar, measure the distance to the car in front. Lane departure warning systems (LDW), based on cameras and image processing, detect and track lanes and alert the driver if these are crossed unintentionally. Whereas ACC can work without any infrastructure, LDW works only if road markings can be reliably detected, which is not the case if the situation is too complicated or they are only partially detectable or not present at all.

Therefore, the use of map data for assistance functions, originally captured for navigation systems, is currently investigated (Blervaque, 2008). Using existing map information has the advantage to (1) provide scene interpretation with suitable background knowledge, (2) amend missing information, caused by missing objects in the real world or failure to detect them, and (3) supply information even beyond the horizon of the car's sensors. However, assistance systems will have requirements beyond today's navigation systems. The 'Enhanced Digital Mapping Project' has classified applications, according to their accuracy requirements, into 'WhatRoad', 'WhichLane', and 'WhereInLane', the latter requiring an accuracy of

±0.2 m for the map (EDMap, 2004). It is concluded that contemporary mapping techniques would be too expensive to provide such maps, even for the 'WhichLane' case.

However, there are nowadays alternatives to traditional map production. *Firstly*, the information required in a map is very dependent on the application. For example, finding the exact position relative to the environment using on-board sensors does not necessarily require a vector map in the usual sense, but rather any features which are stable and can be easily extracted using those sensors. This means that, in order to produce such a map, an interpretation in the classical sense is not required. *Secondly*, measurement systems such as mobile laser scanners are available today, which allow capturing the geometry of the road corridor in great detail, suitable for finding such features (Kukko et al., 2007). *Thirdly*, we are used to compute the relative position (e.g., the position of the vehicle relative to a lane) by an absolute (GPS) measurement and comparison to an absolute geo-referenced map. However, this is not necessary for finding the relative position using relative measurements – in this case, we are relieved from needlessly high (absolute) map accuracy requirements. *Fourthly*, with more and more vehicles using on-board sensors such as cameras and laser scanners, there is the vision that the map can be held up-to-date by feeding this information back to a central server. Similar procedures are in place in the car navigation sector already.

2 Matching and Positioning Based on Overlapping Data

Matching of multiple datasets using overlapping data is a fundamental research topic in photogrammetry, computer vision, and robotics. In the context of 3D scanning, the main interest is usually to register two (and more) independently acquired (3D) point clouds in order to obtain an overall data set in a single coordinate frame. However, since the point clouds are measured relative to a scan head location, after finding the correct transformation of the cloud, the relative orientation of the scan heads is also available. In robotics, on the other hand, there is a strong interest in those standpoints in the first place, in order to track a robot's path. Using the relative orientations, simultaneous localization and mapping (SLAM) can be performed, which incrementally adds all measured data to an overall model (collection of points). In the past, this was often done for mobile robots which were restricted to move on a planar ground, using scanners which operate in a plane parallel to the ground (Thrun et al. 2005). However, it has also been extended to the 3D case (Borrmann et al. 2008).

Closed form solutions are available for finding the required transformation from point correspondences in 2D and 3D (Sansò 1973; Horn 1987). Thus, the main problem is to identify corresponding points in two (or more) point clouds. If an initial alignment is known, which is usually true in case of a scanner on a mobile vehicle, this can be used for establishing correspondences and finding a solution iteratively. Often, the well-known iterative closest point (ICP) algorithm is used for this (Chen & Medioni 1991; Besl & McKay 1992). However, similar to the considerations for the layout of bundle blocks in traditional aerial photogrammetry, a sequence of local matches will bend up quickly. These effects can be reduced by the detection and closing of loops, see e.g. the GraphSLAM algorithm (Thrun et al. 2005). Nevertheless, depending on the size of the loops, additional absolute control will be required.

Such control can be provided by points in the environment with known absolute coordinates. Apparently, it would be unreasonable to provide dense georeferenced point clouds for the entire road network for the sole purpose of positioning a vehicle. Even doing so for a subset, e.g. certain important intersections only, would imply huge amounts of data to be processed and transferred. Thus, instead of iconic matching of point clouds, (symbolic) matching of features, or landmarks, becomes interesting, since they usually allow a very compact representation.

Different geometric features have been proposed in the context of scan matching, like planes, cylinders, spheres and tori (Rabbani et al. 2007). It is important for features to be discriminative, so that map features can be uniquely assigned to scene features. For example, planar faces may be deemed to be good candidates in urban environments. However, it is often the case that the segmentation of the outlines is not very robust so that a single planar patch is not very discriminative (Brenner et al. 2008). Instead, the angles formed by the normal vectors of several planar patches may be used (Brenner & Dold, 2007). However, this requires a suitable constellation of patches and is not a local feature anymore. Several approaches have been proposed to obtain locally unique features, e.g. spin images (Johnson & Hebert 1999) or slippage features (Bokeloh et al. 2008). If features are to be useful for positioning, they should fulfill the following requirements:

- they should be unique in a certain vicinity, either by themselves or in (local) groups,
- their position / orientation should be stable over time,
- few of them should be needed to determine the required transformation with a certain minimum accuracy,

- they should be reliably detectable, given the available sensors and time constraints.

Especially the last point is important in real world applications such as driver assistance systems. Some authors have described scanning systems which use 3D scanners on mobile robots and classify the point cloud in real time, typically using classes such as ground / non-ground and scattered / locally linear (such as wires) / locally planar (such as facades) (Wulf et al. 2004; Lalonde et al. 2006). However, it is probable that it will still take some time until such full 3D scanners are available in vehicles. What can be expected in the near future, though, are scanners which scan in a few planes such as the close-to-production IBEO Lux (IBEO, 2008) scanner, which scans in four planes simultaneously. The Velodyne HDL-64E, used by most finalists of the 2007 DARPA Urban Challenge, scans even in 64 planes simultaneously, but is produced in small numbers only (Velodyne, 2008).

Weiss et al. (2005) describe a system which combines GPS, odometry, and a (four-plane) laser scanner to estimate the position of a vehicle. A feature map is obtained beforehand, which contains the 2D locations of poles of traffic signs and traffic lights. Triangles are defined by selecting triplets of those points, the side lengths of which yield three-dimensional, translation invariant descriptors. Then, when being used, the measurements of a single (horizontal) scan are clustered into segments and segment triples corresponding to feature triples (i.e. matching the descriptor) are searched for.

3 Our Approach and Test Setup

It is our goal to

- explore which features are the most suitable to fulfill the requirements from the list above,
- determine how maps containing such features can be produced reliably and completely automatically,
- augment maps in such a way that for each possible vehicle location, an estimate for the achievable position accuracy, based on the features that can be observed from that location, is available.

In any system where an existing, abstract representation (such as a map or CAD drawing) has to be matched to sensor data, there is a choice as to what kind of abstraction level is used for the actual comparison process. We assume an asymmetry in this process, since the production of maps may use sensors with a high density, accuracy and the possibility for full 3D scans. In contrast, sensors in vehicles may be very low cost, low accuracy and density, and may scan only in a single plane. Therefore, it is important not to burden the online system with too much requirements regarding scene interpretation. After all, the online system is not required to produce a scene reconstruction, but only to match the data with an already existing scene description. On the other hand, the system for map production has practically no computation time constraints and the main objective is to keep processing fully automatic. It is better to produce more data and to consume more computing time, if this guarantees a fully automatic and reliable extraction of the required features in turn.

Fig. 1. The Streetmapper system used to acquire the Hannover scan. Three of the four scanners, the GPS antenna, and the odometer (mounted close to the bumper) can be discerned. The fourth scanner pointing to the left is occluded.

We obtained dense laser scans of a set of roads in Hannover, Germany, acquired by the Steetmapper mobile mapping system, jointly developed by 3D laser mapping Ltd., UK, and IGI mbH, Germany (Kremer & Hunter, 2007). The scan was acquired in November 2007 with a configuration of four scanners, as shown in Fig. 1. Two scanners Riegl LMS-Q120 were pointing up and down at a 20° angle, one is pointing to the right at a 45° angle. Another Riegl LMS-Q140 was pointing to the left at a 45° angle. The LMS-Q120 has a maximum range of 150 m and a ranging accuracy of 25 mm. All scanners were operated simultaneously at the maximum scanning angle of 80° and scanning rate of about 10.000 points/s. Positioning was accomplished using IGI's TERRAcontrol GNSS/IMU system which consists of a NovAtel GNSS receiver, IGI's IMU-IId fiber optic gyro IMU operating at 256 Hz, an odometer, and a control computer which records all data on a PC card for later post processing.

Postprocessing yields the trajectory and, through the calibrated relative orientations of the scanners and the time synchronization, a georeferenced point cloud. As in the aerial case, absolute point accuracy is mostly dependent on the performance of the GNSS/IMU system rather than on the ranging accuracy of the scanners. Single point absolute accuracies of 10 cm and better can be expected under good conditions, however, in cities, with prolonged occlusions from trees and buildings, it can be 1 m and worse. It is important to note that even in this case, the relative accuracy of the points will still be around a few centimeters. As mentioned in the introduction, the relative accuracy is actually the crucial parameter for the purpose of relative positioning.

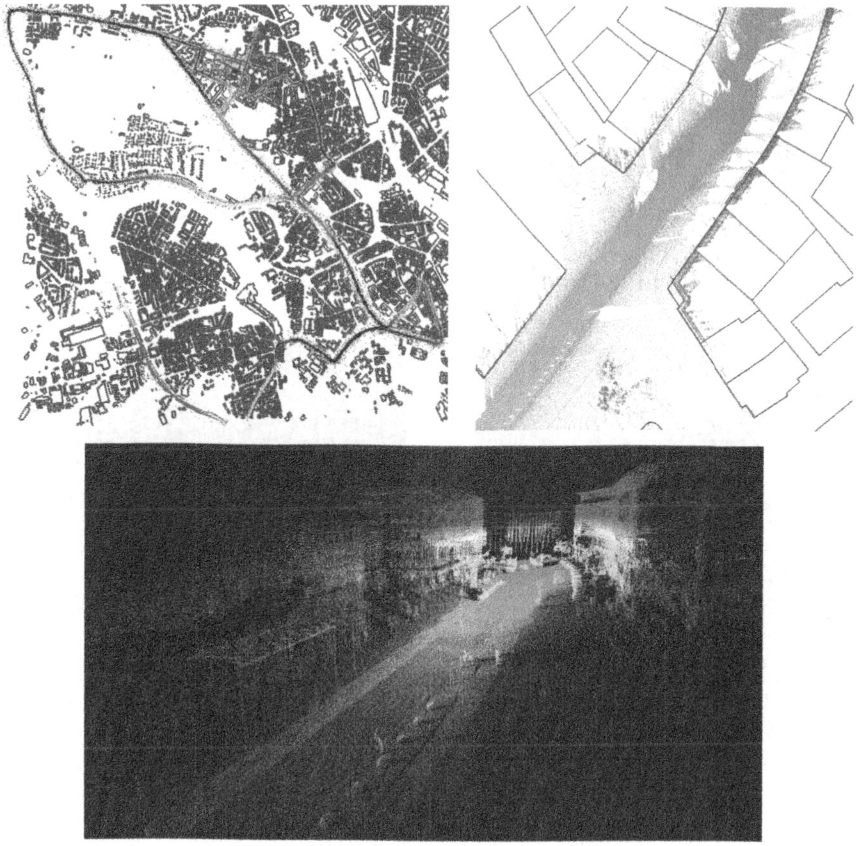

Fig. 2. Top left: entire scan trajectory with ground plans from cadastral map overlaid. Top right: detail, top view. Point color encodes point height using a temperature scale. Bottom: the same detail, shown as 3D view.

Fig. 2 shows an overview and the typical results which can be expected at an inner city street. Note that the scanned area contains streets in densely built up regions as well as highway like roads. It can be seen from Fig. 2 (top right) that the scanned facades (red points in the scan) fit quite well to the ground plan polygons from the cadastral map (magenta), indicating that there were no major errors in GPS/IMU positioning.

The total length of the scanned roads is 21.7 kilometers, captured in 48 minutes, which is an average speed of 27 km/h. During that time, 70.7 million points were captured, corresponding to an effective measurement rate of 24,500 point per second. On average, each road meter is covered by more than 3,200 points.

4 Extraction of Poles

The extraction of poles can be seen as a special case of cylinder extraction, which, as noted in Sec. 2, has been used for scan matching. However, although the 3D point cloud obtained by the Streetmapper system is huge, single poles are not hit by very many points (Fig. 3). Thus, methods which rely on the extraction of the surface, of surface normal vectors, or even of curvature, are not applicable. Also, horizontal slicing to detect circular structures, as used in Luo & Wang (2008), will not work.

Fig. 3. Examples for poles: a traffic light, a street light and two signs, top view (left) and 3D view (right). In the 3D view, two bicyclists can be recognized, standing close to the highest pole.

Fig. 4. Analysis of the local scatter matrix. Temperature scale, red indicates a linear, blue a round or flat structure.

An intuitive approach for the extraction of points with a certain structure in the neighborhood is the evaluation of the local scatter matrix. If eigenvalue analysis yields one large and two small eigenvalues, this indicates a (thin) linear structure. While this works nicely in general (Fig. 4), additional structures mounted at the poles, such as the traffic lights and signs, but also people standing close to the poles, will distort it. Therefore, linking all points classified as being linear to lines may be nontrivial.

Due to this, another model-based approach was investigated. It assumes that the basic characteristic of a pole is that it is upright, there is a kernel region where laser scan points are present, and an outside region where no points are present (Fig. 5). The structure is analyzed in cylindrical stacks. A pole is assumed when a certain minimum number of stacked cylinders is present. In order to be sure that the outside region of each cylinder is empty and not just accidentally hit by no laser ray, an additional ray density analysis is performed. After the stack is identified, the points in the kernel are used for an estimation of the exact position of the pole.

For the entire 22 km scene, a total of 2600 poles was found fully automatically, which is one pole every 8 meters on average. In terms of data reduction, this is one extracted pole per 27,000 original scan points. Although the current implementation is not optimized, processing time is uncritical and yields several poles per second on a standard PC. Of course, poles are not distributed uniformly, and from Fig. 6 it can be seen that there are too few along the highway in the lower left corner. In fact, upon closer examination, those turn out to be trees along the border of a small forest and it will be very dependent on the characteristics of the in-car sensor if they can be used for positioning.

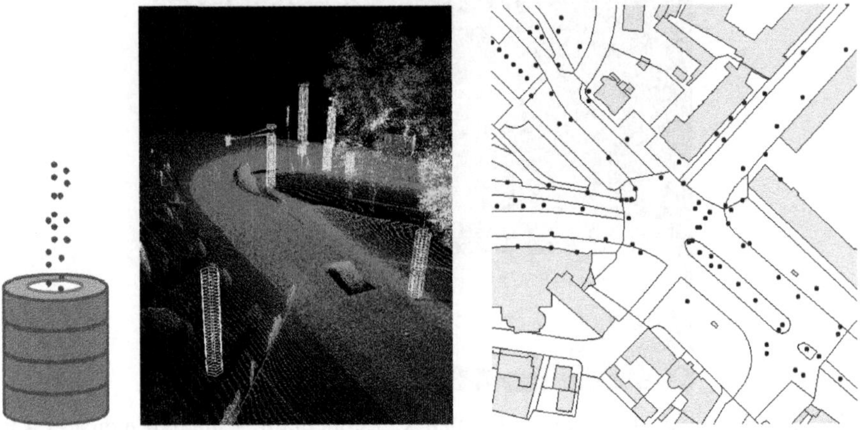

Fig. 5. Analysis in terms of cylindrical stacks. Left: measured laser points in the kernel region (white), no points in the outside region (blue). Middle: identified stacks (white) for the scene in Figs. 3 & 4. Right: Extracted poles (blue points) for the intersection "Königsworther Platz" in Hannover, Germany, with cadastral map overlaid.

Fig. 6. Extracted poles (blue dots) in a larger area with buildings from the cadastral map (green). The thin dotted line is the trajectory of the Streetmapper van.

5 Positioning Based on the Matching of Poles

As a first step towards the goal to augment a map with information regarding the expected positioning accuracy, the following simulation was used. Positions along the Streetmapper trajectory were selected at 10 m distance, yielding 2141 positions. For each of those positions, all visible poles were selected according to a model of the in-car sensor (discussed below). Assuming measurement errors for range and angle as well as inaccurate pole positions, the accuracy for the car positions is derived by error propagation. It is important to note that this simulation covers only an assessment of the achievable accuracy and does not deal with the problem of scene uniqueness and (efficient) assignment of extracted poles to the database, which is a topic on its own.

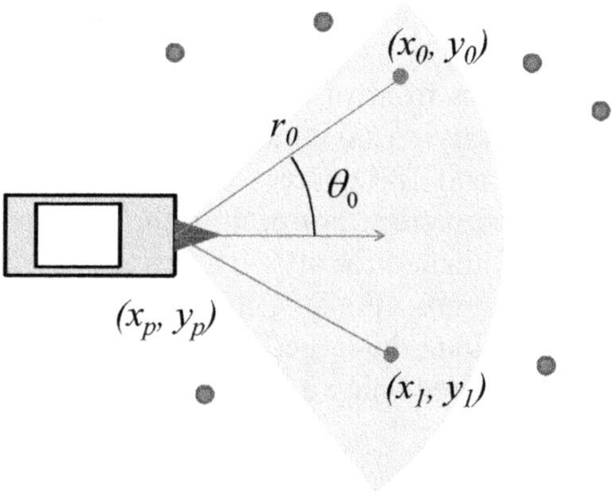

Fig. 7. Sketch of the measurement model. The car observes poles (green) according to its measurement horizon (here, the light green cone). For each pole, four observation equations are set up.

The observation equations are (cf. Fig. 7):

$$
\begin{aligned}
r_i + v_{r_i} &= \left((x_i - x_p)^2 + (y_i - y_p)^2\right)^{1/2} \\
\theta_i + v_{\theta_i} &= \mathrm{atan2}(y_i - y_p, x_i - x_p) - \theta_p
\end{aligned}
$$

where r_i are the measured distances, θ_i are the measured angles, (x_p, y_p) is the car position, θ_p the car orientation, and (x_i, y_i) is the position of the ith pole. All v's are corrections to be minimized. This is the standard polar measurement model which is also popular in robotics (Thrun, 2005). In addition, since the position of the poles (observed beforehand) is also subject to measurement errors, they are introduced in a Gauss-Markoff model as observed unknowns

$$x_{i,0} + v_{x_{i,0}} = x_i$$

$$y_{i,0} + v_{y_{i,0}} = y_i$$

where $(x_{i,0}, y_{i,0})$ are the positions of the poles as extracted by the method described in the previous section. For m observed poles, this leads to 3+2m unknowns and 4m observation equations. Weights are computed from assumed standard deviations σ_r, σ_θ of the in-car sensor and σ_x, σ_y for the extracted poles. Finally, by error propagation, the submatrix of variances and covariances for the car position (x_p, y_p) is obtained. It is visualized below in the form of error ellipses in the plane. Also, the length of the larger axis is used as a scalar measure (which is the standard deviation in this direction).

For the simulations, σ_r = 5 cm, σ_θ = 1°, and $\sigma_x = \sigma_y = 10$ cm was selected. A sensor placed in front of the car was assumed, pointing in the driving direction, with varying (symmetric) opening angle of 20, 40, 60, 80, 100, 120, 140, 160 and 180 degrees and a measurement range of 25, 50, 75, 100, 125 and 150 meters. For each of the 2141 positions, the positioning accuracy was evaluated for all sensor characteristics, i.e., all of the opening angle / range combinations. If the position was undetermined or the standard deviation (along the larger axis of the error ellipse) was larger than 1 m, this was considered being a failure.

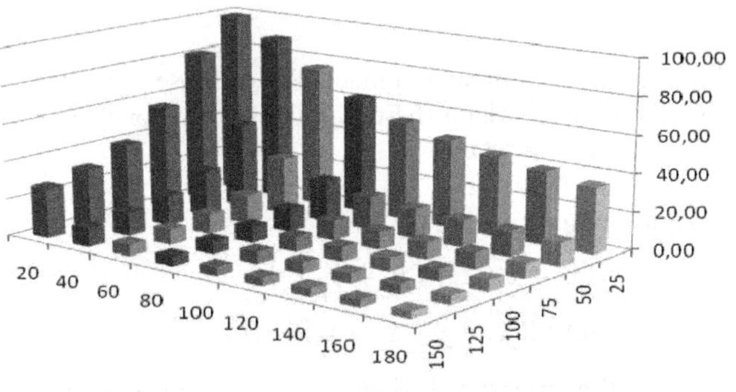

Fig. 8. Percentage of failures to determine the position within 1 m for opening angles of 20° to 180° and measurement ranges from 25 m to 150 m.

The percentage of failures (relative to the total number of 2141 positions) was evaluated (Fig. 8). It can be seen that an opening angle of 20° and a measurement range of only 25 m leads to almost 100% failures. However, with increasing opening angle and measurement ranges this decreases quickly. A 'diagonal' tradeoff between opening angle and range can be observed, for example approximately 10% failures are obtained for 40° / 150 m range, 60° / 100 m range, and 100° / 75 m range. At 150 m range, failures are below 5% for opening angles of 80° or larger.

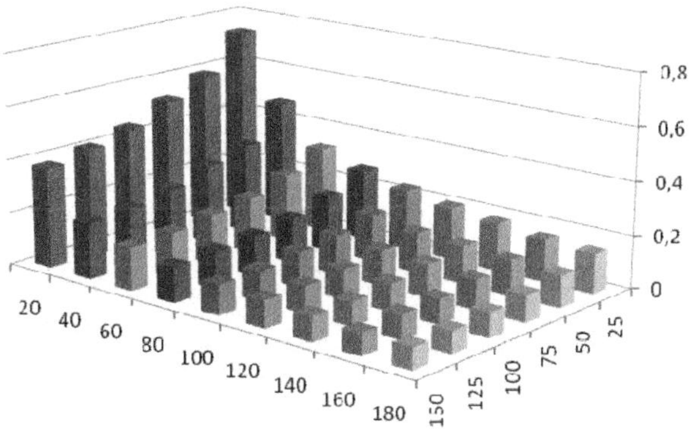

Fig. 9. Average of the positioning accuracy for positions within 1 m accuracy (i.e., positions which did not fail).

The positioning accuracy was evaluated by computing the average of the standard deviation, over all positions which did not fail, i.e., were within a 1 m standard deviation (Fig. 9). Again, an opening angle of 20° is obviously too narrow, with accuracies from 73 cm (at 25 m range) down to 38 cm (at 150 m range). However, for opening angles of 80° and more and ranges of 50 m and more, the error is below 21 cm. At 120° / 100 m, the 10 cm level is reached, which cannot be improved substantially by an increase in angle or range (note 10 cm was also the assumed measurement accuracy for a single pole).

Analyzing the error on a per-point basis, Fig. 10 shows the error ellipses along an inner city street for varying opening angles of 20°, 100° and 180°. Note that the ellipses are scaled by a factor of 100 and in fact the larger axis of any shown ellipse is shorter than one meter. One can see that an opening angle of 20° leads to large errors, whereas the difference between 100° and 180° seems not significant.

All other images show the case of 100° opening angle and varying ranges of 50 m, 100 m, and 150 m. From Fig. 11, one can conclude that the range is not significant in an inner city scene, where lots of poles are present. However, as Fig. 12 shows, in residential areas, the density of poles may be quite low so that a 50 m range will not be enough. However, it can be expected that in this case, other features like extracted façade planes will provide additional information.

Finally, the highway scene in Fig. 13 shows how positioning fails if there are no poles in the scene and how, upon approaching poles, the 150 m range sensor finds a solution first (large blue ellipses), followed by the 100 m and 50 m range sensors.

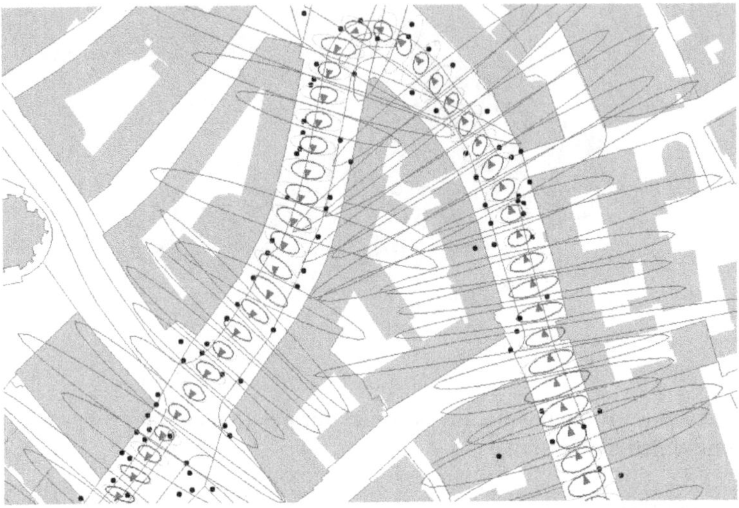

Fig. 10. Error ellipses for a scanner range of 75 m and opening angles of 20° (red), 100° (green), and 180° (blue), for an inner city street. All ellipses are scaled by a factor of 100.

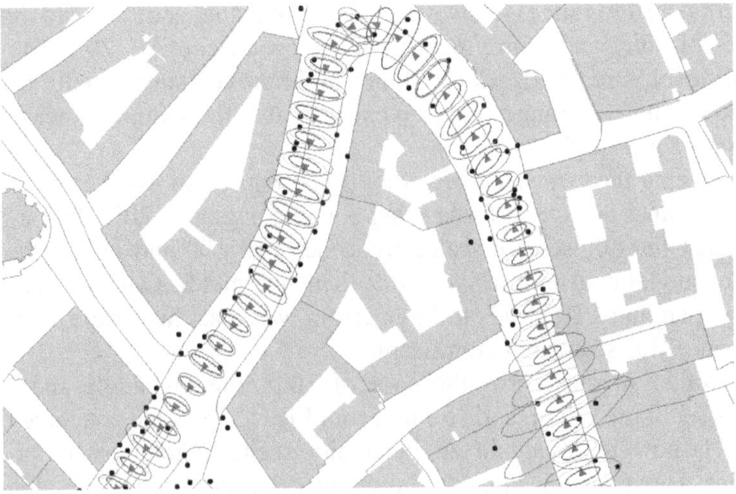

Fig. 11. Same scene as Fig. 10, for a scanner opening angle of 100° and a scan range of 50 m (red), 100 m (green), and 150 m (blue).

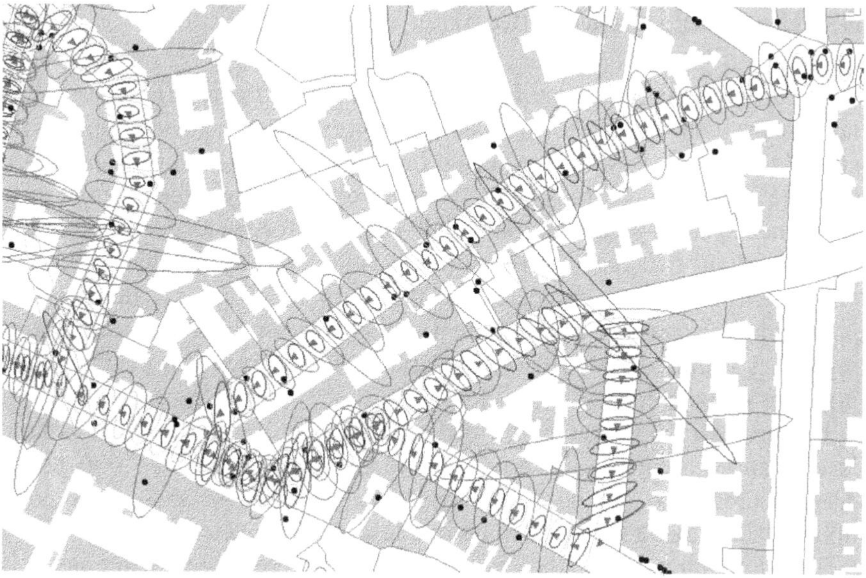

Fig. 12. Residential area, for a scanner opening angle of 100° and a scan range of 50 m (red), 100 m (green), and 150 m (blue).

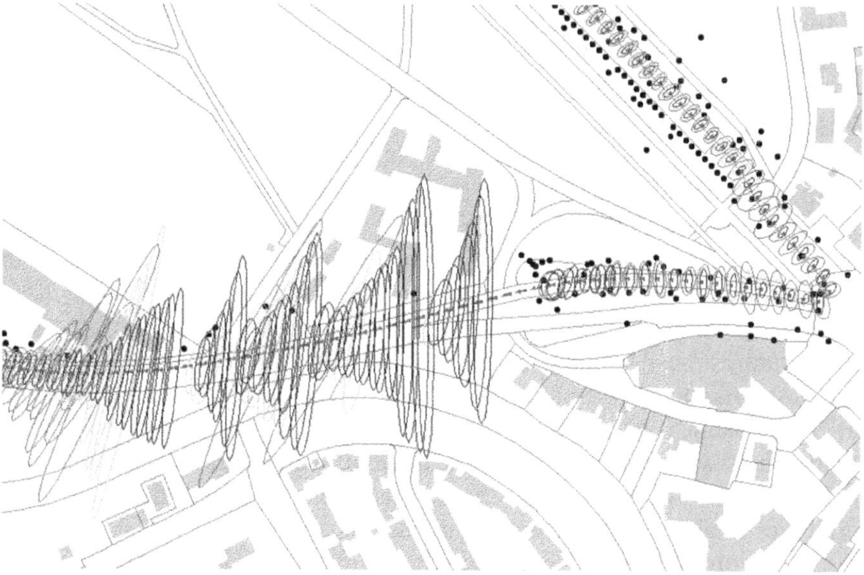

Fig. 13. Change from an inner city scene (right) to a highway scene. Scanner opening angle of 100°, scan range of 50 m (red), 100 m (green), and 150 m (blue).

6 Discussion and Outlook

This paper investigates how features, which are extracted from terrestrial laser scans, can be used for the task of positioning vehicles equipped with suitable range sensors. It is expected that such features will play an essential role in next generation maps, which are especially relevant for driver assistance systems and autonomous driving. As for the feature extraction, it is important that it is carried out fully automatically. In this paper, the extraction of poles was selected for that purpose. To assess the possible positioning accuracy, a measurement model was set up using assumptions regarding a future in-car sensor. The attainable accuracy was evaluated using error propagation.

Feature extraction was performed on a 22 km, 70 million point mobile laser scan of a part of Hannover, Germany, yielding about 2,600 poles. Then, a simulation was used to compute the accuracy for about 2100 positions selected along the original mobile laser scan trajectory.

The simulations show, that in general, relative positioning using poles is feasible and will lead to high positioning accuracies, although the constellation of poles and the exact in-vehicle sensor properties play a large role. Pole constellations of inner city, residential and highway scenes were considered. The influence of different sensor horizons was studied as well by simulation.

For various reasons, the results obtained here will be too optimistic. First, pole extraction still finds some structures which aren't really poles. Even if they are, it was assumed that 100% of them will be detectable by an in-car sensor as long as they are within the sensor's range. Second, visibility has been neglected so far, so that all poles within the range of the in-car sensor were used, even if they are not visible from a certain position. In reality, apart from fixed objects, there will be occlusions from other moving objects. Even for static scenes, this is actually an intricate problem, since future in-car scanners will probably be forward-looking, while mobile laser mapping systems often use different configurations (for example, all four scanners of the Streetmapper system looked backwards). Third, ambiguities were not considered, i.e. it was assumed that the assignment of extracted poles to the poles in the map is correct. Ambiguities may especially result along alleys with trees planted at constant spacing.

On the other hand, in some respect the results are too pessimistic. For example, individual positions were considered, whereas in practice, the position of a vehicle would be tracked (filtered) using additional in-car sensors such as a gyroscope and an odometer. Since tracking is quite stable at relatively short distances, this will have a similar effect as taking a larger set of poles into account, effectively going from a 'visibility cone' to a 'corridor' model. Also, it is obvious that the lateral errors (across the driving direction) could be reduced in many cases by taking additional features into account, especially extracted façade planes.

Our future work will be to extract and use more (different) features, investigate the role of ambiguities and visibility, and to carry out more simulations as well as tests with real in-car sensors.

Acknowledgements

This work has been supported by the VolkswagenStiftung, Germany.

References

Arras, K.O. und Siegwart, R.Y. (1997) Feature Extraction and Scene Interpretation for Map-Based Navigation and Map Building. In: Proc. SPIE, Mobile Robotics XII, volume 3210, pp. 4253-4264.

Besl, P.J., McKay, N.D. (1992) A method for registration of 3-D shapes. IEEE Transactions on Pattern Analysis and Machine Intelligence 14 (2), pp. 239–256.

Blervaque, V. (2008) Prevent Maps & ADAS D12.1 Final Report, ERTICO – ITS Europe, http://prevent-ip.org/ download/ deliverables/ MAPS&ADAS/ PR-12100-DEL-080127-v10-ERT-D12.1 MAPS&ADAS Final Report.pdf, 35p. Last accessed Dec. 8, 2008.

Bokeloh, M., Berner, A. Wand, M., Seidel, H.-P., Schilling, A. (2008) Slippage Features, Tech. Rep. WSI-2008-03, Graphisch-Interaktive Systeme, Wilhelm-Schickard-Institut, Universität Tübingen, 21 p.

Borrmann, D., Elseberg, J., Lingemann, K., Nüchter, A., Hertzberg, J. (2008) Globally consistent 3D mapping with scan matching. Journal of Robotics and Autonomous Systems (JRAS), Vol. 56, Issue 2, pp. 130-142.

Brenner, C., Dold, C. (2007) Automatic relative orientation of terrestrial laser scans using planar structures and angle constraints, Proc. Laser Scanning 2007 and SilviLaser 2007, IAPRS Vol. 36 Part 3/W52, pp. 84-89.

Brenner, C., Dold, C., Ripperda, N. (2008) Coarse orientation of terrestrial laser scans in urban environments. ISPRS Journal of Photogrammetry and Remote Sensing 63 (1), pp. 4-18.

Chen, Y., Medioni, G., (1991) Object Modeling by Registration of Multiple Range Images. International Conference on Robotics and Automation, Vol. 3, pp. 2724–2729.

EDMap (2004), Enhanced Digital Mapping Project Final Report, Technical report, United States Department of Transportation, Federal Highway Administration and National Highway Traffic and Safety Administration, http://www-nrd.nhtsa.dot.gov/pdf/nrd-12/CAMP/EDMap%20Final%20Report/Main%20Report/FinalRept_111904.pdf, 189p. Last accessed Dec. 8, 2008.

Horn, B.K.P., 1987. Closed-form solution of absolute orientation using unit quaternions. Journal of the Optical Society of America 4 (4), pp. 629–642.

IBEO (2008) Ibeo Lux laser scanner, www.ibeo-as.de. Last accessed Dec. 20, 2008.

Johnson, A., Hebert, M., (1999) Using Spin Images for Efficient Object Recognition in Cluttered 3D Scenes. IEEE Trans. PAMI 21(5), pp. 433-449.

Kremer, J., Hunter, G. (2007) Performance of the Streetmapper Mobile LIDAR Mapping System in 'Real World' Projects. In: Photogrammetric Week 2007, Wichmann, pp. 215-225.

Kukko, A., Andrei, C.-O., Salminen, V.-M., Kaartinen, H., Chen, Y., Rönnholm, P. Hyyppä, H., Hyyppä, J., Chen, R. Haggrén, H., Kosonen, I., Čapek, K. (2007) Road environment mapping system oft he finnish geodetic institute – FGI Roamer, Proc. Laser Scanning 2007 and SilviLaser 2007, IAPRS Vol. 36 Part 3/W52, pp. 241-247.

Lalonde, J.-F., Vandapel, N., Huber, D. F., Hebert, M. (2006) Natural Terrain Classification using Three-Dimensional Ladar Data for Ground Robot Mobility. Journal of field robotics 23(10), pp. 839-861.

Luo, D., Wang, Y. (2008) Rapid extracting pillars by slicing point clouds. Proc. XXI ISPRS Congress, IAPRS Vol. 37 Part B3b, pp. 215-218.

Rabbani, T., Dijkman, S., van den Heuvel, F., Vosselman, G. (2007) An integrated approach for modelling and global registration of point clouds. ISPRS Journal of Photogrammetry and Remote Sensing 61 (6), pp. 355–370.

Sansò, F., (1973) An exact solution of the roto-translation problem. Photogrammetria 29 (6), pp. 203–216.

Stiller, C. (2005) Fahrerassistenzsysteme – Von realisierten Funktionen zum vernetzt wahrnehmenden, selbstorganisierenden Verkehr, in: M. Maurer & C. Stiller, Eds., 'Fahrerassistenzsysteme mit maschineller Wahrnehmung', Springer Berlin Heidelberg New York, pp. 1–20.

Thrun, S., Burgard, W., Fox, D. (2005) Probabilistic Robotics, The MIT Press, Cambridge, Mass.

Velodyne (2008). Velodyne scanner, www.velodyne.com/lidar. Last accessed Dec. 10, 2008.

Weiss, T., Kaempchen, N., Dietmayer, K. (2005) Precise ego localization in urban areas using lasedrscanner and high accuracy feature maps. Proc. 2005 IEEE Intelligent Vehicles Symposium, Las Vegas, USA, pp. 284-289.

Wulf, O., Brenneke, C., Wagner, B. (2004) Colored 2D Maps for Robot Navigation with 3D Sensor Data. Proc. IEEE/RSJ Int. Conf. on intelligent Robots and Systems, Sendai, Japan, pp. 2991-2996.

Automatic Revision of 2D Building Databases from High Resolution Satellite Imagery: A 3D Photogrammetric Approach

Nicolas Champion[1] [a,b], Georges Stamon[b], Marc Pierrot-Deseilligny[a]

[a] MATIS – Institut Géographique National, 2-4 avenue Pasteur, 94160 Saint Mandé, France – firstname.lastname@ign.fr
[b] Université Paris Descartes – CRIP5-SIP, 45 rue des Saints-Pères, 75006 Paris, France – firstname.lastname@math-info.univ-paris5.fr

Abstract. Among all the issues involved in Geographic Information Science, automating the update of 2D building databases is a crucial and challenging issue. Such an update usually starts out with a manual change detection process. The main goal of this paper is to present a new method to automate the detection of changes in a 2D building database, starting from satellite images. The workflow of our approach is divided into 2 phases. Primitives, extracted from multiple images or from a correlation Digital Surface Model (DSM), are firstly collected for each building and matched with primitives derived from the existing database to achieve a final decision about acceptance or rejection. A specific algorithm, based on the DSM and a computed Digital Terrain Model (DTM), is subsequently used to extract new buildings. The method is here introduced and tested in two test areas, very different regarding the land use and topography. The outcomes of the method are assessed and show the good performance of our system, especially in terms of completeness, robustness and transferability.

[1] The first author would like to thank the French Space Agency (CNES) for providing Pléiades Simulations

M. Sester et al. (eds.), *Advances in GIScience*, Lecture Notes in Geoinformation and Cartography, DOI 10.1007/978-3-642-00318-9_3, © Springer-Verlag Berlin Heidelberg 2009

1 Introduction

Traditionally, the mapping process is carried out in National Mapping Agencies (NMAs) and comes in the form of a linear workflow. As stated in (Heipke et al., 2008), it is composed of 4 separate stages. Once the database specifications defined, objects of interest (buildings, roads, etc.) are captured by operators on the field (with GPS or from airborne images) then processed and stored in a Geographic Information system (GIS) as a vector database. At the end of the process, the question that immediately arises concerns the update of the database, more specifically the strategy to use for that purpose. This question is a very topical issue in developed countries, as topographic databases, in particular 2D building databases, have been completed during the last decade.

1.1 Map Update Strategies

As shown in Heipke et al. (2008), two strategies can be considered for this update. The first strategy – the participatory way – consists in collecting the information about changes from other public agencies (e.g. town councils, cadastre) and other contracted bodies (e.g. real estate agencies). Even general public's service can be called upon. This last solution – the collaborative way – is considered in an increasing number of Web 2.0 applications, such as the OpenStreetMap[2], Google Map Maker[3] and WikiMapia[4] projects, or the Map Insight[5] tool, developed by TeleAtlas. The second strategy – the centralised way – consists in collecting the information about changes from external data such as airborne images or laser scanning data. This strategy also consists in comparing the existing database to more recent data in order to detect new, demolished or modified objects (Figure 1) in the scene.

The participatory/collaborative way is appealing and could lead to promising applications. The main limitations of this update strategy are related to the inevitable remaining errors in the database that require an ultimate ground truth in order to be eliminated. The centralised way may make up for this deficiency, by introducing a ground truth (more recent external data). The main issue here concerns the change detection step, which is known to be long and costly, when carried out manually Steinnocher and

---

[2] www.openstreetmap.org. Last date accessed 12/12/08
[3] www.google.com/mapmaker. Last date accessed 12/12/08
[4] http://wikimapia.org. Last date accessed 12/12/08
[5] http://mapinsight.teleatlas.com. Last date accessed 12/12/08

Kressler (2006). In addition, update cycles tend to be shorter (from 10 to for instance 3 years), according to a new trend in users' needs. As a consequence, there is a high necessity to introduce a certain degree of automation in the change detection step e.g. to develop expert systems that are able to send an alarm to an operator when a change is detected between the database and more recent external data.

(1) (2) (3)

Fig. 1. Update Problems. (1): Demolished buildings. (2): New buildings. (3): Modified (misregistered) buildings.

Among all possible external data, images – in particular satellite images – have qualitatively advanced to a point where they become interesting input to solve such change detection issues. Thus, most satellite systems offer now a high spatial resolution (the panchromatic resolution is 1m for Ikonos, 60 cm for Quickbird, 50cm for Worldview-I, 50cm for the Pléiades-HR system that shall be ready for launch in early 2010) and have already been used in previous works for mapping purposes, for instance in (Lafarge et al., 2008) for the automatic production of 3D city models. Moreover, satellite systems have had their reactivity considerably upgraded during the last decade and are now capable to acquire a considerable amount of information in a relatively short amount of time, which evidently complies with the reactivity required in a change detection procedure.

1.2 Related Works

Most of the change detection methods found in literature – (Vögtle and Steinle, 2004), (Steinnocher and Kressler, 2006), (Matikainen et al., 2007), (Olsen and Knudsen, 2005), (Rottensteiner, 2008) – are based on a preliminary land cover classification of the scene: buildings are firstly detected in input data by means of geometric, photometric or structural features, and subsequently compared to the outdated database in order to detect changes in the scene.

Thus, Matikainen et al. (2007) proposes to perform a building detection from First and Last Pulse (FP/LP) laser scanning data and aerial colour orthophotos[6]. The DSM, computed from the FP data, is firstly segmented and classified into "ground" and "buildings or tree" objects, according to geometric considerations. Buildings are then separated from trees through a classification tree based approach (Breiman et al., 1984) and are lastly compared to the existing map in order to find out which buildings have been changed, deleted or constructed. Beyond the feasibility to use such a classification-based method in a change building detection process, this paper puts the emphasis on the good performance and the high level of automation of the approach.

Olsen and Knudsen (2005) propose to break the change detection workflow into two related sub-tasks: the verification of existing registrations on the one hand, the estimation of the position of new buildings on the other hand. The proposed approach actually deals with an image-based classification process. CIR (RGB/IR) images are segmented and classified into "building" and "non building" (training sets are derived from the existing map). A rule-based system allows to achieve a final decision about the change status: an existing building is confirmed if it is verified in at least one image (*soft decision*). New buildings are accepted if detected in both RGB and IR aerial images (*hard decision*). This method gives good results, provided roof materials (used in the scene) exist in the training sets and the number of demolished buildings be small with respect to the number of buildings in the scene.

Rottensteiner (2008) proposes to split the change detection procedure into three stages. First, a Dempster-Shafer fusion process is carried out on a per-pixel basis and results in a classification of the input data – here CIR images and a DSM – into one of four predefined classes: buildings, trees, grass land and bare soil. Connected components of building pixels are then grouped to constitute initial building regions and a second Dempster-Shafer fusion process is carried out on a per-region basis to eliminate regions corresponding to trees. The third stage corresponds to the actual change detection process, where the detected buildings are compared to the existing map, which results in a very detailed change map.

Eventually, 2 noteworthy projects are related to change detection and update procedures: the ATOMI project (Switzerland) (Niederöst, 2000), (Baltsavias, 2004) and the WIPKA project (Germany). They are the fruit of a productive cooperation between academic research institutions and

[6] i.e. images geocoded in the same geographic reference as the database to update

National Mapping Agencies. The WIPKA project aims at automatically verifying a topographic database (ATKIS). A specific algorithm is used to update roads and a knowledge-driven approach (Busch et al., 2004) is used to automatically verify area objects contained in the dataset. This project is all the more interesting as it outlines the requirements that a change detection process has to fulfill to be efficient: firstly, the process must be interactive, as it appears to be an appropriate trade-off between fastidious manual work and erroneous fully automatic work. Secondly, a change detection process must be data-dependent i.e. the system design is built with respect to the specifications of the database to check.

1.3 Overview of the Method

The main goal of this paper is to present a new approach in order to automatically detect changes in a 2D building vector database, starting from more recent satellite images. The database to update corresponds to a cadastral map and represents the 2D footprints of buildings. This cadastral map has two special features that are later considered to build our system: on the one hand, buildings are captured manually in the field and their limits are also given by the limits of walls (and not gutters like in satellite images); on the other hand, the building outlines actually correspond to the intangible limits between ownerships (this database is initially built for tax purposes): when shared by 2 different homeowners, one physical building is systematically split into 2 objects in the database.

Fig. 2. Left: a triplet of Pléiades-HR images (simulations). Right: Zoom in. Pléiades-HR Simulations (top) are computed from 25cm airborne images (bottom).

The input of our method is given by recent tri-stereoscopic CIR Pléiades-HR images (Figure 2), with a base-to-height ratio of 0.15 and a spatial resolution (Ground Sample Distance – GSD) of 50cm. These satellite images actually correspond to simulations, obtained by applying Modulation Transfer Functions (MTF) and Signal to Noise Ratio (SNR) to high resolution (25cm) airborne images, as described in (Cantou et al., 2006) and illustrated in Figure 2. A correlation DSM[7] is computed automatically with a stereo-matching algorithm (Pierrot-Deseilligny and Paparoditis, 2006), based on the 2D minimisation of a cost that considers discontinuities and radiometric similarities (image matching scores). Note that the DSM is given in the form of a matrix (regular grid) of height values (raster model). CIR and NDVI (Normalized Difference Vegetation Index) orthophotos are then computed and a vegetation mask is easily obtained by applying a threshold to NDVI orthophotos (this index is high for vegetation because vegetation absorbs visible radiation and reflects infrared radiation). The output of the method is a change map, in which buildings are labeled unchanged, demolished, modified or new.

The method described here is meant to be an alternative to the previously described methods, mostly based on a preliminary land use classification and is based on two key ideas. The first one takes up (Busch et al., 2004) and consists in splitting the change detection procedure (known to be difficult to solve) into two subtasks easier to solve, the verification of the database (phase I) and the detection of new buildings (phase II). The second one consists in using massively 3D geometric primitives in the system.

2 Automatic Verification of the Database (Phase I)

This first phase implements a knowledge-driven approach: the key idea here is to use the prior knowledge provided by the database about the geometric description of buildings and to collect (2D & 3D) information from input sources in order to confirm or not their existence in the database.

[7] A Digital Surface Model (DSM) is a 2.5D representation of the *earth* surface, including any above-ground objects, such as trees and human settlements. A Digital Terrain Model (DTM) is a 2.5D representation of the *terrain* (*bare ground*) surface.

2.1 General Workflow

As illustrated in Figure 3, geometric linear primitives are first extracted from input data (2.2). For each object (building) to verify, the primitives that fulfill specific constraints are selected and used to compute a per-building similarity score (2.3). A final decision about acceptance or rejection is then achieved through a dedicated threshold-based module (2.4). This first phase also leads to a partially updated database, in which buildings are labeled demolished, modified or unchanged.

2.2 Extraction of Primitives

The performance of this phase is obviously related to the pertinence of primitives extracted from input data. Two kinds of primitives are used in our work, firstly 2D contours and secondly 3D segments.

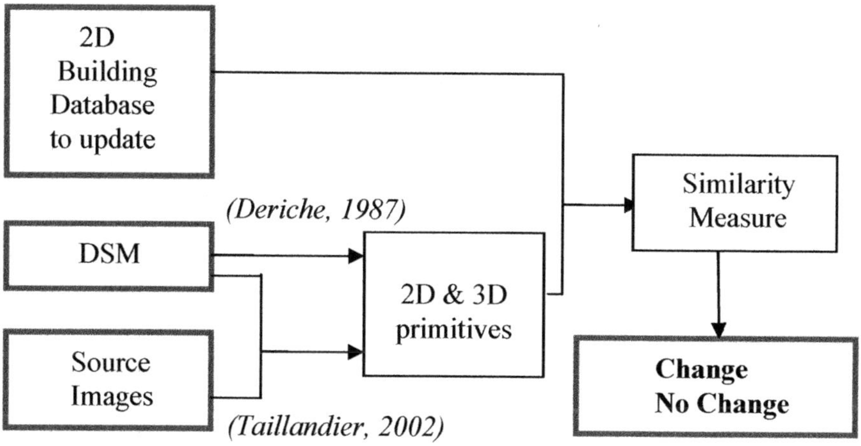

Fig. 3. Workflow of Phase I (Automatic verification of the database).

2.2.1 Extraction of 2D Primitives

The classical gradient operator (Deriche, 1987) is firstly applied to the DSM and a hysteresis detector is used to extract the local maxima in the direction of gradients, which are then chained, polygonized and deliver the 2D contours. These primitives also correspond to the height discontinuities in the DSM. As the DSM is not surprisingly less accurate in shadow and vegetated areas, due to classical drawbacks of stereo-matching algorithms, the 2D contours are less reliable at such locations (Figure 5) and other features consequently need to be extracted.

2.2.2 Extraction of 3D Segments

3D segments are thus injected in our system. They are generated directly from multiple satellite images with (Taillandier, 2002). As sketched in Figure 4, this algorithm can be divided into 4 stages. 2D segments are firstly extracted from source images (stage 1) with (Deriche, 1987). A set of so-called associations (i.e. possible correspondences) is subsequently determined (stage 2), by matching 2D segments from the Object Space through the Sweep Plane technique (Collins, 1995): an association also corresponds to a set of 2D segments (in images), whose projections onto a plane ($z = z_k$) intersect each other in at least one voxel (x, y, z_k) of the Object Space. The set of associations is then pruned (stage 3), firstly under a geometric constraint, secondly under a unicity constraint. Filtered associations are lastly used to reconstruct the 3D segments (stage 4). More details can be found in (Taillandier, 2002).

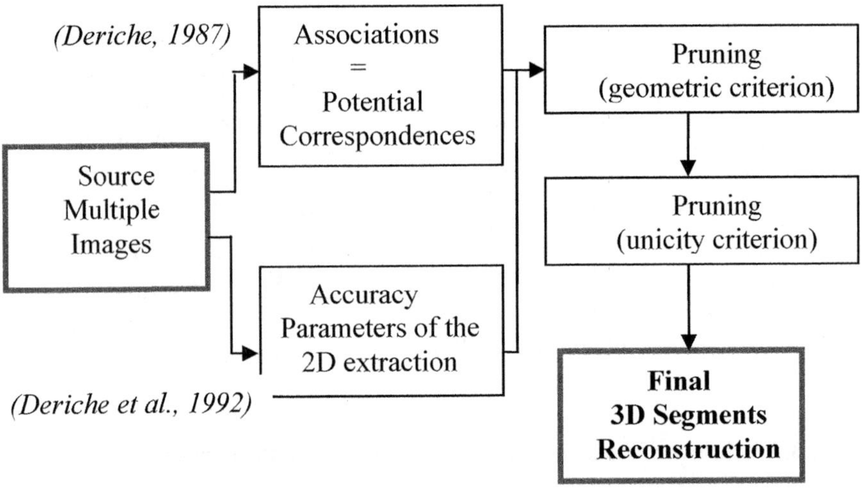

Fig. 4. Workflow of the reconstruction of 3D segments (Taillandier, 2002).

As illustrated in Figure 5, these 3D segments are particularly interesting as they are shown to be accurate in planimetry: it is related to the fact the uncertainty on the determination of image 2D segments is considered during the reconstruction of 3D segments (Deriche et al., 1992). They also represent a good caricature of buildings in an urban environment (they are used as input data in a 3D city modeler).

Fig. 5. 2D contours (in white, left) and 3D primitives (in white, right) are superimposed on the RGB orthophoto. Contrary to 2D contours, the extraction of 3D segments is reliable, even in challenging vegetated areas.

2.3 Matching

2.3.1 Selection of Pertinent Database Primitives

As previously mentioned, the building boundaries correspond to the limits of ownerships in the database. Therefore, as shown in Figure 6, boundaries are split into 2 categories: inner boundaries (i.e. shared by 2 buildings) correspond to the intangible limit between 2 ownerships and only seldom correspond to a physical (height or radiometric) discontinuity; outer boundaries (i.e. belonging to only one building) must match primitive(s) in source data. As a result, only outer boundaries (i.e. verifiable outlines) are kept for the subsequent matching process.

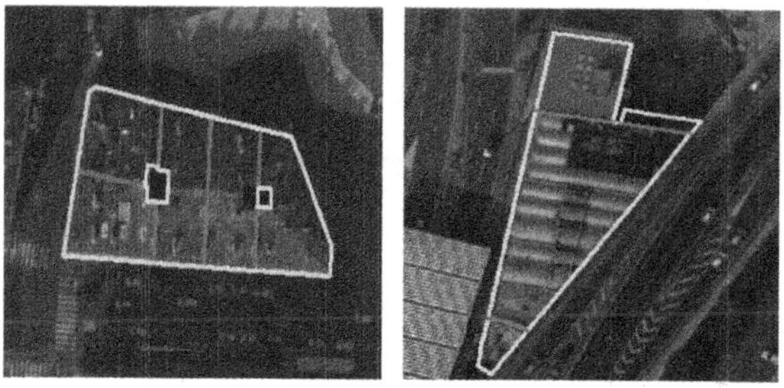

Fig. 6. Inner boundaries (red) and outer boundaries – verifiable outlines (green).

2.3.2 Computing a Similarity Score

In this module, the primitives that are likely to offer a proof about the existence and the correct registration of buildings are selected: firstly, they must be located at a given distance of a building; secondly, they must satisfy a user-defined relative orientation (with respect to the building outlines). A similarity measure SM is then computed for each building: it corresponds to the coverage rate of selected primitives on the building. More formally, let B be a building to be checked and b_j, a verifiable outline (Refer to Figure 7 for one illustration). A Region Of Interest ROI_j is then defined for each verifiable outline b_j, as a buffer given by its width d_0, centered on and aligned with b_j. The similarity measure SM is given by:

$$SM = \frac{\sum_{b_j \in B} \|b_j\| \rho\left(b_j, \left\{p_i : p_i \in ROI_j \; and \left|\theta_i^j\right| \leq \theta_0\right\}_i\right)}{\sum_{b_j \in B} \|b_j\|}$$

Where:

- p_i is an extracted primitive
- Θ_i^j is the relative orientation between p_i and b_j
- Θ_0 is a user-defined orientation
- $\|.\|$ gives the length of b_j
- ρ computes the coverage rate between b_j and selected primitives

2.4 Decision-Making Module

A thresholding procedure lastly classifies objects in three predefined classes (T_L and T_H are user-defined thresholds):

- Demolished if $SM \leq T_L$
- Modified if $SM > T_L$ and $SM < T_H$
- Unchanged if $SM \geq T_H$

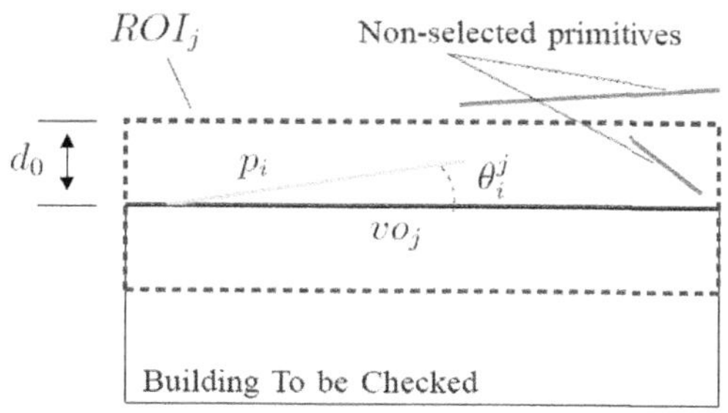

Fig. 7. Similarity Measure (SM) – Sketch.

3 Detection of New Buildings (Phase II)

The goal of this phase is to detect new buildings in the scene. The key idea here is to extract the above-ground objects that neither correspond to a building already present in the database nor a tree.

3.1 General Workflow

A specific methodology is proposed to achieve that goal. As sketched in Figure 8, a DTM is calculated (See section 3.2) and a normalized DSM (nDSM = DSM - DTM) is computed. A geometric threshold (here 2.5m) is then applied to build a new above-ground mask. This binary mask is lastly compared to the initial mask (composed of the vegetation mask and a building mask, easily derived from the database to update) through appropriate morphological tools: the objects that appear in both masks are filtered out and remaining objects are assumed to correspond to new buildings.

3.2 Automatic DTM Generation

The key issues of this second step are the computation of the DTM and its accuracy. In our work, the DTM is directly derived from the DSM and the initial above-ground mask.

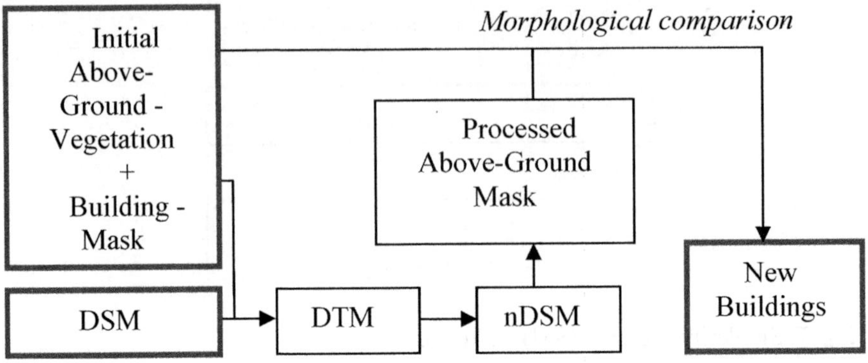

Fig. 8. Workflow of Phase II (Detection of new buildings).

3.2.1 Theory

The DTM generation problem is considered solved when the terrain surface $z = \{z_{c,l}\}_{c,l}$ is determined at the nodes $p_{c,l} = (x_c, y_l)$ of the regular grid, already defined for the input DSM. Note that (c, l) belongs to [1,M] x [1,N], with M and N, respectively the number of columns and rows of the regular grid. Our DTM generation method belongs to the category of deformable models, described for instance in (Montagnat, 2000) and formulates the terrain surface reconstruction problem as the minimisation of a given energy $E(z) = K(z) + \lambda G_{\alpha,k}(z,\sigma)$, where K(z) refers to the regularisation term and $G_{\alpha,k}$ to the data term.

Regularisation Term

The regularisation term corresponds to the discrete approximation of the second derivative and contributes to minimize height variations in the final DTM. That amounts to give internal smoothing properties (i.e. prior knowledge about the surface regularity) to the ground surface to reconstruct.

$$K(z) = \sum_{(c,l) \in [1,M]x[1,N]} \left(\frac{\partial^2 z}{\partial^2 x_c}\right)^2 + \left(\frac{\partial^2 z}{\partial^2 y_l}\right)^2 \qquad (1)$$

Where:

$$\frac{\partial^2 z}{\partial^2 x_c} = z_{c-1,l} - 2z_{c,l} + z_{c+1,l} \ (2) \quad and \quad \frac{\partial^2 z}{\partial^2 y_l} = z_{c,l-1} - 2z_{c,l} + z_{c,l+1} \ (3)$$

Data Term

The data term measures the distance between the model to estimate $z_{c,l}$ and observations $obs_{c,l}$ (i.e. DSM height points out of the above-ground mask).

$$G_{\alpha,k}(z) = \sum_{\substack{(c,l)\in[1,M]x[1,N] \\ p_{c,l} \notin \text{Vegetation Mask} \\ p_{c,l} \notin \text{Initial Building Mask}}} \rho_{\alpha,k}\left(\frac{z_{c,l} - obs_{c,l}}{\sigma}\right) \quad (4)$$

The main problem concerns the outliers i.e. above-ground points present in the DSM but not in the above-ground mask. They may correspond to cars, kiosks, other street furniture, etc., inaccuracies in the vegetation mask (omitted trees) or inaccuracies in the correlation DSM (false correlation). As such outliers systematically make the final surface deviate from the true terrain, they have to be discarded from the data term calculation.

A method to filter outliers and to keep only inliers (i.e. true ground points) is then envisaged: it is based on the M-estimator theory (Zhang and Faugeras, 1992). A robust norm is used in the data term and enables to downweight measurement points not close enough to the estimated model (typically outliers). In an iterative process, a coarse DTM is also firstly computed. The residuals (i.e. the oriented distance between the computed model $z_{c,l}$ and the corresponding observations $obs_{c,l}$) are secondly calculated and a new distance is assigned for each observation, according to the robust norm and the corresponding residual. The surface runs nearer to inliers along the minimisation process, as illustrated in Figure 9.

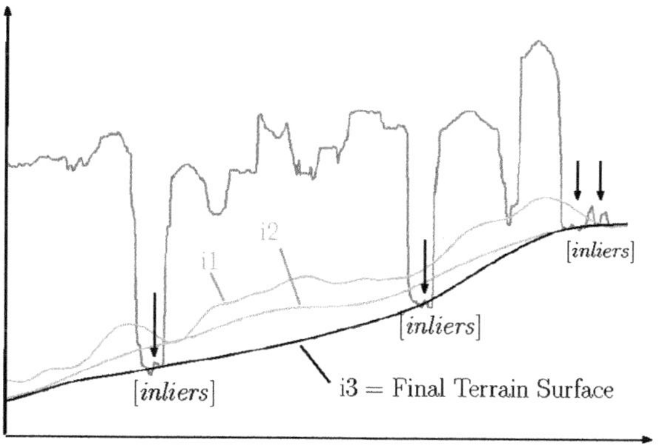

Fig. 9. Illustration of the minimisation process. The DTM, computed at three different iterations (i1 < i2 < i3) shows that the surface is attracted by inliers and discards outliers along the process.

However, as stated in (Kraus and Pfeifer, 1998), the problem is not symmetric: points with a positive residual correspond to lower points and are likely to be to inliers; conversely, points with a negative residual are likely to be outliers. To favour lower points and to penalize higher points, the norm attributed to positive residuals has to be more increasing than the norm attributed to negative residuals. In our work, a Tukey norm (Equation 5) is chosen for negative residuals and a (more increasing) Huber norm (Equation 6) is chosen for positive residuals.

The Tukey and Huber norms are fully determined as soon as the difference between the model and observations follow a Gaussian normal law. In our context, this difference follows a Gaussian law, centered but not normalized: a standard deviation σ needs to be computed in a "clean" (without any outliers) horizontal area. Mathematically, this standard deviation measures the residual noise between our model and an ideal sample of inliers observations. In other words, this standard deviation measures the correlation noise in our correlation DSM.

$$\rho_{\alpha,k}(x)=\begin{cases}\dfrac{\alpha^2}{6}\ \mathit{if}\ x<0\ \mathit{and}\ |x|\geq\alpha\\ \dfrac{\alpha^2}{6}\times\left[1-\left(1-\left(\dfrac{x}{\alpha}\right)^2\right)^3\right]\ \mathit{if}\ x<0\ \mathit{and}\ |x|<\alpha\end{cases}\qquad(5)$$

$$\rho_{\alpha,k}(x)=\begin{cases}\dfrac{x^2}{2}\ \mathit{if}\ x>0\ \mathit{and}\ |x|<k\\ k\times\left(|x|-\dfrac{k}{2}\right)\ \mathit{if}\ x>0\ \mathit{and}\ |x|\geq k\end{cases}\qquad(6)$$

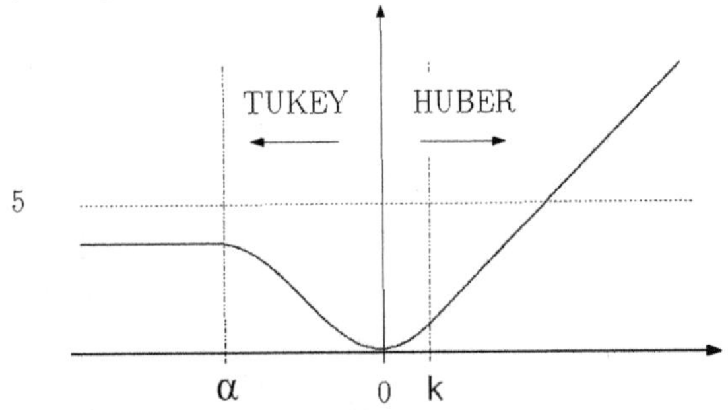

Fig. 10. The dissymetric Tukey-Huber norm used in the robust outliers rejection process.

Balance Coefficient

Our model also behaves like a membrane floating on the DSM and clinging to inliers. The coefficient λ is used in order to balance both terms. The higher λ is, the better the model fits observations. The smaller λ is, the smoother the model is. Once λ fixed, the final DTM is a trade-off between internal properties and its closeness to ground DSM points.

3.2.2 Practical Aspects

Implementation

The Fletcher-Reeves conjugate gradient (Ueberhuber, 1997) is used to minimize the energy E. This algorithm is chosen for its high stability and because it allows a limited storage of height values in memory. It requires an initial solution that here corresponds to the DSM so that all potential inliers are preserved in the initial solution. The required stopping criterion is here given by a constraint on the gradient norm (the gradient of the energy goes to zero at a minimum). Lastly, a paving strategy is considered here, as the minimization process is shown to be long. A mosaicking step is thereafter necessary.

Remark about the Input Building Mask

The DTM is all the more accurate as inliers are present in initial observations. As a consequence, the buildings, considered demolished in the first phase i.e. corresponding to potential ground points are removed from the building mask during the generation of the DTM: here, the splitting of the change detection procedure appears to be crucial to improve the accuracy of the DTM and, at the end, the detection of new buildings.

4 Experiments

4.1 Description of the Dataset

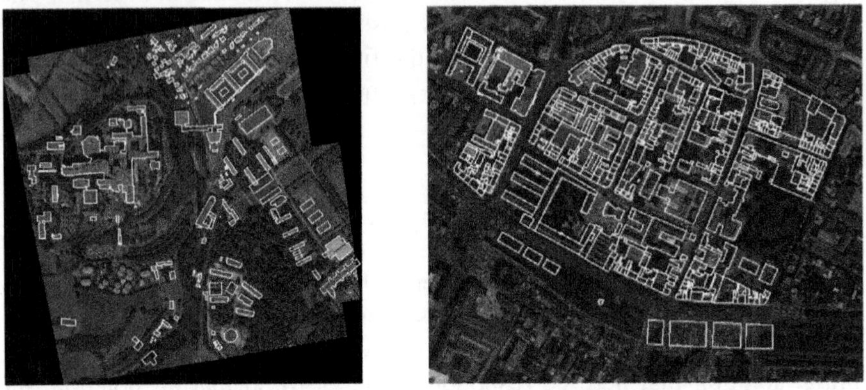

Fig. 11. The database to update, superimposed on the RGB orthophoto. Left: Toulouse; Right: Amiens.

Two test areas are used for this study: Toulouse and Amiens, both located in France (Figure 11) but very different regarding the kind of land use and topography. Toulouse has an area of about 1 km^2 and mostly features a suburban area, composed of detached buildings, very different to each other with respect to the size, height, shape and roofing material (tile, concrete, gravel…). The test area is hilly (the difference in altitude is about 100m) and vegetated (wooden areas). Amiens has an area of 0.5 km^2 and features a densely built-up urban land use, composed of adjacent small houses, generally covered by slate. The terrain is relatively flat (the difference in altitude is smaller than 10m) and vegetated areas only concern treed streets and public gardens.

Regarding the input data, most inaccuracies appear in the DSM. The automatic production of DSM from multiple images is actually known to be a very challenging topic. Advances, made in this field in the last decade, have led to great improvements, especially in favorable configurations, i.e. with very high resolution aerial images (GSD=10-25cm), with a large overlapping. The context here is very challenging (GSD=50cm, in a tri-stereocopic configuration): the low spatial resolution mainly leads to smoothed building outlines and classic problems in shadow areas (where the matching scores are not reliable) systematically lead to overestimated height areas in the DSM.

Lastly, as usually carried out in a change detection test, simulated changes are performed to better control what to be detected. Therefore, a reference (up-to-date) database is firstly (manually) edited; an outdated database is subsequently derived by simulating new, modified and demolished buildings: 51 changes (25 new constructions, 11 destructions, 15 modifications) are simulated in Toulouse; and 38 changes (10 new constructions, 11 destructions, 17 modifications) in Amiens.

4.2 Evaluation Method

The evaluation procedure consists of a comparison of the change map delivered by the method to the ground truth that here corresponds to the initial reference up-to-date database. 2 quality measures (Heipke et al., 1997) are used for that purpose: the completeness and the correctness.

$$Completeness\ (TPR) = \frac{TP}{TP + FN} \in [0;1]$$

$$Correctness = \frac{TP}{TP + FP} \in [0;1]$$

where TP (True Positive), FP (False Positive), FN (False Negative), TN (True Negative) are denoted in Table 1 (confusion matrix).

From a practical point of view, the completeness refers to errors kept in the final database, once the change detection (and update) ended. The correctness refers to unchanged objects, unnecessarily checked by an operator. The optimal value for both quality measures is 1.

Table 1. Confusion Matrix

Ground Truth / Algo	Change	No Change
Change	TP	FP
No Change	FN	TN

4.3 Results

Figure 12 illustrates the workflow of Phase II for the Toulouse test area. Regarding the DTM (Figure 12-3), two profiles, along a North-South and a West-East axes, are plotted in Figure 13. They give a visual qualification of the DTM produced with our method. The accuracy of our DTM is also estimated with ground truth data and appears to be very good: the height standard deviation, between a sample of reference height points (captured by an operator) and corresponding values in the DTM is 1.47m, which is in line with the theoretical value (ranging from 0.9m to 1.8m) expected for a tri-stereoscopic satellite system such as Pléiades-HR. Lastly, the change map is depicted in Figure 14: confirmed buildings are highlighted in green, demolished buildings in red, changed buildings in orange and the estimated positions for new buildings in cyan.

4.4 Evaluation and Discussion

4.4.1 Evaluation of the Results

The evaluation outcomes are illustrated in figure 15 (TP are depicted in green, FP in orange, FN in red and TN in blue). Table 2 gives the completeness and correctness rates for both areas, first computed on a per-building and then on a per-pixel basis. Note that the parameters set (T_L, T_H) used to present the results – (T_L, T_H) = (0.1, 0.57) for Toulouse and (T_L, T_H) = (0.1, 0.52) for Amiens – optimizes the TP (benefit) to FP (cost) rate for each area.

Table 2. Per-building and per-pixel evaluation

Test Area / Quality Measures	Toulouse	Amiens
Per-Building Evaluation		
Completeness	96,2%	98,3%
Correctness	41,6%	34,5%
Per-Pixel Evaluation		
Completeness	97,5%	97,7%
Correctness	75,3%	69,4%

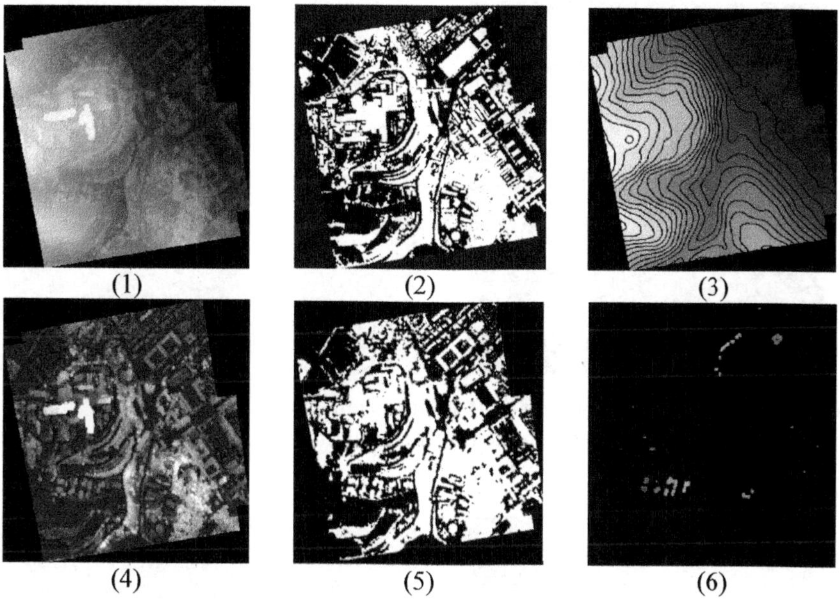

Fig. 12. Detection of new buildings in Toulouse. **(1)** : Initial DSM (lighter = higher). **(2)** : Initial above-ground Mask. $\{\text{Trees} \cup \text{Buildings}\}$. **(3)** : Automatically processed DTM (with contour lines superimposed). **(4)** : nDSM. **(5)** : Processed above-ground mask. **(6)** : New building mask.

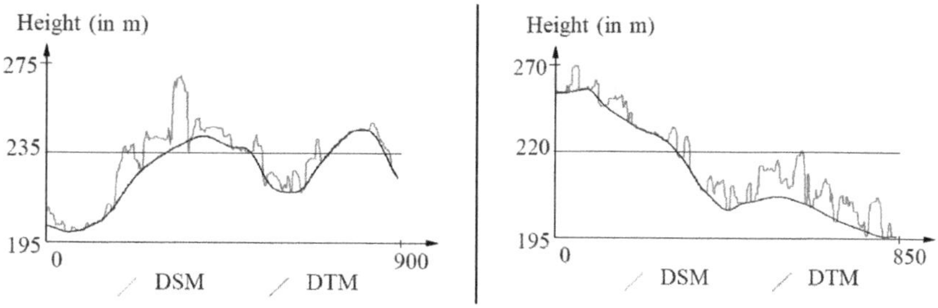

Fig. 13. Profile along a North-South (left) and West-East (right) axes. The DTM obtained with our method perfectly clings to inliers.

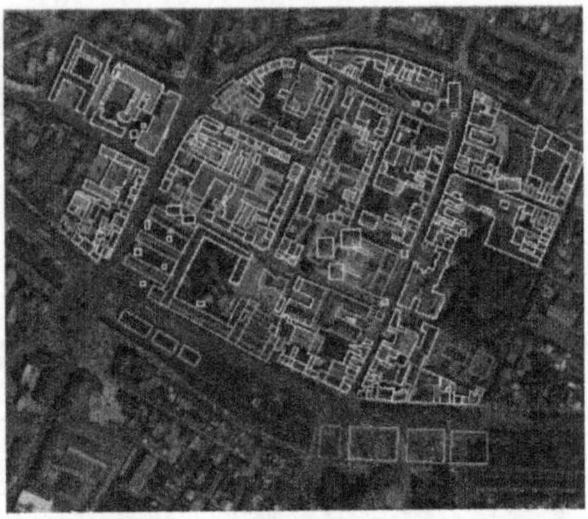

Fig. 14. Change Map obtained in Toulouse (left) and Amiens (right). Confirmed buildings are depicted in green, demolished buildings in red, changed buildings in orange and new buildings in cyan.

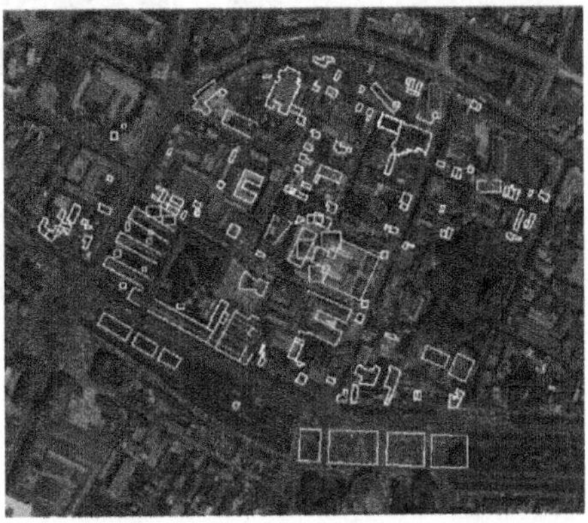

Fig. 15. Evaluation in Toulouse (left) and Amiens (right). TP cases are depicted in green, FN cases in red, FP in yellow and TN in blue.

4.4.2 Discussion

In a similar way as the other change detection approaches found in literature, our method delivers a lot of false alarms (the correctness is 41.6% for Toulouse and 34.5% for Amiens). The nature of these false alarms depends on the phase.

In Phase I, false alarms are almost entirely related to the building size: a lot of confusions occur with small buildings, which explains why the correctness is significantly higher, respectively 75.3% and 69.4%, when computed on a per-pixel basis. In Amiens (Figure 16-1), buildings are small and generally connected to each other. Therefore, there are few verifiable outlines (outer boundaries): at difficult locations, typically in the north shadow side of buildings, the extraction of primitives is poor and an alert is wrongly sent. A similar problem appears in Toulouse (Figure 16-2), especially in the northern part of the scene that contains a lot of small buildings, with small recesses and overhangs. In such a configuration, the DSM is not accurate enough to allow a correct extraction of 2D contours. Small 3D segments could overcome the problem but the pruning procedure (used at the end of the 3D reconstruction – cf. Section 2.2.2) intrinsically gives the priority to longest segments: as a result, no primitives are extracted and an alert is wrongly sent.

In Phase II, most FP cases are related to inaccuracies in input data: non masked trees in the vegetation mask and overestimated height areas in the DSM are systematically alerted as new buildings (Figure 16-3).

This relative drawback might appear to be a high limitation of our system, as it would entail in an operational context a pointless verification of unchanged building by an operator. But this view has to be put into perspective. Firstly, the problem related to false alarms appears to be hard to avoid as it mainly concerns small buildings, for which pertinent information is known to be difficult to extract. Secondly, this problem does not alter the quality of the final database, which is guaranteed by a high completeness rate (96.2% in Toulouse and 98.3% in Amiens) i.e. a good detection of the changes present in the scene (illustrated for instance in Figure 16-3,4,5). Thirdly, the system clearly and drastically speeds up the update procedure, as it gives only one quarter of the database to verify, with almost all the changes included. Lastly, the method appears to be easily transferable from one test area to another one.

The demonstration is achieved by plotting the Receiver Characteristic Curve (ROC) (Fawcett, 2004) that gives the True Positive Rate according to the False Positive Rate. This ROC curve shows that a parameters set $(T_L,T_H) = (0.1, 0.54)$, used for both areas, would give similar results. Compared to the previously used parameters sets, only one additional FN case appears in Toulouse and only 9 additional FP cases in Amiens. The good performance of the system is also preserved. Such a feature is evidently related to the fact invariant 3D geometric primitives are preferred to radiometric primitives in our system: from a radiometric point of view, buildings differ a lot between Toulouse and Amiens and a colour-based method could not overcome such differences; by contrast, from a geometric point of view, buildings are quite similar in both areas, which explains the high transferability of a geometric approach like the one presented here.

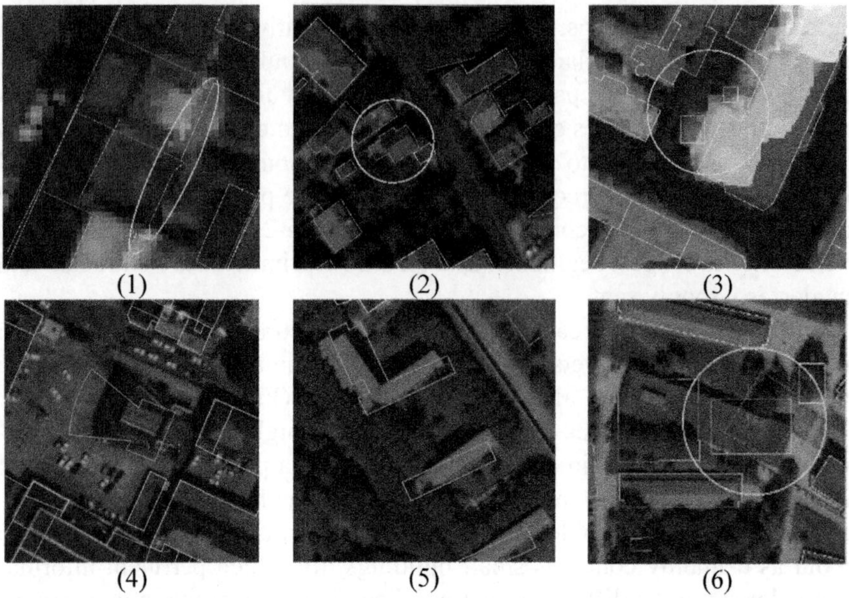

Fig. 16. Evaluation Details. **(1) (2):** (FP) False alarms in Amiens and Toulouse. **(3):** (FP) False alarms related to (height) inaccuracies in source data (DSM). **(4) (5):** (TP) Building misregistrations detected by the algorithm. **(6):** Detection of a difficult configuration, here a demolished (red) and a new building (cyan) located at the same place.

4 Conclusion and Future Trends

In this paper, we first show that the mapping procedures are nowadays mostly related to update and change detection issues and that satellite imagery has a major role to play in this context. We then present an original method to detect the changes between a 2D building database and more recent high resolution satellite images. Two main contributions are presented here. First, we propose a semi-recursive approach: the objects present in the database are first verified (phase I) and the outcomes are used to improve the detection of new buildings (phase II). Second, the 2D database to update is plunged into a 3D environment by using invariant 3D geometric primitives (3D segments and DTM), reconstructed with tri-stereoscopic Pléiades-HR capabilities. These two contributions, combined together, give good results, both in terms of completeness and transferability (robustness). In the future, other experiments will be carried out on larger areas to confirm the robustness of the method. A special concern will also deal with the extraction of more pertinent primitives (especially for smaller buildings) and with considering the inaccuracies of input data (more specifically the DSM and vegetation mask) more reliably in the system. Beyond mapping purposes, the method could also be applied in other fields, typically in damage assessment after disastrous events like earthquakes.

References

Baltsavias, E.P. (2004), Object extraction and revision by image analysis using existing geodata and knowledge: current status and steps towards operational systems, in: International Journal of Photogrammetry and Remote Sensing, Vol. 58, pp. 129–142.

Breiman, L., Friedman, J., Stone, C.J. andOlshen, R.A. (1984), Classification and regression trees, The Wadsworth Statistics/Probability Series.

Busch, A., Gerke, M., Grünreich, D., Heipke, C., Liedtke, C. and Müller, S (2004), Automated verification of a topographic reference dataset: System design and practical results, in: International Archives of Photogrammetry and remote Sensing, vol. XXXV-B2, pp.735–740.

Cantou, J. P., Maillet, G., Flamanc, D. and Buissart, H. (2006) Preparing the use of Pleiades images for mapping purposes : preliminary assessments at IGN-France, in: International Archives of Photogrammetry and Remote Sensing, Vol. XXXVI.

Collins, R. (1995), A space-sweep approach to true multi-image matching, in: Computer Science Department, University of Massachusetts, Technical. Report.

Deriche, R. (1987), Using Canny's criteria to derive a recursively implemented optimal edge detector, International Journal of Computer Vision, Vol.1, pp.167-187

Deriche, R., Vaillant, R. and Faugeras, O. (1992), From noisy edges points to 3d reconstruction of a scene : A robust approach and its uncertainty analysis, in: Theory and Applications of Image Analysis. World Scientific Series in Machine Perception and Artificial Intelligence, pp. 71–79.

Fawcett, T. (2004), ROC graphs: Notes and practical considerations for researchers, in: Technical Report, HP Laboratories, USA

Heipke, C., Mayer, H., Wiedemann, C. and Jamet, O. (1997), Automated reconstruction of topographic objects from aerial images using vectorized map information, in: International Archives of Photogrammetry and Remote Sensing, vol. XXXII, pp. 47–56.

Heipke, C., Cygan, M., Sester, M., Renlinag, Z. and Jie J. (2008), GIS Updating from imagery and collateral data sources, Tutorial jointly organized by ISPRS TC II & IV.

Kraus, K. and Pfeifer, N. (1998), Determination of terrain models in wooded areas with airborne laser scanner data," ISPRS Journal of Photogrammetry and Remote Sensing, vol. 53, pp. 193–203

Lafarge, F., Descombes, X., Zerubia, J., Pierrot-Deseilligny, M. (2008), Building reconstruction from a single DEM, Proc. IEEE Computer Vision and Pattern Recognition (CVPR)

Matikainen, L., Kaartinen, K. and Hyyppä, J. (2007), Classification tree based building detection from laser scanner and aerial image data, in: Proceedings of ISPRS Workshop Laser Scanning 2007.

Montagnat, J.,Delingette, H., Scapel, N., Ayache, N. (2000), Representation, shape, topology and evolution of deformable surfaces. Application to 3D medical image segmentation. Technical Report 3954, INRIA.

Niederöst, M. (2000), Reliable reconstruction of buildings for digital map revision, in: International Archives of Photogrammetry and Remote Sensing, vol. XXXIII-B3, pp. 635–642.

Olsen, B. and Knudsen, T. (2005), Automated change detection for validation and update of geodata, in: Proceedings of 6th Geomatic Week.

Pierrot-Deseilligny, M. and Paparoditis, N. (2006), An optimization-based surface reconstruction from Spot5-HRS stereo imagery, in International Archives of Photogrammetry and Remote Sensing, vol. XXXVI.

Rottensteiner, F., Trinder, J., Clode, S. and Kubik, K. (2005), Using the Dempster-Shafer method for the fusion of lidar data and multi-spectral images for building detection, Information Fusion, pp. 283–300.

Rottensteiner, F. (2008), Automated updating of building data bases from digital surface models and multi-spectral images, in:International Archives of the Photogrammetry, Remote Sensing and Spatial Information Sciences, Vol.XXXVII B3A, pp.265-270.

Steinnocher, K. and Kressler, F. (2006), Change detection, EuroSDR Technical Report.

Taillandier, F. and Deriche, R. (2002), A reconstruction of 3D linear primitives from multiple views for urban areas modelisation, in: ISPRS Commission 3 Symposium on Photogrammetric Computer Vision (PCV02).

Ueberhuber, C. (1997), Numerical Computation, volume 2 – chapter 14 (Minimization Methods). Springer.

Vögtle, T. and Steinle, E. (2004) Detection and recognition of changes in building geometry derived from multitemporal laserscanning data, in: International Archives of the Photogrammetry, Remote Sensing and Spatial Information Sciences, Vol. XXXV-B2, pp. 428–433.

Walter, V. and Fritsch, D. (2000), Automated revision of GIS databases, in: Proceedings of the Eighth ACM Symposium on Advances in Geographic Information Systems, pp. 129–134.

Zhang, Z. and Faugeras, O. (1992), 3D scene analysis: a stereo based approach, in: Image and Vision Computing Journal.

Accuracy of High-Resolution Radar Images in the Estimation of Plot-Level Forest Variables

Markus Holopainen[1], Sakari Tuominen, Mika Karjalainen, Juha Hyyppä, Mikko Vastaranta, Hannu Hyyppä

[1] Department of Forest Resource Management, University of Helsinki, Latokartanonkaari 7, Helsinki, Finland
markus.holopainen@helsinki.fi

Abstract. In the present study, we used the airborne E-SAR radar to simulate the satellite-borne high-resolution TerraSAR radar data and determined the accuracy of the plot-level forest variable estimates produced. Estimation was carried out using the nonparametric k-nearest neighbour (k-nn) method. Variables studied included mean volume, tree species-specific volumes and their proportions of total volume, basal area, mean height and mean diameter. E-SAR-based estimates were compared with those obtained using aerial photographs and medium-resolution satellite image (Landsat ETM+) recording optical wavelength energy. The study area was located in Kirkkonummi, southern Finland. The relative RMSEs for E-SAR were 45%, 29%, 28% and 38% for mean volume, mean diameter, mean height and basal area, respectively. For aerial photographs these were 51%, 26%, 27% and 42%, and for Landsat ETM+ images 58%, 40%, 35% and 49%. Combined datasets outperformed all single-source datasets, with relative RMSEs of 26%, 23%, 33% and 39%. Of the single-source datasets, the E-SAR images were well suited for estimating mean volume, while for mean diameter, mean height and basal area the E-SAR and aerial photographs performed similarly and far better than Landsat ETM+. The aerial photographs succeeded well in the estimation of species-specific volumes and their proportions, but the combined dataset was still significantly better in volume proportions. Due to its good temporal resolution, satellite-borne radar imaging is a promising data source for forest inventories, both in large-area forest inventories and operative forest management planning. Future high-resolution synthetic aperture radar (SAR) images could be combined with airborne laser scanner data when estimating forest or even tree characteristics.

Keywords: Forest inventory, forest planning, radar imaging, E-SAR, TerraSAR, aerial photographs, Landsat

M. Sester et al. (eds.), *Advances in GIScience*, Lecture Notes in Geoinformation and Cartography,
DOI 10.1007/978-3-642-00318-9_4, © Springer-Verlag Berlin Heidelberg 2009

1 Introduction

1.1 Radar Images in Forest Resource Estimation

Medium-resolution satellite imagery of the optical wavelength region has long been applied in forest inventory (e.g. Kilkki and Päivinen, 1987; Tokola, 1989; Tomppo, 1990). Despite their great value in large-area inventories, they have been a disappointment in small-area mapping and operative forest planning. The main reason is that after numerous efforts, the accuracy of estimates derived from medium-resolution satellite image interpretations has not been high enough at the stand or field plot level - the mean volume root mean squared error (RMSE) has typically been between 55% and 60% at the stand level (e.g. Pussinen, 1992; Holopainen and Lukkarinen, 1994; Hyyppä et al., 2000). At the plot level the RMSEs are even higher: 65-80% (e.g. Poso et al., 1999; Franco-Lopez et al., 2001; Tuominen and Poso, 2001). When the size of the target area increases, the RMSEs decrease relatively rapidly, being 10-20% at the typical forest estate level in Finland (over 25 ha; Tokola and Heikkilä, 1995; Tomppo et al., 1998).

In addition to the lack of accuracy, another factor prohibiting the use of satellite images in recording optical wavelength energy is cloudiness. For instance, under the conditions encountered in Finland, a full countrywide, cloud-free mosaic of satellite images cannot be acquired annually. In tropical forests the situation is even worse. This is a problem with current requirements for up-to-date forest inventory estimates. A major advantage of radar images, compared with optical region satellite images, has been their good availability (temporal resolution) under all imaging conditions. This makes radar imaging, especially the synthetic aperture radar (SAR) carried by satellites, an intriguing option in developing methods for operational inventory of forest resources - in case the accuracy would be suitable for plot- and stand-level estimation. However, the spatial resolution of satellite radar images has so far been lacking.

The Seasat satellite produced as early as in 1978 the first radar images obtained from a satellite, with 25-m spatial resolution. In the 1980s and 1990s the development of SAR satellite radars proceeded towards the aims set at the global scale: large-area applications without detailed information. In the inventory applications based on SAR satellite radar images, the emphasis was on the estimation of large forest areas (e.g. Rauste, 1990) and biomasses (e.g. Rauste et al., 1994; Rauste, 2006). During the last decade, SAR images have been used in plot- and stand-level forest variable estimation, as well (e.g. Hyyppä et al., 2000; Kellndorfer et al., 2003). The results by Hyyppä et al. (2000) showed that with SAR images accuracies similar to those of medium-resolution satellite images (Landsat Thematic Mapper, TM) can be obtained.

Promising results at the plot and even tree levels were obtained with profiling airborne radars in the 1990s (e.g. Hyyppä, 1993; Hyyppä et al., 2000). Hyyppä et al. (2000) compared remotely sensed materials available at that time in the estimation of forest variables and stated that profiling radar was the only means to obtain stand-level volume estimates accurate enough for the requirements of operative forest inventory. In their study, the mean volume RMSE obtained by using the profiling radar was 34%, while that of the traditional ocular field inventory method was 26%. However, due to its small scanning area the profiling radar is not practical for operative forest inventory.

The biggest jump in forest inventory technology in recent years has been in applications based on helicopter- and airplane-based laser scanners (ALS = airborne laser scanning) operating in near-infrared (NIR) wavelength area. Research results have shown that ALS is at least as accurate as traditional ocular field measurements in estimating the stand mean volume at plot level with area-based methods (e.g. Naesset, 1997, 2004a, 2004b; Holmgren, 2003; Suvanto et al., 2005; Packalén and Maltamo 2006, 2007; Holopainen et al., 2008) or via single-tree characteristics (e.g. Hyyppä and Inkinen, 1999; Maltamo et al., 2004; Hyyppä et al., 2008; Kaartinen et al., 2008). Area-based ALS-method or single-tree detection provide accuracies ranging between 10 to 20% for mean volume RMSE.

ALS is carried out at relatively low altitudes, which consequently makes it relatively expensive per area unit. Other remotely sensed data will still be needed, especially when updated information is required e.g. several times per year. Of special interest are inexpensive images with good temporal resolution that can be utilized in multiphase sampling and change detection in addition to the ALS-measurements.

Simultaneously, SAR-measurements experienced a break-through similar to that in the ALS method, when in the early 2000s satellite radar imagery with spatial resolutions as high as 1-3 m were developed. In addition to the improved spatial resolution, one of the central improvements in the new SAR-satellite images has been their ability to utilize interferometry and polarimetry.

Interferometry utilizes the interferogram generated from the phase differences of two SAR images taken from slightly differing positions. With the interferogram, a coarse surface model of the landscape is obtained. In forests, this surface model is located somewhere between the ground and tree canopies, depending on the frequency used. On the other hand, a so-called coherence image that describes the quality of the interferogram can be calculated between two interferometric SAR images. In the case of multitemporal image pairs, even small changes in the target, such as movements of branches or needles, can reduce the interimage coherence. Coherence information can be used in the estimation of growing stock volume, as well, because with larger volumes the coherence between two images decreases.

Polarization means the direction of orientation of the electric field vector of the electromagnetic wave transmitted by the radar. In SAR instruments the vibration direction of the transmitted or received radio wave can be either horizontally (H) or vertically (V) polarized in relation to the antenna orientation. In full polarimetric imaging, all four combinations of transmit or receive (HH, HV, VH and VV) are simultaneously recorded. Multiple polarizations can be used in image interpretation in ways similar to the multiple bands of a satellite image. The backscatter intensity of the cross-polarization bands (HV and VH) is a good estimator of the volume of growing stock: the greater the biomass, the greater is the backscatter at the cross-polarization band (Henderson et al., 1998). SAR polarimetry and interferometry allow the use of extra features, such as scattering mechanisms and height differences of scatterers.

The first high-resolution radar satellite, the Japanese Advanced Land-Observing Satellite (ALOS), was launched in 2006. It carries a phased array-type L-band synthetic aperture radar (PALSAR), which uses the L wavelength area (23.6 cm). The spatial resolution of PALSAR is 10-30 m (for single- or full-polarization imaging modes, respectively) (Rosenqvist et al., 2007). The Canadian Radarsat-2 was launched in 2007. It uses the C wavelength area (5.6 cm). The most important update is full polarimetric imaging (with Radarsat-1 only HH polarization was possible). The spatial resolution is approx. 2 m in single-polarization imaging mode. The German TerraSAR-X satellite, launched in 2007, uses the X wavelength area (3.1 cm). Compared with earlier SAR satellites open to civilians, the most important advance of TerraSAR-X is the so-called Spotlight imaging mode, which allows the targets to be viewed for a longer time, resulting in higher spatial resolution (approx. 1 m). There is a full polarimetric imaging mode, but it is not open for civilians (Düring et al., 2008).

Rauste et al. (2008) reported, that the estimation of growing stock volume is slightly more accurate with full-polarimetric ALOS than with the earlier Japanese Earth Resources Satellite 1 (JERS-1), but the estimates still saturate at 150 m^3/ha. Results obtained with the RadarSat-2 or TerraSAR-X have not yet been published. However, in the TerraSAR an airborne sensor, Expeimental Synthetic Aperture Radar (E-SAR), owned by the German Aerospace Centre (DLR) has been used to simulate the results obtainable with TerraSAR.

1.2 Objective of this Study

The objective of this study was to determine the usability of bitemporal (spring and autumn) E-SAR images in the estimation of forest variables at the plot-level. These variables included mean volume, tree species-specific volumes and their proportions of the total volume, basal area, mean height and mean diameter. The accuracy of estimates obtained using the airborne E-SAR can be considered an upper limit for those obtainable later with the TerraSAR, since the latter operates in the X-band only while the E-SAR contains both X and L bands.

2 Material and Methods

2.1 Field Data

The study area is located in Kirkkonummi, southern Finland (24.45° E and 60.22° N). It covers approx. 1000 ha of managed boreal forest. Relascope sample plots were measured in the field. In seedling stands, circular sample plots with a radius of 5.64 m were measured. Stratified sampling, based on Landsat Enhanced Thematic Mapper Plus (ETM+) satellite image, aerial photographs and old forest management plan, were used in selecting field sample plot locations. Stands with large mean volumes received higher weight in the field sample, thus the mean forest attributes are somewhat higher than generally in the area (Table 1). The final locations of the measured field plots were registered with a Differential Global Positioning System (DGPS).

Table 1. Means and maximums of the variables examined in the field dataset.

	Mean	Maximum
Mean diameter, cm	25.1	46.9
Mean height, m	19.1	32.1
Basal area, m^2/ha	17.2	63.0
Mean volume, m^3/ha	160.2	547.5
Mean volume of Scots pine, m^3/ha	50.6	364.1
Mean volume of Norway spruce, m^3/ha	72.4	483.7
Mean volume of deciduous trees, m^3/ha	37.2	241.0
Proportion of Scots pine, %	37.1	100
Proportion of Norway spruce, %	32.1	100
Proportion of deciduous trees, %	28.3	100

2.2 Remotely Sensed Data and Feature Extraction

2.2.1 E-SAR

The E-SAR images were captured on August 31, 2000 and May 2, 2001. Both flights included L-band (HH, HV, VV and VH polarizations) and X-band (HH and VV polarizations). In all, 12 E-SAR image bands, all georeferenced using a digital elevation model (DEM), were thus available. Radiometric calibration was carried out by the DLR, with the help of corner reflectors located in the study area.

Features extracted from the E-SAR image included mean and standard deviation of the backscatter intensity and band ratios. The features were extracted from areas corresponding to the relascope sample plots. In comparison to satellite radar imaging, the variation of incidence angle of the transmitted radiation in airborne SAR imaging is significantly larger within the imaging area. In the E-SAR data employed, the near-range incidence angle was 25° and that of the far-range approx. 56°. Since the incidence angle greatly impacts the intensity of the backscatter received (especially in open terrain), the E-SAR incidence angle was calculated for each field plot using the plot location, flight altitude and imaging geometry of the antenna. Furthermore, the terrain slope at the plot centre towards the viewing direction of the antenna was calculated for each plot, because it may impact the backscatter received, as well. In dense forests, the impacts of the incidence angle and terrain slope are smaller, especially in the X band, which cannot illuminate the ground.

2.2.2 Landsat ETM + Satellite Image

The Landsat ETM+ image used in this study was acquired on April 21, 2001 (path 188, row 18). The ground resolution was 30 m in bands 1-5 and 7, 60 m in thermal band 6 and 15 m in the panchromatic (PAN) band. The image was georeferenced into the Finnish Uniform Coordinate System with ground-control points collected from a base map. The satellite image features used in this study were pixel values of bands 1-7 (pixel containing the field plot centre), and average and standard deviation extracted from a 2 x 2 pixel window around the field plot centre in the PAN band.

2.2.3 Aerial Photographs

False-colour aerial photographs with NIR, red (R) and green (G) bands from summer 2000 were used. The images were captured with a film camera and digitalized by scanning the negatives. The digitalized images were orthorectified, using base map and a DEM to a ground resolution of 0.5 m.

The aerial photographic features used in this study are presented below. These were calculated separately for the three bands, from 40 x 40 pixel windows around the field plot centre, except in the case of standard texture.

1. Band averages
2. Band standard deviations
3. Standard texture derived from a 32 x 32 pixel window, which was further divided into 1 x 1, 2 x 2, 4 x 4 and 8 x 8 pixel blocks. The features were calculated as standard deviations of the means of pixel blocks. Finally, the standard deviation of these four calculated standard deviation features was calculated.
4. Textural features derived from co-occurrence matrices: angular second moment, contrast, entropy and local homogeneity (e.g. Haralick et al., 1973).

2.3 Feature Normalization and Feature Selection

There were large differences in the ranges of the features derived from the three remotely sensed data sources. For this reason, the features were normalized before estimation in such a way that the mean of each band corresponded to 0 and the standard deviation to 1. Without standardization, those features having the highest ranges would have predominated in the k-nearest neighbour (k-nn) search (see Section 2.4 below), whether correlating with the forest variables or not.

Sequential forward selection was applied in the feature selection. This method adds features iteratively to the model. In the first iteration round, the feature producing the best estimate (lowest RMSE) for the forest variable examined (e.g. mean volume of growing stock) is selected. In the following iteration rounds, those features that produced the best estimation result with those already present in the feature set were selected, one by one. Iteration continues until the RMSE is no longer lowered. Sequential forward selection may stop after one or two features, if none of the rest further improves the accuracy when combined with the first features. This does not mean that the selected subset would be the best of all existing subsets. Stopping at a local minimum was a problem when unstandardized, raw features were used in the preliminary tests.

All three sets were subjected to each feature selection at a time. Additionally, a dataset containing all three sources was created and feature selection was applied to this combined dataset, as well.

2.4 Estimation of Plot-Level Forest Variables

The k-nn method was used in the forest variable estimation (e.g. Kilkki and Päivinen, 1987; Tokola et al., 1996; Franco-Lopez et al., 2001 (Eq. 1)). A central assumption is that field plots (or stands) that are similar in reality will be similar in the space defined by remotely sensed data features, as well. The forest variables of any image pixel can then be estimated with the help of reference field plots measured in the field by calculating the averages of the nearest neighbours. In the present study, similarity was determined by the Euclidean distances in the image feature space. The nearest neighbours were weighted with inverse distances (Eq. 2).

$$\hat{y} = (\sum_{i=1}^{k} w_i y_i) / k \tag{1}$$

where
$\hat{y}$ = estimated value for variable y
y_i = measured value for variable y at the *i*:th nearest field plot
w = weight of field plot *i* in the estimation

$$w_i = \frac{1}{d_i^2} / \sum \frac{1}{d_i^2} \tag{2}$$

d_i = euclidian distance to the *i*:th nearest field plot (measured in the feature space)
k = number of neighbours used in the estimation

An essential parameter affecting the results obtained with the k-nn method is the number of neighbours, k, for which values 3,4 and 5 were tested in this study. These were deemed suitable in relation to the number of field plots. Selecting the value for k is always a compromise: a small k increases the random error of the estimates, while a large k results in averaged estimates and reduces the variation available in the original dataset.

3 Evaluation of Results

Evaluation of the estimation accuracy was carried out using cross-validation. In the process, each field plot at a time is left out of the reference dataset and the forest variable estimates are calculated using the remaining field plots. The estimates are then compared with the values observed in the field. The RMSE (Eq. 3) and relative RMSE (Eq. 4) were derived from the comparisons.

$$RMSE = \sqrt{\frac{\sum_{i=1}^{n}(\hat{y}_i - y_i)^2}{n}} \tag{3}$$

$$RMSE\% = 100 * \frac{RMSE}{\bar{y}} \tag{4}$$

where
y_i = measured value of variable y at plot i
$\hat{y}_i$ = estimated value of variable y at plot i
$\bar{y}_i$ = mean of measured values of variable y
n = number of field plots

The estimation accuracies obtained using the three remotely sensed datasets are presented in Tables 2 and 3. The best value of k varied, depending on the data and target variable, but four neighbours generally gave the best results, thus the results presented were obtained with this k value. The lowest RMSEs were obtained with the combined dataset, except in the case of mean diameter. Of the single data sources, E-SAR performed especially well in the mean volume estimation, where the RMSE was significantly lower than those obtained with aerial photographs or the Landsat ETM+. In the case of other general forest variables (mean diameter, mean height and basal area), E-SAR and aerial photograph results were relatively alike, and far better than Landsat ETM+. Two types of tree species-specific variables were studied: volumes of spruce (*Picea* A. Dietr.), pine (*Pinus* L.) and deciduous trees and their proportions of total volume in the field plot. Estimation success was generally better in the case of proportions of mean volume (Fig. 1), thus these results are presented in Table 2. The estimation accuracies of both variable types were poorer than those of the plot mean variables shown in Table 1. Of the single data sources, aerial photographs performed best.

Table 2. Estimation results for general stand variables with the three remotely sensed datasets and the combined set. Results were obtained with four nearest neighbours. If using three or five neighbours resulted in a lower RMSE, this is shown in brackets. The combined set was tested with four neighbours only.

	RMSE, % of mean			
	Mean diameter	Mean height	Basal area	Mean volume
Landsat ETM+	40.45 (38.51)	34.51	48.66 (46.86)	58.33
Aerial photographs	25.69	26.54 (26.48)	42.14 (40.61)	50.71 (49.33)
E-SAR	29.01 (27.82)	28.44 (27.46)	38.28	44.82
Combined set	26.30	23.15	32.87	38.30

Table 3. Estimation results of tree species-specific proportions of the mean volume with the three remotely sensed datasets and the combined set. Results were obtained with four nearest neighbours. If using three or five neighbours resulted in a lower RMSE, this is shown in brackets. The combined set was tested with four neighbours only.

	RMSE, % of mean		
	Proportion of pine	Proportion of spruce	Proportion of deciduous trees
Landsat ETM+	106.5 (105.99)	77.23 (76.79)	102.6 (100.26)
Aerial photographs	76.17 (70.97)	77.64	74.92
E-SAR	81.77	84.61 (83.22)	86.15
Combined set	64.28	58.79	68.66

Features from all data sources were selected from the combined set, while those selected from single data sources are presented below. Of the ETM+ satellite image bands, bands 5 and 6 were suited best for the estimation of general stand variables (mean volume, mean diameter, mean height and basal area), while the volume and proportion of conifers were estimated best with bands 3 and PAN, and the volume and proportion of deciduous trees with bands 2 and 4.

In aerial photographs, the best features for forest variable estimation were the textural features based on standard deviations of pixel blocks (standard texture) and co-occurrence matrices, augmented with the average of R band for basal area and mean volume. Estimates of conifer volumes and properties were most accurate when features based on co- occurrence matrices (and standard deviations of pixel blocks) were used together with averages for the G band. This was the case with deciduous trees, as well, but quite surprisingly, the NIR band was not among the features that first entered the model. Overall, the textural features proved to be more important than the averages of aerial photograph pixel values.

Of the E-SAR features, tree dimensions (mean height, mean diameter) were best estimated with ratios of the polarization bands of same-date images. For basal area, intensity of the HV polarization and standard deviations of backscatter of different polarizations of the L band performed best. The intensity of the L-HH polarization and band ratio L-HH/L-HV described the mean volume of growing stock most accurately. The standard deviations of the backscatter intensities for the X-VV and L-VV polarizations were suited for the estimation of volumes and proportions of conifers, while the volume of deciduous species was estimated most accurately with the standard deviation for the backscatter of the X-HH polarization and the proportion of deciduous species with band ratios L-HH/L-VV and L-HH/X-VV derived from the autumn image.

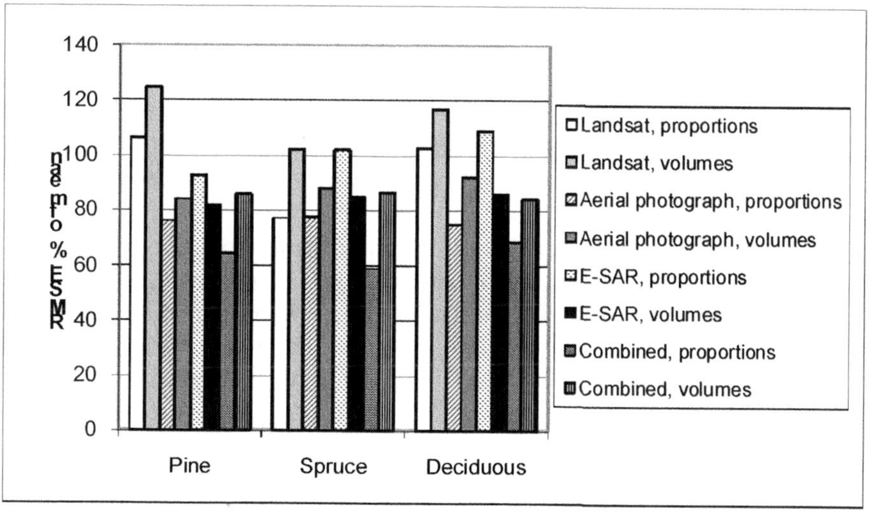

Fig. 1. Estimation results of the two types of tree species-specific variables (proportions of mean volume and mean volume m^3/ha) with the three remotely sensed datasets and the combined set. Results were obtained with four nearest neighbours.

4 Discussion and Conclusions

In the present study, we analysed the usability of E-SAR radar imagery in the estimation of plot-level forest variables. E-SAR is an airborne instrument, that simulates the data obtained with the high-resolution satellite radar TerraSar. The accuracies were compared with results obtained using aerial photographs and a medium-resolution Landsat ETM+ satellite image recording optical wavelength energy. A combined dataset was created and tested, as well. The dimensionality of all datasets was reduced with sequential forward selection of features. The lowest RMSEs were obtained with the combined dataset. Compared with single data sources, the improvements were largest in mean volume, basal area and species-specific volumes. Of the single data sources, E-SAR was able to produce the most reliable results concerning general forest variables, especially the mean volume of the plot, while Landsat ETM+ produced the overall poorest estimates.

The mean errors of traditional ocular forest inventory used in operational forest management planning vary from 16% to 38% (Poso, 1983; Laasasenaho and Päivinen, 1986; Pussinen, 1992; Haara and Korhonen, 2004; Saari and Kangas, 2005). This means that the approx. 45% error level reached with E-SAR imagery at the field plot level closely resembles that of ocular field inventory (the RMSEs are somewhat lower at the stand level). The 38% error obtained with the combined dataset is even more promising.

A central problem in ocular field inventory is the poor accuracy of species-specific estimates. For example Haara and Korhonen (2004) obtained relative RMSEs of 29%, 43% and 65%, for pine, spruce and birch (*Betula* L.), while the biases were -5.5%, 4.4% and 5.7%, respectively. During the next few years, the Finnish operational forest management-planning will change to an ALS-based system. With the planned area-based method, mean volumes can be estimated more accurately than with ocular field inventory, but the species-specific estimations of RMSEs are similar. Packalén and Maltamo (2007) showed that the plot-level RMSEs for species-specific volumes varied between 51% and 102%. Holopainen et al. (2008) obtained plot-level tree species-specific accuracies between 70% and 72% for pine, spruce and deciduous trees by combining ALS and aerial photographic data. In the present study the aerial photographs outperformed the E-SAR in the species-specific estimation success by 6-11 %, the RMSEs being 75%-68%, which is in line with those obtained with ocular field inventory or area-based ALS data.

The E-SAR images were tested to determine the capacity of the new high-resolution TerraSAR satellite images, which are currentry available, as well. Concerning the high-resolution SAR satellite images, full polarimetric imaging has recently become operational. In combining data from several satellite types, information from different wavelength areas can be obtained. These factors should improve the estimation accuracies in forest applications, compared with previous instrument generations. Furthermore, spatial resolution has been improved and is now at the scale of 1 m (single-polarization imaging). Current expectations are high, especially for utilization of SAR-interferometry and polarimetry (Krieger et al., 2005).

A central task for future forest resource inventories will be detection of changes, i.e. updating the forest inventory data. In addition to the traditional forest variables, more interest will be placed on changes in biomass, bioenergy and carbon balance. Climate change will probably increase forest damage, creating a demand for monitoring methods, as well. Our results suggest, that high-resolution SAR images outperform medium-resolution optical wavelength region satellite images in the task of monitoring growing stock volumes and biomasses.

The next steps will include testing with actual TerraSAR-X imagery to determine whether the results simulated with airborne E-SAR will hold true with satellite-borne instruments. Further research areas will include the use of SAR image segments as feature extraction units, developing the methods of feature selection and combining the high-resolution SAR data with ALS data.

Acknowledgements

The study was partly supported by the Finnish Academy for the project Improving Forest Supply Chain by Means of Advanced Laser Measurements (L-Impact).

References

Düring, R., Koudogbo, F.N. & Weber, M. (2008) TerraSAR-X and TanDEM-X Revolution in Spaceborne Radar, Proceedings of the ISPRS XXI Congress, Beijing, China.

Franco-Lopez, H., Ek, A.R. & Bauer, M.E. (2001) Estimation and mapping of forest density, volume and cover type using the k-nearest neighbors method. Remote Sensing of Environment 77: 251 - 274.

Haara, A. & Korhonen, K. (2004) Kuvioittaisen arvioinnin luotettavuus. Metsätieteen aikakauskirja, 2004, 489-508.

Haralick, R. M., Shanmugan, K. & Dinstein, I. (1973) Textural features for image classification. IEEE Transactions on Systems, Man and Cybernetics, 3, 610-621.

Henderson, Floyd M., Lewis & Anthony, J. (editors). (1998) Principles & Applications of Imaging Radar, Manual of Remote Sensing, Third Edition, Volume 2, John Wiley & Sons, Inc., ISBN 0-471-29406-3.

Holmgren, J. (2003) Estimation of forest variables using airborne laser scanning. PhD Thesis. Acta Universitatis Agriculturae Sueciae, Silvestria 278, Swedish University of Agricultural Sciences, Umeå, Sweden.

Holopainen, M. & Lukkarinen, E. (1994) Digitaalisten ilmakuvien käyttö metsävarojen inventoinnissa. Helsingin yliopiston metsävarojen käytön laitoksen julkaisuja 4. (In Finnish).

Holopainen, M., Haapanen, R., Tuominen, S. & Viitala, R. (2008) Performance of airborne laser scanning- and aerial photograph-based statistical and textural features in forest variable estimation. In Hill, R., Rossette, J. and Suárez, J. Silvilaser 2008 proceedings:105-112.

Hyyppä, J. (1993) Development and feasibility of airborne ranging radar for forest assessment. Helsinki University of Technology, Laboratory of Space Technology, 112 pp. ISBN 951-22-1888-7.

Hyyppä, J. & Inkinen, M. (1999) Detecting and estimating attributes for single trees using laser scanner. The Photogrammetric Journal of Finland, 16:27-42.

Hyyppä, J., Hyyppä, H., Inkinen, M., Engdahl, M., Linko, S. & Zhu, Y-H. (2000) Accuracy comparison of various remote sensing data sources in the retrieval of forest stand attributes. Forest Ecology and Management, 128:109-120.

Hyyppä, J., Hyyppä, H., Leckie, D., Gougeon, F., Yu, X. & Maltamo, M. (2008) Review of methods of small-footprint airborne laser scanning for extracting forest inventory data in boreal forests. Internationa Journal of Remote Sensing 29:1339-1366.

Kaartinen, H, Hyyppä, J. Liang, X., Litkey, P., Kukko, A., Yu, X., Hyyppä, H. & Holopainen, M. (2008) Accuracy of automatic tree extraction using airborne laser scanner data. In Hill, R., Rossette, J. and Suárez, J. Silvilaser 2008 proceedings:467-476.

Kellndorfer, J. M., Dobson, M. C., Vona, J. D. & Clutter, M. (2003) Toward precision forestry: Plot-level parameter retrieval for slash pine plantations with JPL AIRSAR. IEEE Transactions on Geoscience and Remote Sensing 41(7): 1571-1582.

Kilkki, P. & Päivinen, R. (1987) Reference sample plots to combine field measurements and satellite data in forest inventory. Department of Forest Mensuration and Management, University of Helsinki. Research notes, 19:210-215.

Krieger, G., Papathanassiou, K. & Cloude, S.R. (2005) Spaceborne Polarimetric SAR Interferometry: Performance Analysis and Mission Concepts, EURASIP Journal on Applied Signal Processing 2005(20):3272-3292

Laasasenaho, J. & Päivinen, R. (1986) Kuvioittaisen arvioinnin tarkistamisesta. Folia Forestalia 664. 19 s.

Maltamo, M., Eerikäinen, K., Pitkänen, J., Hyyppä, J. & Vehmas, M. (2004) Estimation of timber volume and stem density based on scanning laser altimetry and expected tree size distribution functions. Remote Sensing of Environment, 90, 319-330.

Næsset, E. (1997) Estimating timber volume of forest stands using airborne laser scanner data. Remote Sensing of Environment, 51, 246-253.

Næsset, E. (2004a) Practical large-scale forest stand inventory using a small footprint airborne scanning laser. Scandinavian Journal of Forest Research, 19, 164-179.

Næsset, E. (2004b) Accuracy of forest inventory using airborne laser-scanning: Evaluating the first Nordic full-scale operational project. Scandinavian Journal of Forest Research, 19, 554-557.

Packalén, P. & Maltamo, M. (2006) Predicting the plot volume by tree species using airborne laser scanning and aerial photographs. Forest Science, 56, 611-622.

Packalén, P. & Maltamo, M. (2007) The k-MSN method in the prediction of species specific stand attributes using airborne laser scanning and aerial photographs. Remote Sensing of Environment, 109, 328-341.

Poso, S. (1983) Kuvioittaisen arvioimismenetelmän perusteita. Silva Fennica 17:313-343.

Poso, S., Wang, G. & Tuominen, S. (1999) Weighting alternative estimates when using multi-source auxiliary data for forest inventory. Silva Fennica 33:41-50.

Pussinen, A. (1992) Ilmakuvat and Landsat TM-satelliittikuvat välialueiden kuvioittaisessa arvioinnissa. Aerial photos and Landsat TM -image in compartmentwise survey. University of Joensuu, Faculty of Forestry. Thesis for the Master of Science in Forestry Degree. 48 s.

Rauste, Y. (1990) Incidence-angle dependence in forested and non-forested areas in Seasat SAR data. International Journal of Remote Sensing, 11:1267-1276.

Rauste, Y. (2006) Techniques for Wide-Area Mapping of Forest Biomass Using Radar Data. Dissertation for the degree of Doctor of Science in Technology. VTT Publications 591.

Rauste, Y., Häme, T., Pulliainen, J., Heiska, K. & Hallikainen, M. (1994) Radar-based forest biomass estimation. International Journal of Remote Sensing, 15:2797-2808.

Rauste, Y., Lönnqvist, A., Molinier, M., Ahola, H. & Häme, T. (2008) ALOS Palsar Data in Boreal Forest Monitoring and Biomass Mapping, Proceedings of 1st Joint PI Symposium of ALOS Data Nodes, Kyoto, Japan, 19 - 23 Nov. 2007.

Rosenqvist, A., Shimada, M., Ito, N. & Watanabe, M. (2007) ALOS PALSAR: A Pathfinder Mission for Global-Scale Monitoring of the Environment, IEEE Transactions on Geoscience and Remote Sensing 45(11), 3307-3316.

Saari, A. & Kangas, A. (2005) Kuvioittaisen arvioinnin harhan muodostuminen. Metsätieteen aikakauskirja 1/2005:5-18.

Suvanto, A., Maltamo, M., Packalén, P. & Kangas, J. (2005) Kuviokohtaisten puustotunnusten ennustaminen laserkeilauksella. Metsätieteen aikakauskirja, 2005:413-428.

Tokola, T. (1989) Satelliittikuvien käyttö koealaotantaan perustuvassa suuralueiden inventoinnissa. Joensuun yliopisto, metsätieteellinen tiedekunta. Metsätalouden suunnittelun syventävien opintojen tutkielma. 72 s.

Tokola, T. & Heikkilä, J. (1995) Satelliittikuvainventoinnin puuston tilavuusestimaattien luotettavuus tilatasolla. Research meeting of Forest Research Institute, North Karelia, Finland. Metsäntutkimuslaitoksen tiedonantoja 568:23-35.

Tokola, T., Pitkänen, J., Partinen, S., & Muinonen, E. (1996) Point accuracy of a non-parametric method in estimation of forest characteristics with different satellite materials. International Journal of Remote Sensing, 12:2333-2351.

Tomppo, E. (1990) Designing a satellite image-aided national forest inventory. Proceedings from SNS/IUFRO workshop in Umeå 26-28 Feb. 1990. Remote Sensing Laboratory, Swedish University of Agricultural Sciences, Report 4:43-47.

Tomppo, E., Katila, M., Moilanen, J., Mäkelä, H. & Peräsaari, J. (1998) Kunnittaiset metsävaratiedot 1990-1994 (Forest resources per municipalities 1990-1994). Metsätieteen aikakauskirja 4B. (In Finnish).

Tuominen, S. & Poso, S. (2001) Improving multi-source forest inventory by weighting auxiliary data sources. Silva Fennica 35(2):203-214.

Assessment of Solar Irradiance on the Urban Fabric for the Production of Renewable Energy using LIDAR Data and Image Processing Techniques

Cláudio Carneiro[1], Eugenio Morello[2,3], Gilles Desthieux[4]

[1] Geographical Information Systems Laboratory (LASIG), Ecole Polytechnique Fédérale de Lausanne (EPFL), Switzerland, claudio.carneiro@epfl.ch
[2] SENSEable City Laboratory, MIT, Cambridge, MA, USA, eugenio@mit.edu
[3] Human Space Laboratory, DIAP, Politecnico di Milano, Milano, IT, eugenio.morello@polimi.it
[4] Haute Ecole du Paysage, d'Ingénierie et d'Architecture (HEPIA), University of Applied Sciences Western Switzerland (HES-SO), gilles.desthieux@leea.ch

Abstract. A general understanding of the solar admittance and solar gains incident on the urban fabric is very useful to assess the potential implementation of renewable energies at the scale of the city. The authors propose a tool that uses Light Detection and Ranging (LIDAR) data to automatically derive this information in a fast and accurate way with no need to refer to the construction of complex models of the urban layout. In particular, a complete methodology from the extraction of LIDAR data to the environmental analysis of urban models and the visualization of results is presented. Aim of the work is to establish a process to investigate digital urban models integrating cross-disciplinary competences, like remote sensing, GIS, image processing and urban and environmental studies. Toward this goal, working on several interfaces, tools and datasets was necessary to provide a consequent structure to the introduced methodology.

Case study for application was two districts in Geneva (a historical district, and a modern one) where LIDAR data are available. The use of a hybrid approach from raw LIDAR data and vectorial digital maps (GIS data) of roofs footprints derived from a 3-D urban model of the city of Geneva allowed to interpolate a 2.5-D

M. Sester et al. (eds.), *Advances in GIScience*, Lecture Notes in Geoinformation and Cartography, DOI 10.1007/978-3-642-00318-9_5, © Springer-Verlag Berlin Heidelberg 2009

urban surface model (DUSM) of roofs buildings, with a resolution grid of 0.50 by 0.50 metres. This step reveals itself as fundamental for processing the environmental analysis of the urban texture. In particular, the implemented tool calculates solar radiation and solar accessibility on buildings' roofs, in order to investigate the potential for implementing photovoltaic panels.

Keywords: 3D data analysis, LIDAR, 2.5-D urban surface models, urban morphology, image processing, solar radiation, 2-D and 3-D visualization

1 Introduction

Recently, the increasing attention to environmental issues in urban studies has opened up many questions about how planners should manage them in the design process. In fact, numerous authors and architects are convinced that cities play a leading role in controlling sustainability: strategies for redefining more efficient cities in terms of energy performance and environmental quality were the centre of attention in seminal work by Richard Rogers in defining policies for UK cities (Rogers, 1997; Urban Task Force, 1999) and supported the debate around the promotion of more compact cities (Jenks et al., 1996, 2000). Anyway, a comprehensive and reliable toolkit for sustainable urban design is lacking among practicing professionals.

Today's availability of 3-D information about cities offers the possibility to analyse the urban fabric in a very innovative way. In fact, even if LIDAR data permit to derive precious and precise information about the physical layout of cities, still very few applications have been implemented in order to process these data for the environmental analysis and to get a quick understanding about the performance of the urban form. For instance, the increasing interest in the quantification of energy-based indicators at the scale of the city, strongly suggests the integration of 3-D geography and urban studies in order to provide useful applications for the urban planning. In particular, we introduce a novel cross-disciplinary approach that covers the whole procedure from data acquisition from Airborne Laser Scanning (ALS) to the environmental analysis through the image processing of digital urban models. The tool presented here can be intended just as one tile of a larger mosaic aiming at the quantification of environmental parameters at the urban scale. Using the capability of LIDAR data in order to construct an accurate 3-D urban surface model, a tool for calculating solar radiation incident on urban fabric is implemented.

The investigation of solar radiation in architecture is not new and there are already several tools that calculate radiation performance of buildings very accurately. Nevertheless, these tools are very useful at the micro-scale of architecture (environmental performance software) or at the macro-scale of landscape and regional geography (GIS tools), but the focus on the urban district and tools for automatically calculating irradiation on a whole urban site are mostly lacking.

The calculation of solar radiation gives us some clues about the energy-performance of the urban fabric concerning a general understanding of the different contributions of vertical and sloped surfaces in the urban context. Hence, some critical questions in urban design and planning arise: can we build denser cities without decreasing the potential for passive solar architecture? Which is the incidence of beam versus diffuse solar irradiance contributions in typical urban textures? Those questions could benefit from comparative studies between different urban design solutions using the methodology introduced here.

The two areas selected for the analysis present very different characteristics in terms of extension and building typologies. The first pilot zone is a square 300 metres wide near the Rhone River and the old town of Geneva, in Switzerland, such as presented in shadowed white zone of figure 1 (left image). The urban texture is quite compact and densely built, presenting 36 different buildings with average height of 18.5 metres counted on site. The second pilot zone is a square 600 metres wide near the lake of Geneva, on a residential area, such as presented in shadowed white zone of figure 1 (right image). The urban texture presents a low density urbanization characterized by two storey's tall buildings (294 were counted on site) in average and by generous open areas. The reconstruction of the 2.5-DUSM, as explained later in text is shown in figure 2.

Fig. 1. Two pre-defined areas (pilot zones) of the case study in Geneva's city; left: dense built area near the Rhone River and the old town; right: residential area near the lake of Geneva

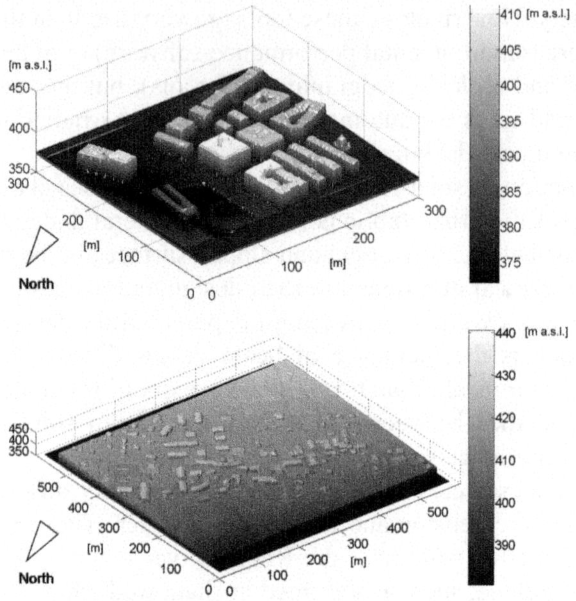

Fig. 2. The 2.5-DUSM visualized on Matlab for the first (image above) and second (image below) pilot zones. Heights are expressed in meters above sea level

2 Related Work

Reference to this work is on one hand 3D GIS applied to urban model building, on the other hand environmental analysis, in particular solar irradiance, in urban area. Both fields are well established in the literature, but most often separately.

In the field of 3-D geography, previous literature on interpolation of LIDAR point clouds is vast. The advantages and disadvantages of several interpolation methods, such as triangle-based linear interpolation, nearest neighbour interpolation and kriging interpolation were presented by Zinger et al. (2002). The most accurate surfaces are created using a grid with a sampling size that relates as close as possible to the LIDAR point density during the acquisition phase (Behan, 2000). For applications where high level of accuracy is demanded, control techniques that analyse the quality of digital terrain models can be carried out (Menezes et al., 2005).

A method to interpolate and construct a 2.5-DUSM (incorporating the geographical relief), based on LIDAR and GIS buildings data, has been proposed by Osaragi and Otani (2007). As related research, there are some semi-automatic methods available to create 3-D GIS data from LIDAR data, such as the Virtual London at UCL (Steed et al., 1999) and the Map-Cube at Tokyo CadCenter Corporation (Takase et al., 2003).

In the field of environmental analysis, the literature proposes various tools to compute solar irradiance on urban area. A lot of them are based on Computer Aided-Design (CAD) in architectural domain and consist in simulating solar access: RADIANCE lighting simulation model (Compagnon, 2004), TOWNSCOPE II (Teller and Azar, 2001), SOLENE (Miguet and Groleau, 2002) and other works presented by Ward (1994) and Robinson and Stone (2005). However, such tools require contextual data specific to a given district and can with difficulty generalize and automate calculation on several districts. For this reason, Batty et al. (1999) stress the need to couple such CAD tools with 3D GIS so as to include data processing and spatial analysis systems and to provide automatic or systematic analysis on urban area.

In this perspective of integrated approach, Rylatt and al. (2001) developed a solar energy planning system on urban scale using GIS-based decision support. Two different studies concerning the analysis of solar potential of roofs have been recently presented (Kassner et al., 2008; Beseničara et al., 2008).

Pioneers in the use of image processing techniques to analyse environmental indicators and morphology of digital urban models was a group of researchers at the Martin Centre, University of Cambridge (Ratti, 2001; Ratti and Richens, 2004; Ratti et al., 2005). Today's increasing availability of 3-D information from user generated contents and remote sensing surveys, makes this technique promising and very useful for a general understanding of the performance of our cities.

The technique is based on the use of very simple raster models of cities, called 2.5-DUSMs. These models reproduce the geometry of the urban fabric and are produced by regularly spaced matrices of elevation values, which contain 3-D information on 2-D digital support, stored in Bitmap format. Developing software algorithms derived from image processing, it is possible to develop efficient strategic tools for analysing and planning the sustainable urban form, measuring geometric parameters and assessing radiation exchange, energy consumption, wind porosity, visibility, spatial analyses, etc. Results are fast and accurate.

On the basis of this literature, our work will propose a conceptual and methodological framework that formalises the introduction of solar irradiance and morphological analysis into 3D building models using GIS in a consistent way.

3 Dataflow Process

The process for structuring the proposed methodology is based on five major steps as represented in the dataflow of figure 3: (1) the construction of the 2.5-D urban surface model, (2) the segmentation procedure for laser scanning data to derive slopes and orientations of roofs, (3) the analysis of suitable building roofs in terms of surface area for photovoltaic purpose, (4) the solar irradiation on building roofs through the image processing of urban models, (5) the visualizations of results. All these steps will be further detailed any analyzed on the next three sections of this paper.

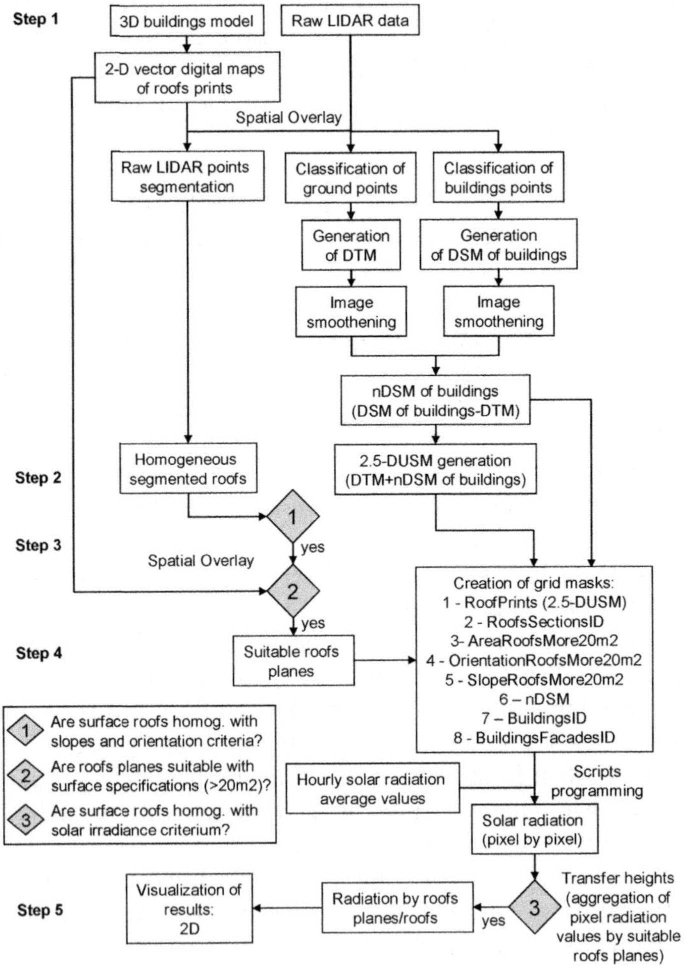

Fig. 3. Dataflow summarizing the process used to calculate irradiation on roof surfaces

4 Presentation of Methodology: the Implementation of the 2.5-D Urban Surface Model from LIDAR Data to the Calculation of Irradiation on Roof Surfaces

4.1 The Implementation of the 2.5-DUSM from LiDAR Data to Environmental Analysis

For the 2.5-DUSM interpolation of the case study of the city of Geneva here presented two data sources are required: raw LIDAR data and 2-D vectorial digital maps of roofs prints of buildings derived from a 3-D urban model. The existing 3-D urban models were reconstructed by the enterprise CIBERCITY, combining a hybrid approach that uses ortophotos and raw LIDAR data.

First, we interpolate a digital terrain model (DTM) by classifying the LIDAR points according to the following sequential operations:

- Using a GIS software, LIDAR points contained within building polygons and in the 1 metre buffer generated from building polygons are eliminated.
- Using the classification tools provided by TerraScan software, LIDAR points whose elevation value vary significantly from surrounding points are considered to be points indicating features such as aerial points (for example, if the laser beam touches a bird), trees and vehicles, and thus are removed.

After eliminating the points indicating all these features, a DTM can be interpolated only from ground points – with good density of LIDAR points (such as in our testing areas, located in Geneva city, where LIDAR points were acquired with a density of 4 to 6 points per square metre) there is no great difference among some of the existing gridding interpolation methods that can be employed, such as the nearest neighbour binning, inverse distance weighting, triangulation with linear interpolation, minimum curvature, kriging and radial basis functions (Gonçalves, 2006). All these interpolation methods are accessible in common GIS software available on the market.

For its generalised use by the scientific community for DTM interpolation, the triangular interpolation was chosen. Secondly, we interpolate (using only the LIDAR points classified as being contained within vectorial roofs prints) a value for each grid cell corresponding to a roof value for all the existing buildings on the areas of study. Thus, a triangulation with linear interpolation is also applied to each one of the roofs of the buildings. It is important to note that along edges of roofs, LIDAR points whose elevation value vary significantly from surrounding points are considered to be points indicating features such as low points and are removed (using the classification tools provided by TerraScan software). For each building, and more specifically for each grid cell contained within, the building height (also defined as nDSM of buildings) is taken to be the value of subtraction of the terrain elevation (calculated in the DTM interpolated) from the building elevation. Lastly, each building is added to the DTM as a column (whose borders are defined from the vectorial roofs prints), using the building height found previously for each cell contained within, as described in last paragraph. The final result allows the construction of a 2.5-DUSM of roofs, which is composed of only terrain and buildings heights information (DTM + nDSM). Data source and parameters needed for generating the 2.5-DUSM of roofs are shown in figure 4.

Finally, in order to complete the image enhancement of the model, we have to refine the facades of buildings that are sloped because of interpolation. In order to achieve this goal, we applied an image smoothing using a 3 by 3 low-pass filtering. Each building's contour pixel was deleted and then expanded again, in order to assign more constant values to roofs edges (figure 5).

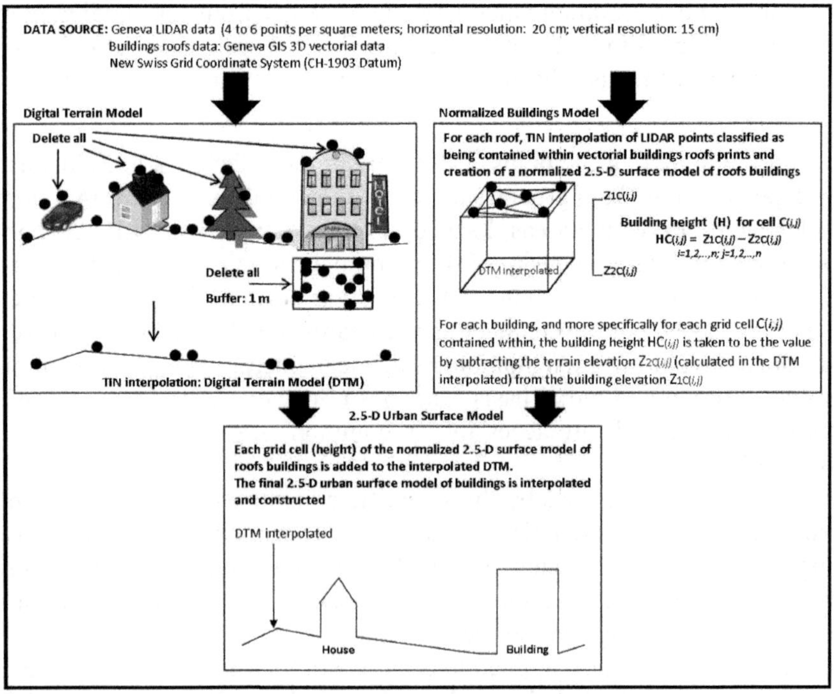

Fig. 4. Interpolation and construction of 2.5-D Urban Surface Model (2.5-DUSM)

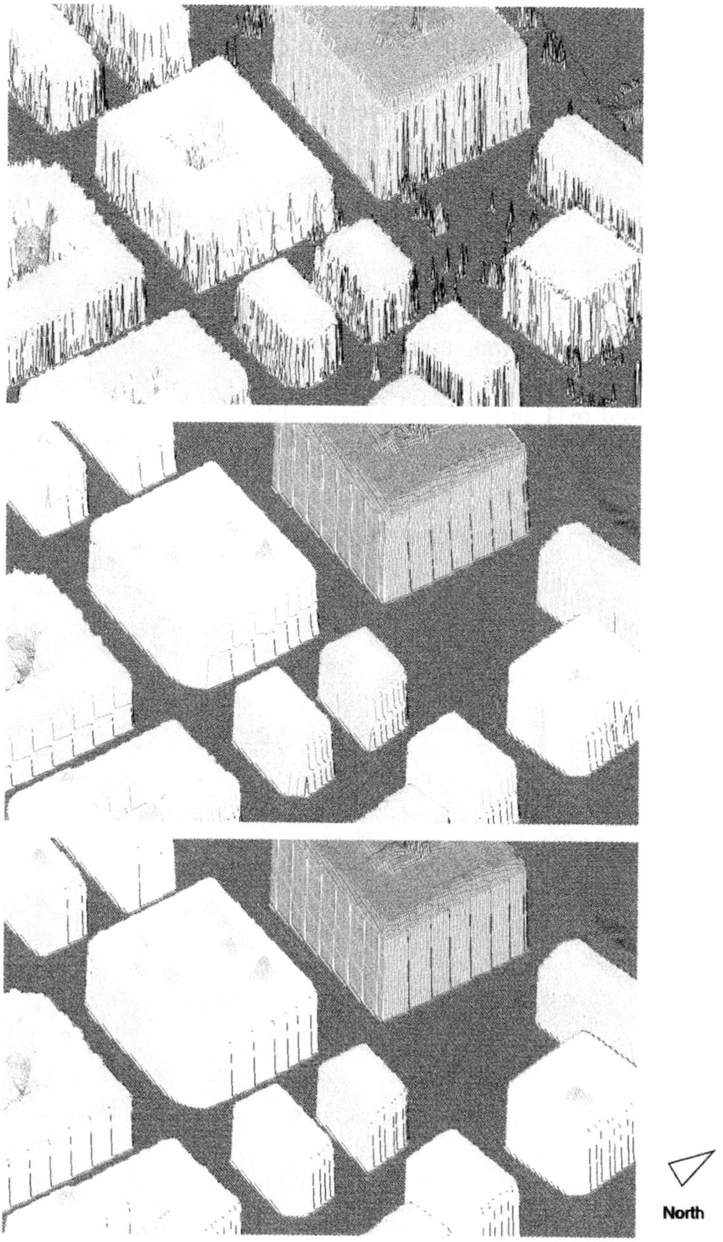

Fig. 5. The enhancement of the urban model; from the top: (a) the raw LIDAR data, (b) the original DEM, (c) the reconstruction of roofs edges and buildings facades. With this last step, the facades become vertical

4.2 Segmentation Procedure for Laser Scanning Data: Derivation of Slope and Orientation of Roofs

A segmentation procedure for laser scanning data was implemented to search for planar faces in order to define more accurately slope and orientation of each roof section, when compared to slope and orientation automatically calculated using 2.5-DUSM. It follows a region growing principle that takes the deviation from a plane in 3D into consideration, as initially described by Quint and Landes (1996) and later on enhanced to application on LIDAR data by Vögtle and Steinle (2000). Figure 6 shows the result of the algorithm for an area of the second pilot zone using all pulse laser scanning data.

As proposed by Lemp and Weidner (2005) the algorithm has been also applied using only last pulse laser data but results were not satisfying, especially along roofs edges.

Fig. 6. Example of segmentation procedure (areas in red) for laser scanning data on part of the second pilot zone

4.3 Surface Roof Requirements and Creation of Masks for Suitable Roofs

The segmentation procedure enables create homogeneous roof planes in terms of orientation and slope. On the other hand, the 2-D vectorial digital map of roofs prints contains such an accurate geometry of roof but with no information on slope or orientation.

Therefore, through the overlay between the layers of segmented roof and the 2D map of roofs prints, it is possible to add, in the second layer, attribute data on orientation and slope and to calculate from slope real area of each roof.

In historical city centres, building roofs are often of complex structure and split into numerous roof sections. A lot of them are too small to implement PV panels. So it is matter of making a pre-selection of roofs that meet some minimal requirements in terms of surface area. Let consider that an installed power capacity of 1 KW requires 8 m^2 of PV panel (PACER, 1996). In practise, the installed power of a PV panel should not be inferior to 2 KW, for cost-effectiveness purpose. This means that the minimum surface area of PV panel unit is equal to 16 m^2. If we consider a need of some distance margin between the border of the roof and the border of the PV panel, in our model, we will select roofs with surface area superior or equal to 20 m^2.

For flat roofs, additional slope is provided to panels to increase their efficiency. This result in decreasing the part of the roof area being used as a inclined panel provokes shadow on the panel located behind it (figure 7). For instance, following the equation 5 given in PACER (1996), for panels with slope of 30° (Beta), the ratio A_{panels} / A_{roof} decreases to 40% (only 40% of the roof area can be used). Therefore, high electrical production involves an optimum between the total PV area on a roof and the efficiency of each panel. Given the recent technological developments of PV panels and the increase of efficiency, a low slope of about 5° enables to meet the optimum, which corresponds to a use of 80% of the flat roof. So, in our model, we select flat roofs where 80% * real area $\geq$ 20 m^2.

$$\frac{L}{P} = \frac{1}{\cos\beta + \sin\beta / \tan\theta} \tag{1}$$

Where:

L: length of the inclined panel
P: distance between panels
β: slope of the panel
θ: shadow angle

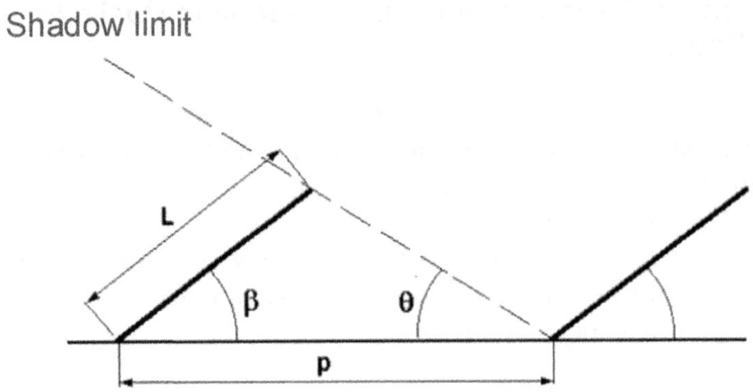

Fig. 7. Implementation of inclined PV panels on flat roof (adapted from PACER, 1996)

Finally, after having selected suitable roof planes in terms of surface area requirement, grid pixel masks of roof footprints, orientation and slope are created from 2D vector map of roofs prints, with a resolution 0.5 meter by 0.5 meter. Such masks serve as input for the calculation of irradiation.

4.4 Solar Irradiation Calculation on Buildings Roofs

The calculation is made with MatLab scripts from hourly irradiation average values, pixel by pixel. Theoretical background about calculating solar radiation in urban area and its application to Geneva sectors are respectively presented in details in sections 5 and 6 of this paper.

4.5 Meteorological Data Sources

The solar irradiation data we use in our application come from METENORM® that is a meteorological reference incorporating a catalogue of meteorological data and calculation procedures at several locations in the world. The database offers, among others, average data on global and diffuse radiation incident on horizontal surface, for a given time period and scale. By subtraction we can deduce beam irradiation. From the monthly values (station data), METEONORM calculates hourly values of all these parameters using a stochastic model. The resulting time series correspond to "typical years".

The version (4.0) we used provides statistical data of a typical year that corresponds to the period 1970-1990 for the Geneva location[1]. With such data we would like to calculate yearly balance of irradiation on each roof pixel in a given area in Geneva. However, calculating irradiation for each hour of a typical year and for each pixel of a high resolution 2.5-DUSM model would result of several days of computer time simulation. Consequently, we reduced our solar irradiation dataset by averaging hourly values for each month. The yearly dataset contains thus 288 (24hours x 12months) hourly values instead of 8'760. The solar geometry has to be considered for the day the most representative of each month, which is the day 15th.

5 Theory on Solar Radiation: Calculating Hourly Global Radiation Incident on an Inclined Plane (Roof)

Raw data on solar irradiation are generally derived at hourly time scale in most meteorological databases. In the following, we therefore consider such a time scale in the calculation of irradiation. As we will see, calculating solar radiation incident involves describing interactions between the sky model, solar geometry and 2.5D buildings model (Compagnon, 2004).

Let assume the case of a roof or section of a roof inclined at an angle β from the horizontal position and with an azimuth exposition angle γ. At locations where the hourly global and diffuse radiation on horizontal surfaces are measured and known, the global radiation on an inclined surface can be written, by referring to Iqbal (1983, p. 320), as:

$$I_{\beta\gamma} = I_{b\beta\gamma} + I_r + I_s \quad (1)$$

Where:

$I_{\beta\gamma}$: hourly *global* radiation incident on an inclined plane at an angle β and with an azimuth angle γ

$I_{b\beta\gamma}$: hourly *beam* radiation incident on an inclined plane

I_r: hourly *ground-reflected* radation incident on an inclined plane

I_s: hourly sky *diffuse* radiation incident on an inclined plane

The expression of the three components of the global irradiation and the way of calculating them are detailed in Iqbal (1983).

[1] The most recent version 6.1 (2008) offers radiation data for the time period 1980-2000

However, in urban area, it should be taken into account shadowing effects due to obstructions in the surrounding environment of the irradiated surface, like other buildings, neighbour building structures (chimney), trees, etc.

For the beam component, by combining the 2.5D-model of urban area and geometric position of the sun hour per hour, it is possible to calculate the shaded surface area of a roof section during a given hour (Glenn and Watson, 1984; Incropera and DeWitt, 1960).

Assuming an isotropic model, the amount of sky diffuse radiation is to be linked with the sky visibility from the surface roof of interest. Therefore, obstructions on diffuse radiation and on sky visibility more in general can be analysed through the Sky View Factor (SVF); it expresses the relationship between the visible area of the sky and the portion of the sky covered by surrounding buildings and other obstacles viewed from a specific point of observation (Souza et al., 2003). The SVF encompasses reduction of visibility due to the slope of an inclined plane and obstacles in the surrounding built environment. The diffuse radiation incident is thus obtained through the product of SVF (%) by available diffuse radiation on unobstructed horizontal surfaces. An example of calculating SVF with 3D GIS is proposed by Souza et al. (2003) through their 3DSkyView tool.

The ground reflected (albedo) component being negligible, we do not consider the shadowing effect.

6 Image Processing of Raster Images for the Study of Solar Radiation

In this study the reconstruction of the 2.5-D urban model was applied for implementing specific tools for calculating the solar irradiance intercepted by urban roofs according to figure 4. Aim of the proposed application is first to determine the rate of accessibility of urban roofs to direct solar radiation and second to estimate the potential usage of those surfaces for the production of renewable energy from the sun. To fulfil the analysis results require being stored in different modalities. Outputs of the proposed tools are both numerical data and visualizations. Numerical data of solar radiation is collected pixel by pixel on the roofs and also stored in synthetic tables containing the list of the identified roofs' sections. Visualizations, on the other hand, make numerical results visible on the 2-D plans of the case-study areas.

The technique used in this application is based on the image processing of the provided models that are interpreted as raster images, i.e. 2-D arrays where every intensity value represents the height of the pixel in metres. The density[2] and accuracy[3] of the LIDAR data used are relatively high. Thus, a 2.5-DUSM (grid) of 0.50 by 0.50 metres, which has a sampling size that relates as close as possible to the LIDAR point density during the acquisition phase (Behan, 2000), was interpolated and constructed. The technique computes irradiance values on all target points being visible from a viewpoint location represented by the position of the sun.

Two types of input data are required for running the tool: first, geographical data that inform the location of the case-study areas (geographical coordinates of the case-study area, the height above sea level and the physical extension of the site in metres) and second the statistical irradiance values (see section 4). From these letter irradiances it is possible to derive irradiance values for every orientation and inclination of surface, according to solar geometry formulae (see section 5).

A list of several masks is the next necessary set of inputs to run the model. These are the following:

- The digital elevation model (DEM) mask, here also called 2.5-D urban surface model, whereby intensity values of pixels represent the height above the sea level. This input image is the main 2.5-D information to run the script;
- The masks including labels of buildings and roofs' sections. Hence, it is possible to determine the exact identification of each building and roof's section.
- The masks containing the IDs of buildings facades and the nDSM of buildings were, respectively, used to compute the areas of the vertical surfaces and the volume of buildings (applied on the morphological analysis shown in section 7.3). A technique of image enhancement was used to define the exact height of each perimeter pixel of buildings.
- The masks for describing roofs are essentially two and inform about orientations and inclinations (slope) of roof pixels and consider also constraints due to the potential use for the installation of PV modules (refer to Section 4).
- The mask of areas of roofs' sections bigger than 20 m^2. All roofs' sections not included on this mask were excluded from computation.

Solar geometry formulae allow to derive the beam and the diffuse components of hourly radiations just starting from the previous mentioned inputs. In the computation of the time distribution of radiations during the

[2] 4 to 6 points per m^2

[3] Planimetric precision of 20 centimetres and altimetric precision of 15 centimetres

day, however, we can just assess values representative for clear sky conditions, even if they tend to produce conservative estimates (Duffie, 1980, p.77). Therefore, even if accurate, results of the application should be used mostly in comparative studies to evaluate differences of solar caption in time.

Once the irradiance arrays are calculated and the input images are defined, the core of the computation can run. The shadow casting routine first introduced by Ratti (2001) is applied here at a specific hour of the day and is used to detect which pixels on roofs are in shadow and which collect direct sunlight. The procedure calculates the shadow volume. First, the three components of the vector pointing towards the sun are defined. Then, we compute the components of an opposite vector, scaled so that the larger of the x and y components is just one pixel, and the z component is adjusted to the image calibration. If we translate the model by the x and y components, and simultaneously reduce its height by subtracting the z component, we get part of the shadow volume. If we continue translating and lowering by multiples of this vector, and take the maximum of this volume with that previously calculated, we build up the whole shadow volume. The process can be stopped when all levels are zero or the translation has shifted the volume right off the image. To reduce the shadow volume to an actual map of shadows on the roofs and ground level of the city, the original model is subtracted from the shadow volume. Pixels with negative or zero values are in light; positive values are in shade (please refer to Morello and Ratti, 2009 for details). The same procedure can be applied also for the calculation of the SVFs on the model. In that case, multiple iterations of shadow casting are processed starting from simulated source points homogeneously distributed on the sky vault.

Hence, when we are able to determine for every pixel its shadowing condition, its SVF, its orientation, its inclination and its linear extension, we can assign the incident solar irradiance calculated in Watt per square metre. Moreover, we need to define the linear length of pixels of roofs' sections: depending on their inclination, pixels can assume different areas. This can be easily calculated applying trigonometric formulae.

As final result we store a set of irradiation values in dedicated arrays. First we accumulate all radiation maps in a 4-D matrix (length of the site x width of the site x 24 hours of the typical day of the month x 12 months), so that it is possible a posterior to reproduce the visualization of solar admittance on an hourly basis. The second type of array is a 2-D table where the rows represent the ID of roofs' sections and columns store hourly, daily and monthly irradiance values.

7 Results of the Analysis

7.1 Presentation

The analysis aims to reveal the behavior to solar admittance on two case-study areas that differ in terms of size and land occupation. In order to produce a significant comparative study, it is fundamental to derive useful urban indicators that allow us to correlate the incident solar radiation in an objective way. To achieve this goal we propose for instance a morphological study to normalize specific constraints due to the diverse urban contexts.

7.2 Solar Admittance on the Case-Study Areas

In this investigation we extract information concerning the solar admittance of two portions of urban fabrics. In doing so, we analyze a limited number of buildings and discard objects beyond the selected area of investigation. This constraint inevitably produces some edge effects, because some obstructions produced by the neighboring discarded urban fabric are not taken into account. Anyway, for these particular urban sites, we are confident that the impact on end-results is negligible.

The first step was to estimate the suitable areas of roofs eligible for the production of energy from the sun through solar panels or PV modules. This analysis can be conducted just considering urban geometry, in particular assuming the following constrains (results shown in figure 8, left image): (a) some orientations of roofs' slopes are discarded a priori (E, NE, N, NW, W); (b) as explained in Section 4.3., areas smaller than 20 m^2 are not taken into consideration, since their energy production is not convenient; (c) only 80% of horizontal surfaces of roofs are computed due to the inclination of panels (see section 4.3.); (d) slopes higher than 60 degrees are inefficient and thus eliminated; (e) finally, as a more restrictive condition (figure 8, right image) we discard also areas that have a SVF smaller than 60% (refer to the SVF maps of figure 9), in order to consider the effects generated by the mutual shadings casted by buildings that otherwise would not be addressed.

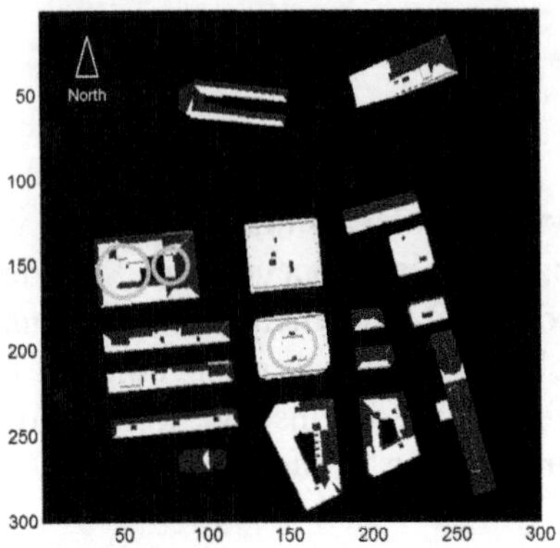

Fig.8. Estimation of suitable areas (in white) and unsuitable roofs (in gray) for the installation of solar panels or PV modules on urban roofs. On the map on the right the authors propose a more restrictive condition, discarding areas where SVF values are smaller than 60%. Hence, areas like terraces facing courtyards or behind chimneys (highlighted with circles) are eliminated

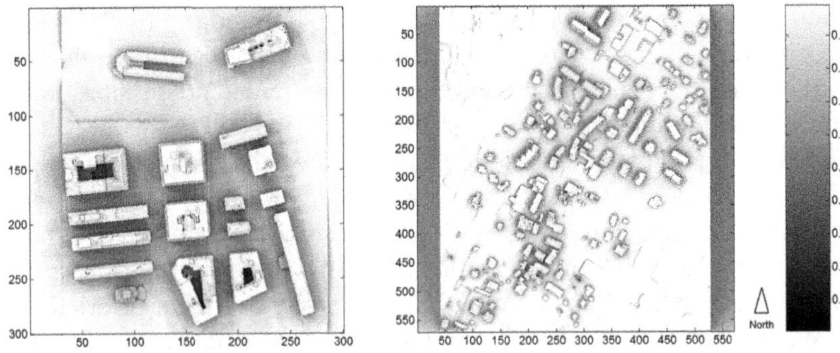

Fig. 9. The SVF computed on the two case-study areas. The average SVF on the roofs is 0,82 for the area on the left and 0,86 for the area on the right

The outputs of this analysis are displayed as visual maps with irradiance values represented on the roofs, such as in the examples presented in figures 10, 11 and 12. Values are aggregated on each roofs' section to highlight in a synthetic representation which are suitable surfaces for installing solar or PV panels. Daily, monthly or annual values can be displayed.

If we compare the two pilot zones some interesting characteristics emerge. The second pilot zone collects almost twice as much solar energy on sloped roofs as the first pilot zone as shown in figure 12, also because the second area has a double amount of roofs' surface available. An example of collected hourly solar irradiances over the first pilot zone on roofs in typical days for December, March and June (MWh) is shown in figure 13.

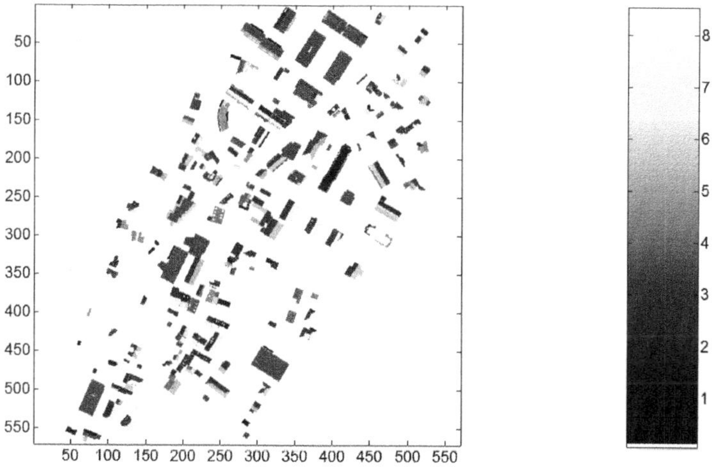

Fig. 10. Daily solar gains collected on the roofs of the second pilot zone on the 15th of December (kWh/ m^2).

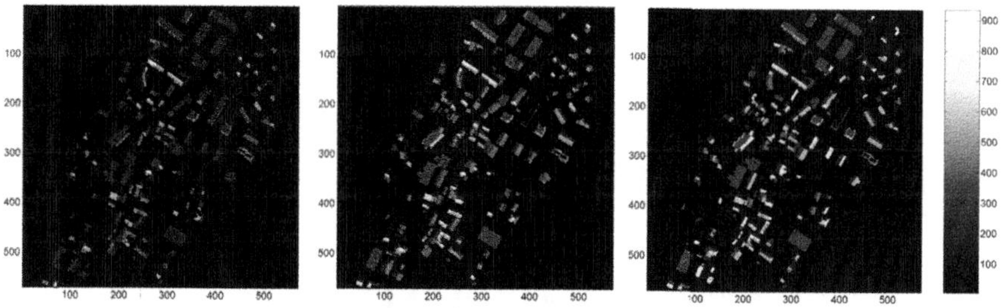

Fig. 11. Hourly maps of irradiances (W/m^2) collected on the roofs for the second pilot zone on the 15th of March. From left: the map at 9 AM, 11 AM and 1 PM

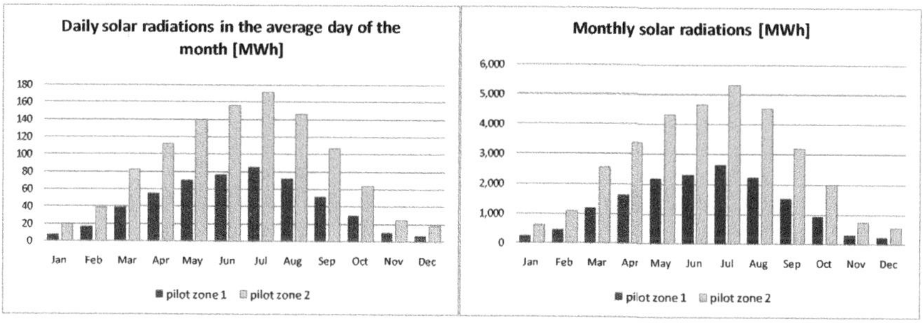

Fig. 12. Solar radiation values collected on roofs on the two pilote zones in Geneve. From the left: daily solar irradiances for the average day of the month (MWh) and monthly irradiances (MWh)

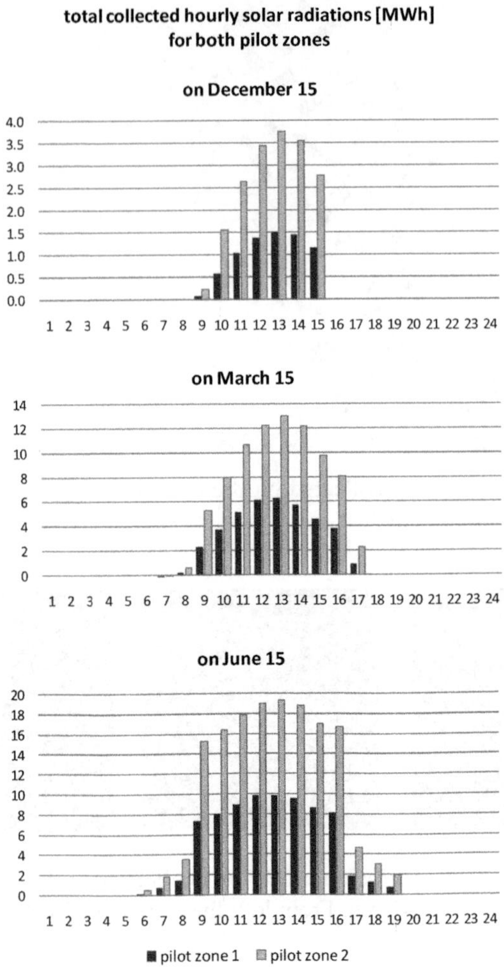

Fig. 13. Collected hourly solar radiation from 6 AM to 8 PM over the roofs of both pilot zones on typical days for December, March and June (MWh)

7.3 Morphological Analysis of the Urban Texture in Relation to Solar Radiation

Aim of the morphological study is to better understand how different urban models behave in terms of renewable energy production, in this case solar energy.

Since the two sites that we are investigating present very different characteristics in terms of extension and building typologies, it is very important to find indicators to conduct a morphological comparative study. The first pilot zone is a more central urban area with a dense fabric, tall buildings and few open spaces. On the contrary, the second zone presents a low density urbanization characterized by two storey's tall buildings in average and by generous open areas. In order to calculate plausible and comparable measures of urban density we discard non urban land (agricultural areas, big parks, and rivers) and we refer to the area defined by the gray contour in Table 1.

Beside more traditional urban morphological indicators (urban land, covered area, built volume, estimation of floor area, mean built height, theoretical population) and urban density measures (urban built density, ground coverage index), we assess the surface to volume ratio, as an indicator of compactness. In fact, the first pilot zone is more compact than the second (S/V ratios are respectively 0,20 and 0,30). It was possible to calculate this parameter, since we provided the necessary masks to distinguish between roofs and the footprints of buildings.

Moreover, we define a set of parameters that address the solar admittance of the urban fabric. These are: (a) the area of roofs suitable for installing PV modules, also expressed as a ratio to the total area of roofs; (b) the irradiance density, i.e. the incident energy per m^2; (c) an urban irradiance density, defined as the total net incident irradiance on urban roofs divided by the population. This last parameter does not represent a measure that can be converted into useful energy and for that reason is not shown in Table 1.

From the analysis of the total incident radiation on two areas, interesting considerations emerge (figure 14, image above and Table 1). In fact, if we normalize the data considering the total areas of roofs (figure 14, image below), then the irradiance on the second area is slightly higher than on the first one. Hence, results are very similar even if the two urban morphologies are very different. Therefore, compact urban areas can perform as well as less dense suburbia in terms of solar admittance.

Results of this analysis show that the production of solar energy on low density areas is easier due to two main parameters: urban obstructions and population density. First, urban obstructions have a higher impact on the compact urban area (the mean sky view factor SVF computed on roofs is 0,82 in the first pilot zone versus a value of 0,86 in the second).

Second, the building typologies in the second case-study area allow to obtain a higher degree of 'potential solar roof' area per person. In fact, if we refer to the theoretical population of the two areas (derived as a standard value from the built volume), then the second zone shows better results (figure 14 above). This is due to the lower densities and consequently to the larger amount of roofs' areas available *pro capite*. For instance, considering the population hosted in both sites, the mean irradiance per person in the second pilot zone is almost three times as much as in the first one (see Table 1). We define this latter index as the 'urban irradiance density' (kWh/person) with the aim to assign to each person the average solar irradiance incident on the roofs during the entire year. It is important to observe that this irradiance density takes into account the total incident solar radiation and not the fraction that can be converted into useful energy alone. Further research will convert this value into electrical energy production.

As a general remark, it is important to outline that this study does not want to promote low density as an urban model. According to the literature and trends in sustainable urban design practice (Burchell et al., 2005; Jenks et al., 1996, 2000; Urban Task Force, 1999), the compact city model is preferrable since it permits sinergies and green policies that require minimum population densities. Anyway, if we only have to analyse the performance of different urban textures in terms of solar admittances, than the actual city model does not optimize its performance. At the same urban density, a better design should take into account the orientation of roofs and facades, the slopes of pitched roofs, mutual obstructions between buildings.

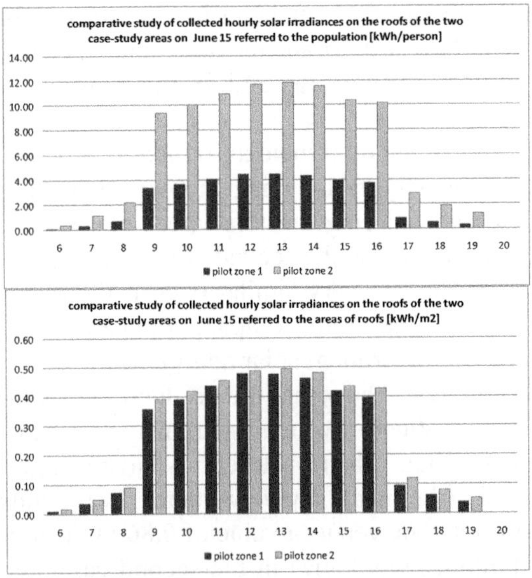

Fig. 14. Comparison of solar irradiance (kWh) collected on the 15th of June from 6 AM to 8 PM on the two case-study areas. Above the values weighted by the population (kWh/person) and below the irradiances referred to the areas of the roofs (kWh/m^2)

Table 1. Morphological indicators computed on the two case-study areas

	PILOTE ZONE 1	PILOTE ZONE 2
MORPHOLOGICAL INDICATORS		
urban land [m²] (red boundary)	42711.00	150670.00
covered area [m²] (black footprint)	17903.00	34592.00
built volume [m³]	331520.00	244600.00
estimation of floor area [m²]	110510.00	81535.00
Mean height [m]	18.52	7.07
population [person]	2210.13	1630.67
area of roofs [m²]	20660.44	39109.46
area of facades [m²]	46535.00	34751.00
surface to volume ratio [1/m]	0.20	0.30
POTENTIAL AREAS FOR INSTALLING "SOLAR ROOFS"		
suitable areas of roofs for the installation of PV panels [m²]	11438.05	20757.35
percentage of roofs suitable for the installation of PV panels [%]	55.36	53.08
DENSITY MEASURES		
urban built density [m³/m²]	7.76	1.62
ground coverage index [m²/m²]	0.42	0.23
Total irradiance per year [MWh/year]	16.05	32.88
gross irradiance density on the urban land [MWh/m²/year]	0.38	0.22
net irradiance density computed on roofs [MWh/m²/year]	0.78	0.84
mean irradiance density per person [MWh/person/year]	7.26	20.16

8 Visualization of Results: 2-D Displays for Communicating Purposes

According to Nielsen (1993), the acceptability of any visual exploratory system is strictly related to its utility (feasibility of the information to be visualized) and its usability (cognitive visual interpretation of the 3-D urban models proposed).

This combination between utility and usability determines the level of acceptability among the different users (in particular for architects and urban planners) of the proposed 3-D urban models (Reichenbacher and Swienty, 2007).

User requirements concerning the utility and usability of 3-D urban models for communication and visualization purposes have not yet caused much attention within the world of researchers and developers. Thus, for different users and applications, it is fundamental to clearly distinguish which levels of detail (LOD) should be implemented and, based on this classification, which urban objects should be or should not be visualized. This filter of criteria is essential, in order to avoid too dense and confusing urban scenes.

The 2-D visualization of results here proposed is based on the user requirements study undertaken for the city of Geneva, such as presented by Carneiro (2008). For both pilot zones, figures 15 and 16 show the percentage of suitable for solar panels and figures 17 and 18 show the average monthly irradiance values in KWh/m^2.

Fig. 15. Percentage of suitable area for solar panels on the first pilot zone with a Sky View Factor (SVF) of 60%

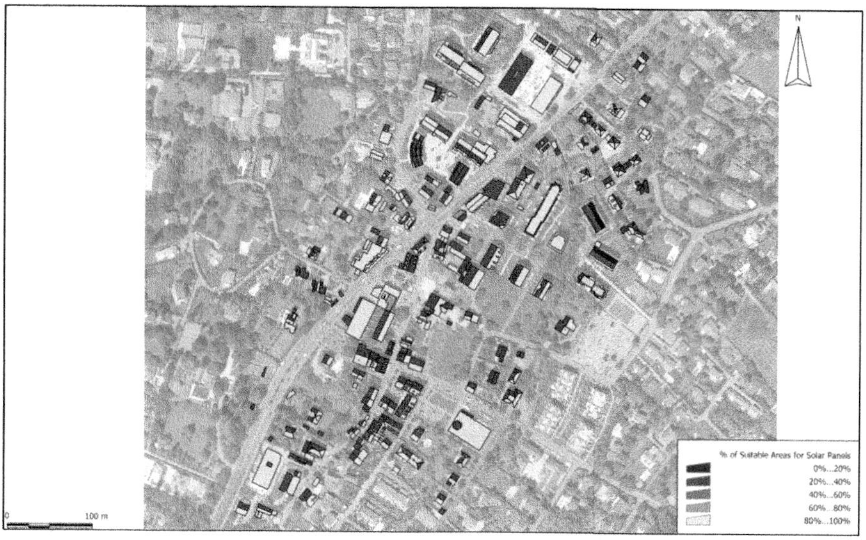

Fig.16. Percentage of suitable area for solar panels on the second pilot zone with a Sky View Factor (SVF) of 60%

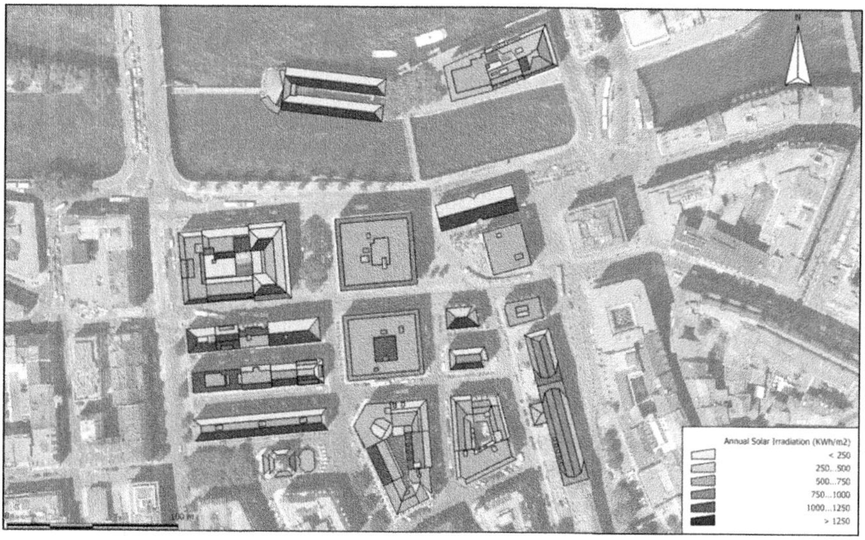

Fig. 17. Annual solar irradiance values (KWh/m^2) for the first pilot zone

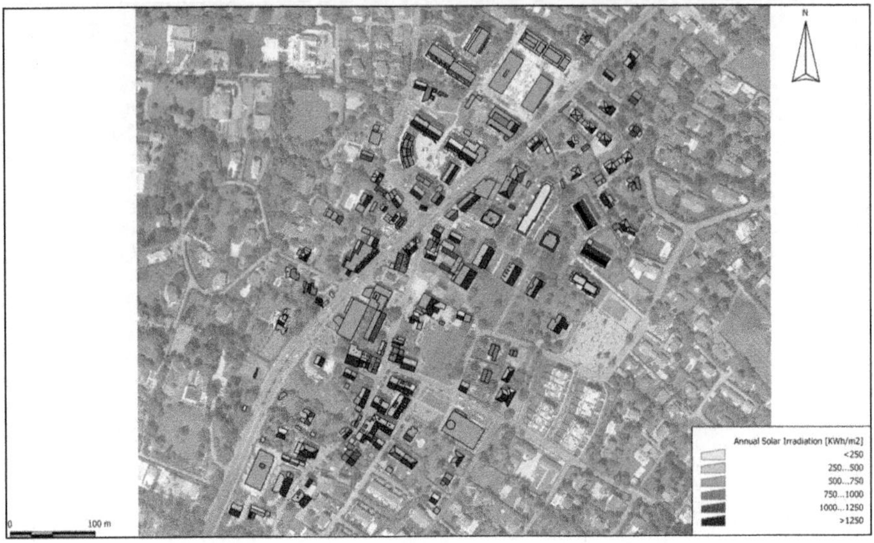

Fig. 18. Annual solar irradiance values (KWh/ m^2) for the second pilot zone

9 Conclusions

In this paper a complete methodology from the extraction of LIDAR data to the analysis of solar admittance over urban models and the visualization of results was introduced. Combining different disciplines interfaces and datasets reveal constrains of today's applicability of LIDAR images for the environmental prediction of the urban form. Hence, a first result is to bring 3-D geography and urban studies together in a process that goes from data acquisition and processing to urban modelling and urban design application.

Applications of this methodology in urban design and planning are very promising. In our case study we limit the analysis to the physical built environment, but we could extend the investigation to assess the impact of new buildings in the city and use the technique for improving design schemes based on an evaluation of quantitative indicators before and after changes are introduced. If applied on more urban typologies once LIDAR data are available, the morphological analysis could lead to interesting indicators that could be used by urban planners in future for predicting the environmental behaviour of different urban textures. In fact, guidelines that refer to comprehensive environmental indicators are still missing in the urban design literature.

Besides urban planning strategies, also environmental policies could be promoted if a comprehensive mapping of solar irradiances on buildings is provided. In fact, a database of solar admittances on buildings is important for two main reasons: first, it could affect the real estate market and assess the values of the urban fabric also according to its energetic performance; second, the community could program interventions and define specific incentives able to take into account more precisely the real potential energy production strategies of each building.

10 Future Work

Future work will focus on the calculation of the potential electric power produced through PV modules as a natural continuation of this work. With the obtained radiation values it is already possible to derive general indications about the potential production of electric power, considering simple rule of thumbs related to the efficiency of PV panels (for instance, about 15% of the incident irradiance can be converted into useful electricity).

Moreover, some improvements concerning the technical implementation of the presented methodology are needed. The accuracy of results depends mainly on two issues that we tried to overcome: the quality of input LIDAR data and the constraints of the image processing technique of 2.5-D urban surface models. In order to enhance the precision of LIDAR data, a process of image reconstruction was implemented and a hybrid approach that integrates the use of vectorial buildings roofs prints data was proposed in order to reconstruct the perimeters of buildings roofs.

As here presented, existing vectorial digital maps (GIS data) can be used if available and updated, but outlines of buildings roofs from this source of information should be always handled with special care. In fact, the 2-D outlines of buildings footprints do not have to represent the outline of the building roof. Modifications between GIS data and laser data can have numerous reasons which can not automatically be recognized. Proposals using vectorial digital maps as input for 2.5-D urban surface model interpolation and construction should be attentive of the fact that in some cases map information might not give the correct hints about 3-D buildings shapes.

Improvements in the ways we store the synthetic data are needed, since our visualizations refer now to relative values aggregated on roofs' sections. In so doing, we do not store the absolute values of irradiance pixel by pixel on the roofs, thus averaging results that are sensitive to the effective characteristics (areas, orientation and slope) of the considered roof's sections.

Finally, future work should refine some procedures presented in the methodology, in particular:

- ameliorate the quality of the urban model with the integration of trees, which remains one of the most delicate aspects concerning environmental analysis using 2.5-D urban surface models;
- provide a statistical analysis on 2.5-D urban surface models;
- investigate the utility and usability of the proposed visualizations outputs through users' evaluation.

Acknowledgements

The authors are definitely indebted to *Thomas Vögtle,* from the *University of Karlsruhe* for his support on the segmentation process of LIDAR data, *Peter Gallinelli,* from *Geneva Institute of Technology*, *André Mermoud* from the *University of Geneva,* for their support on solar radiation theory *Laurent Niggeler and Adrien Vieira de Mello* from the *Direction Cantonale de la Mensuration Officielle* (DCMO) of the canton of Geneva, Switzerland, for their vast contribution to this project. Also, the authors would like to thank the *Fondazione Rocca* and its president, *Gianfelice Rocca,* for providing generous funding for this research.

References

Batty, M., Dodge, M., Jiang, B., Smith, A. (1999) Geographical information systems and urban design, Edited by Stillwell, J., Geertman, S., Openshaw, S., Geographical Information and Planning, Published by Berlin Springer, pp. 43-65.

Behan, A. (2000) On the matching accuracy of rasterised scanning laser altimetre data, The International Archives of the Photogrammetry, Remote Sensing and Spatial Information Sciences, Vol. XXXIII, Part B2, pp. 75-82.

Beseničara, J., Trstenjak, B., Setnikac, D. (2008) Application of Geomatics in Photovoltaics, The International Archives of the Photogrammetry, Remote Sensing and Spatial Information Sciences, Vol. XXXVII, Part B4, pp. 53-56.

Burchell, R., Downs, A., McCann, B., & Mukherji, S. (2005) Sprawl Costs: Economic Impacts of Unchecked Development. Washington, Covelo, London: Island Press.

Carneiro, C. (2008) Communication and visualization of 3-D urban spatial data according to user requirements: case study of Geneva, The International Archives of the Photogrammetry, Remote Sensing and Spatial Information Sciences, Vol. XXXVII, Part B2, pp. 631-636.

Compagnon, R. (2004) Solar and daylight availability in the urban fabric, Energy and building, Vol. XXXVI, pp. 321-328.

Duffie, J.A. (1980) Solar engineering of thermal processes, John Wiley & Sons, New York.

Glenn, T., Watson, I. (1984) The Determination of View-Factorsin Urban Canyons. Journal of Climate and Applied Meteorology, Vol. XXIII, pp. 329-335.

Gonçalves, G. (2006) Analysis of interpolation errors in urban digital surface models created from LIDAR data, Proceedings of the 7th International Symposium on Spatial Accuracy Assessment in Resources and Environment Sciences. Edited by M. Caetano and M. Painho, pp. 160-168.

Incropera, F., DeWitt, D. (1960) Introduction to heat transfer. New York: John Wiley & Sons.

Iqbal, M. (1983) An Introduction to Solar Radiation, Academic Press, New York.

Jenks, M., Burton, E., Williams, K. (1996) The Compact City: A Sustainable Urban Form?, E & FN Spon, London, UK.

Jenks, M., Burton, E., Williams, K. (2000) Achieving Sustainable Urban Form, E & FN Spon, London, UK.

Kassner, R., Koppe, W., Schüttenberg, T., Bareth, G. (2008) Analysis of the Solar Potential of Roofs by Using Official LIDAR data, The International Archives of the Photogrammetry, Remote Sensing and Spatial Information Sciences, Vol. XXXVII, Part B4, pp. 399-404.

Lemp, D. and Weidner, U. (2005) Improvements of roof surface classification using hyperspectral and laser scanning data, Proceedings of the URBAN 2005 Workshop, Arizona, USA.

Menezes, A.S., Chasco, F.R., Garcia, B., Cabrejas, J., González-Audícana, M. (2005) Quality control in digital terrain models, Journal of Surveying Engineering, Vol. CXXXI, pp. 118-124.

Morello, E. and Ratti, C. (2009) SunScapes: 'solar envelopes' and the analysis of urban DEMs, Computers, Environment and Urban Systems, Vol. XXXIII, Part 1, pp. 26-34.

Miguet, F. and Groleau, D. (2002) A daylight simulation tool for urban and architectural spaces: Application to transmitted direct and diffuse light through glazing, Building and Environment, Vol. XXXVII (8/9), pp. 833-843.

Nielsen, J. (1993) Usability Engineering, Morgan kaufmann-Academic Press, London, United Kingdom.

Osaragi, T., Otani, I. (2007) Effects of ground surface relief in 3-D spatial analysis on residential environment, The European Information Society: Lecture notes in Geoinformation and Cartography, Edited by Sara Irina Fabrikant and Monica Wachowicz, Published by Springer Berlin Heidelberg, pp. 171-186.

PACER-Energies renouvelables (1996) Centrales photovoltaïques. Guide pour le dimensionnement et la réalisation de projets. A l'usage des bureaux d'ingénieurs, Office fédérale des questions conjoncturelles, Bern, Switzerland.

Quint, F. and Landes, S. (1996) Colour aerial image segmentation using a Bayesian homogeneity predicate and map knowledge. International Archives of Photogrammetry and Remote Sensing, Vol. XXXI, Part B3.

Ratti, C. (2001) Urban analysis for environmental prediction, unpublished Ph.D. dissertation, University of Cambridge, UK.

Ratti, C., Richens, P. (2004) Raster analysis of urban form, Environment and Planning B: Planning and Design, Vol. XXXI.

Ratti, C., Baker, N., Steemers, K. (2005) Energy consumption and urban texture, Energy and Buildings, Vol. XXXVII (7), pp. 762-776.

Reichenbacher, T., Swienty, O. (2007) Attention-guiding geovisualization, Proceedings of the 10th AGILE International Conference on Geographic Information Science, 8th-11th May, Aalborg University, Denmark.

Robinson, D., Stone, A. (2005) A simplified radiosity algorithm for general urban radiation exchange, Building Services Engineering Research and Technology, Vol. XXVI, No. 4, pp. 271-284.

Rogers, R. (1997) Cities for a small planet, Faber & Faber, London, UK.

Rylatt, M., Gadsden, S., Lomas, K. (2001) GIS-based decision support for solar energy planning in urban environments, Computers, Environment and Urban Systems, Vol. XXV, pp. 579-603.

Steed, A., Frecon, E., Pemberton, D., Smith, G. (1999) The London Travel Demonstrator, Proceedings of the ACM Symposium on Virtual Reality Software and Technology, December 20-22, pp. 50-57, ACM Press.

Souza, L., Rodrigues, D., Mendes, J. (2003) A 3D-GIS extension for sky view factors assessment in urban environment, The 8th International Conference on Computers in Urban Planning and Urban Management "CUPUM'03 Sendai", 27-29 May, Japan.

Takase, Y., Sho, N., Sone, A., Shimiya, K. (2003) Automatic generation of 3D city models and related applications, International Archives of the Photogrammetry, Remote Sensing and Spatial Information Sciences, Vol. XXXIV-5/W10.

Teller, J. and Azar, S. (2001) TOWNSCOPE II—A computer system to support solar access decision making, Solar Energy, Vol. LXX, pp. 187-200.

Urban Task Force (1999) Towards an urban renaissance: final report of the Urban Task Force - Chaired by Lord Rogers of Riverside, E & FN Spon, London.

Vögtle, T. and Steinle, E. (2000) 3D modelling of buildings using laser scanning and spectral information, The International Archives of the Photogrammetry, Remote Sensing and Spatial Information Sciences, Vol. XXXIII, Part B3.

Ward, G. J. (1994) The RADIANCE Lighting Simulation and Rendering System, Proceedings of the 21st annual conference on Computer graphics and interactive techniques, pp. 459-72.

Zinger, S., Nikolova, M., Roux, M., Maître, H. (2002) 3-D resampling for airborne laser data of urban areas, The International Archives of the Photogrammetry, Remote Sensing and Spatial Information Sciences, Vol. XXXIV, Part 3B, pp. 55-61.

Enhancing 3D City Models with Heterogeneous Spatial Information: Towards 3D Land Information Systems

Lutz Ross[1], Jannes Bolling[2], Jürgen Döllner[3], Birgit Kleinschmit[1]

[1] Department of Landscape Architecture and Environmental Planning, Berlin Institute of Technology, Straße des 17. Juni 145, 10623 Berlin {lutz.ross, birgit.kleinschmit}@tu-berlin.de
[2] Department for Geodesy and Geoinformation Science, Berlin Institute of Technology, Straße des 17. Juni 145, 10623 Berlin jannes.bolling@tu-berlin.de
[3] Hasso Plattner Institute for Software Systems Engineering, Prof.-Dr.-Helmert-Str. 2-3, 14482 Potsdam, juergen.doellner@hpi.uni-potsdam.de

Abstract. Spatial and georeferenced information plays an important role in urban land management processes such as spatial planning and environmental management. As many of the processes are increasingly coined by participation of and collaboration between multiple stakeholders, a common medium capable of integrating different types and sources of spatial information is necessary. It is argued that 3D city models provide such a framework and medium into which heterogeneous information can be integrated. Therefore, the main research question of this contribution is to identify and develop methods for integrating heterogeneous spatial and georeferenced information into 3D city models in the context of urban land management. We present a prototype 3D Land Information System and a use case for the city centre of Potsdam, Germany. In addition, constraints within administrations regarding the systematic, sustainable use of such a system are discussed.

M. Sester et al. (eds.), *Advances in GIScience*, Lecture Notes in Geoinformation and Cartography, DOI 10.1007/978-3-642-00318-9_6,

1 Introduction

Urban land management encompasses all actions, strategies and plans a city or community undertakes to maintain and develop the city's infrastructure, to monitor and protect its natural resources, to build communities, and find a balance between environmental, economic and social needs. It covers a variety of administrative tasks such as city planning, land-use planning, environmental planning and monitoring, public property management, business promotion, city marketing, and technical infrastructure maintenance. These administrative tasks rely on and produce spatial information relevant for decision-making. However, authorities are not the only stakeholders, nor are they the only users or owners of spatial data used in land management processes. Private companies, such as planning and engineering offices, infrastructure providers, geo-data providers, and the public also use, create, analyse, and provide spatial information for urban land management. Thematically the data covers, amongst other things, plans, environmental data, thematic maps, utility network data, transportation network data, environmental assessment studies, and noise emission maps. Consequently, the overall quantity of spatial data relevant for urban land management increases continuously and, because of the differentiated needs and capabilities of data users, modelling concepts and data structures are often process-, application- and scale-dependent.

The underlying thesis of this contribution is that *semantic 3D city models* (Kolbe 2009) provide an innovative and intuitive framework and medium into which spatial and georeferenced information can be integrated to effectively support communication processes in urban land management. The research is motivated by two main themes: developments in 3D city modelling and the utilization of interactive 3D models in landscape and urban planning. In the scope of 3D city modelling developments in sensor technologies as well as in processing the acquired data have resulted in methods to (semi-) automatically process and derive 3D city objects and 3D city models, respectively (e.g., Rottensteiner et al. 2005; Haala and Brenner 1999; Richman et al. 2005). Consequently, the costs for generating 3D city models have dropped continuously, and, for example, in Germany many communes and administrations have added 3D city models to their local data infrastructures or are planning to in the near future. Parallel to this development, two data models, the City Geography Markup Language (CityGML, Gröger et al. 2008, Kolbe 2009) and the Keyhole Markup Language (KML, Wilson 2008), have evolved as Open GIS standards, which can be used for the storage and exchange of 3D city models. In this contribution the *Level of Detail* (LOD) definitions from the

CityGML specifications are adopted to differentiate between simple block-buildings (LOD1), buildings with differentiated geometries including roof structures (LOD2), and architectural models of buildings (LOD3). Furthermore, CityGML is used to model city objects and plan information.

The second theme, the use of interactive 3D models in spatial and environmental planning, is related to the development of 3D city models. Spatial planning has been one of the drivers for developing tools and methods for the creation and visualization of interactive 3D city and 3D landscape models. Research in this field covers case studies (e.g., Danahy 2005, Lange and Hehl-Lange 2005), the question of the right degree of realism of (geo-)virtual environments and 3D visualizations for planning issues (e.g., Appleton and Lovett 2003, Cartwright et al. 2005), and the development and assessment of technologies and methods (e.g., Ranzinger and Gleixner 1997, Doyle et al. 1998 Counsell et al. 2006). Several recurring observations can be made: In the past the preparation of the interactive 3D models usually required extensive and time-consuming data pre-processing, there is often a trade-off between realness and interactivity (Appleton and Lovett 2003), and although a high potential is seen in the technology for e-participation and e-government applications (e.g., Wang et al. 2007), it only plays a marginal role in practice. With the increasing availability of 3D city models this is likely to change in the future. Planning and land management applications in the urban environment can now make use of the existing 3D city models, which considerably reduces implementation effort and costs. A key issue in this context is to research how 3D city models can be enhanced in order to support communication processes, decision-making, information of the public, and 3D analysis. Besides methods for the integration of spatial and georeferenced information, the usefulness and usability of the enhanced 3D city models has to be examined as well. To accomplish this a thorough cooperation and continuous exchange between research team and stakeholders in practical urban land management is necessary and is ensured through meetings, workshops and the utilization of a prototypic 3D Land Information System in planning processes within the city centre of Potsdam, Germany.

2 Study Area and System Specifications

The prototype 3D Land Information System presented was applied for a use case based in the city centre of Potsdam, Germany (cp. Figure 5), and covers an area of about five square kilometres. The 3D city model used as a base model is composed out of 1,304 buildings in LOD2, 50 buildings modelled in LOD3, and a digital terrain model with 3 m ground resolution. An aerial image with 25 cm ground resolution is used as terrain texture and trees and road lights are integrated as 3D symbols in .3ds format. Into this basic 3D city model, whose specifications correspond to several other 3D city models built in Germany in the recent years, further spatial and geo-referenced information (e.g., environmental data, master plans, development plans and construction plans) has been integrated.

The system development is based on LandXplorer Studio technology from Autodesk, a 3D geo-visualization solution with capabilities to create very large 3D landscape and city models from geo-data and 3D models from computer-aided design software (CAD). In addition to LandXplorer Studio Professional, which was used for authoring and managing the prototypic 3D Land Information System, ArcGIS from ESRI was used for geo-processing, SketchUp 6 Professional for 3D modelling, and Adobe Photoshop CS3 for image processing. A workstation with an Intel Xeon quad-core processor, 4 GB RAM, and a GeForce 8800 GTX graphic processing unit (GPU) was used for processing the data and assembling the 3D Land Information System and a Dell XPS Laptop with dual core Intel processor, 2 GB RAM, and a GeForce 7950 GTX GPU was used for presentations and collaborative planning meetings.

3 Methods

Stakeholder meetings with administration officers, architects, investors, and land owners were held regularly to select relevant spatial information for integration into the 3D city model. The integration of spatial information was implemented by the use of established methods in 3D geovisualization (e.g. terrain textures, 3D symbols, 3D modelling) and development of new methods for the transformation of plans to 3D plan representations. The primary aim was to research methods for creating visual representations of the selected spatial information within the 3D city model. Moreover, methods for the integration and access of further information and data assigned or related to the spatial information were researched as well. The resulting enhanced 3D city models were used and evaluated in further meetings. During these meetings, which included formal meetings with decision-makers and informal workshops and presentations with administration officers, architects, investors, and land-owners, comments and discussions on usability issues, data processing needs, potential applications, and data representation were recorded.

3.1 Integration of Geodata

In general two types of geodata are distinguished: raster data and vector data. Both data types can be integrated into the 3D city model by draping them over the digital terrain model (e.g., Döllner 2005). Besides this method, further methods for the integration of vector data exist, such as visualizing point features as 3D symbols or extruding polygons to 3D blocks. The methods and workflow applied for the integration of geodata are described for exemplary geodata sets.

3.1.1 Integration of Data on Protected Areas

Data on protected areas for nature conservation or groundwater protection are stored as geo-vector data in the environmental department of the city administration of Potsdam. The attribute tables hold information on the protection status, the name of the protected area, the date of designation, the legal basis for the designation, and more. Figure 1 depicts three approaches used to visually represent data on protected areas within the 3D city model.

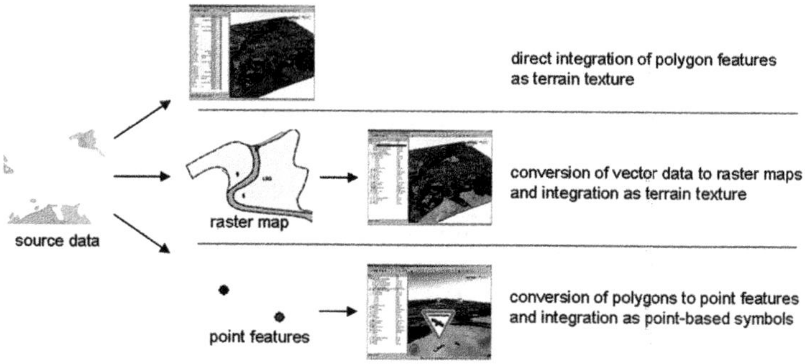

Fig. 1. Integration methods for geo-vector data (from top to bottom): direct integration of geo-vector data, integration as raster map, and integration as 3D symbols positioned inside the source polygons.

The first method is the *direct integration* of the geo-vector data into the 3D city model as interactive terrain texture by projecting the vector features onto the terrain. Interactive in this context means that rule-based and interactive queries can be used to access attribute information and create selections.

The second method, *integration as raster map*, uses geo-coded maps processed from the original vector data by using methods from digital cartography (e.g., colouring features, using signatures and text labels). The geo-coded raster maps derived from this are integrated as terrain textures.

A third approach, *integration as 3D symbols*, was used to integrate 3D symbols by converting the source data (polygon features) into point features. Icons showing official signs used in Germany were applied as 3D symbols placed at the point locations to visually communicate what type of protected areas is depicted. The three methods were applied to nature protection areas, protected biotopes and water protection areas.

3.1.2 Integration of Water Areas

Land-use data from the digital cadastre map was used to select water areas and integrate them as CityGML *WaterObjects* into the 3D city model. Therefore, the selected features were transformed and written to a CityGML file. The support of the CityGML specification enables the 3D city model authoring system to interpret the data and to use a *water shader* on it. A shader in the field of computer graphics is a software instruction, which is used by the GPU to create advanced rendering effects, such as simulating a realistic water surface (Kegel and Döllner 2007). To avoid visual artefacts from z-buffer fighting the digital terrain model was modified and areas covered by water were lowered.

3.2 Integration of Plans from Urban Planning

Several planning processes took place and are still continuing in the study area. So far, three different kinds of plans were examined: master plans, development plans, and construction plans. Although these plans can be differentiated with respect to their content and scale, they are similar in that they describe proposed / possible changes in the cityscape. Thus, the integration of visual representations into 3D city models will have to include changes in the three-dimensional model space. Therefore, methods for the creation of 3D plan representations are examined.

In contrast to geodata, most plans selected for integration were not georeferenced and it was not possible to integrate them directly into the 3D city model. Moreover, information about plan objects, such as the number of floors of a proposed building, is not encoded in attribute tables but in the plan graphics. For this reason, a number of pre-processing steps were necessary to create 3D plan representations from the plans examined. In the worst case, where only image files were available as source data, they had to be geo-coded first and plan features had to be digitized before further models could be made.

3.2.1 Integration of Master Plans

The integration of master plans is exemplified with the master plan '*Speicherstadt*'. The Speicherstadt is an old warehouse and industry complex located at the river Havel in the city centre of Potsdam. Key issues in the planning process were the height concepts and the building density of the plan proposals. The master plan and plan versions were continuously integrated into the 3D Land Information System to provide visual simulations during the planning process.

Initially *extrusion-based modelling* was used to create interactive block models from building footprints and building height information as shown in Figure 2. To ensure an accurate height representation, the absolute building height above sea level is encoded in the geometry of the building footprints. With the in-build import functions of the authoring system block models were processed from this data, which can be interactively and rule-based queried, coloured, and textured. Moreover, the height of the block models can be manipulated and attributes can be edited. In order to include the proposed land-use concept in the 3D plan representation, the geo-coded raster plan was masked with the planning area and integrated into the 3D city model as terrain texture (cp. Figure 2).

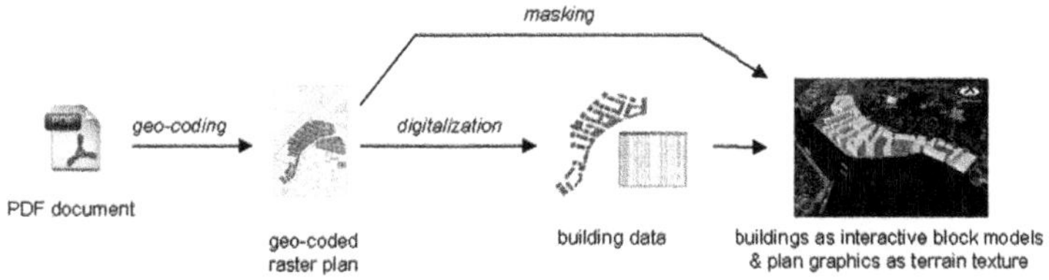

Fig. 2. Extrusion-based modelling approach used for the representation of master plans through block models and terrain textures.

The second method, *3D modelling*, was used to create 3D plan representations with more geometric and appearance detail in external applications. 3D modelling is an established method for creating architectural models and visualizations and it is very flexible in respect of geometry and appearance modelling. For this reason, it is possible to create realistic and comprehensive 3D plan representations which comprise not only the buildings but also the space around them including green spaces and trees, streets, open space, and city furniture. To facilitate 3D modelling, the plan features were classified into categories (buildings, transportation objects, and vegetation objects), and the height of the building above ground and the base height were added to the building footprints as attributes in the GIS environment. After this preparatory work, the features were exported from the GIS to the 3D modelling software. In-built functions of the export plug-in were configured to process block models from the building features and to level features based on attribute information. From this basic 3D model, a detailed model was created. The 3D models were imported via .3ds format and proper positioning was ensured using the centre point of the bounding box as a positioning vector. Figure 3 illustrates the 3D modelling approach.

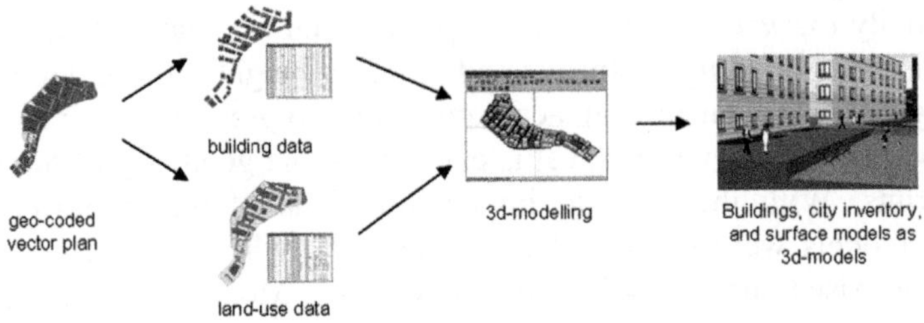

Fig. 3. 3D modelling approach used for the creation of detailed plan representations as 3D models.

A third modelling method, *CityGML-based modelling*, was developed to create a CityGML-based 3D plan representation. The aim behind this approach was to: a) develop a method for the (semi-)automated conversion of plans into CityGML-based 3D representations in order to b) store plans in a CityGML-based 3D geo database. The building data and the land-use data prepared for the 3D modelling method were used as source data. To represent the CityGML object properties in the data, the object classification is refined and additional attributes were specified for the land-use data as shown in Table 1.

The conversion of the land-use data was implemented through an interpolation with a triangulated irregular network followed by a data transformation from ESRI multipatch features to CityGML objects, as illustrated in Figure 4. Citygml4j, a Java class library developed by the Institute for Geodesy and Geoinformation Science at the Berlin Institute for Technology (IGG 2008), and GeoTools, an open source Java code library (geotools.codehaus.org), were used to implement the data transformation. In combination with building models in CityGML format, which were derived by exporting the building models created earlier by the use of the extrusion functions, a CityGML-based 3D plan representation was generated.

Table 1. Object classification used for the transformation of land-use data to a CityGML-based representation

Class	Attributes	Values
Transportation		
	transportation complex name	string
	transportation complex class	string
	transportation complex function	string
	function	string
	usage	string
	surface material	string
	colour	float [0…1], float [0…1], float [0…1]
Vegetation		
	function	string
	average height	float

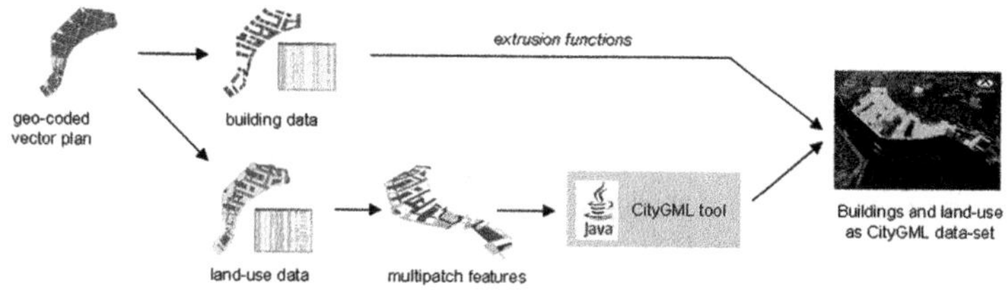

Fig. 4. Illustration of the CityGML-based modelling approach using data transformation and extrusion functions to convert plan features to a CityGML-based plan representation.

3.2.2 Integration of Development Plans

The second category of plans examined was development plans. In Germany, development plans are legally binding planning documents which specify the future land use as well as the building density and building functions. As the content and graphics of development plans are specified by law, the integration of the plan graphics as terrain texture was chosen as primary method.

To further communicate the potential effect of development plans on the cityscape, experiments were made to depict 3D building plots. Borders defining building plots were digitized and regulations about the maximum ground area, gross floor area, number of floors, maximum building height, and building function were added as attributes to the features. This data was used to create block models via extrusion functions, which represent

3D building plots. Transparency was deliberately added to 3D building plots to indicate that the visualization does not show actual or planned buildings, but the 3D space within which buildings can be constructed.

3.2.3 Integration of Construction Plans

Construction plans contain all necessary information to create detailed architectural or even in-door 3D models. Despite this fact, it was decided that these plans should be integrated as relatively simple LOD3 models to visually communicate the planning character and to keep the 3D modelling effort low. The workflow utilized is analogous to the 3D modelling approach employed earlier. The only difference was that in this case geo-coded site plans in Drawing Exchange Format (.dxf) could be used as source data for the modelling process.

The site plans only contained poly-lines and points, however. Thus, it was necessary to topologically correct the data and create a feature data set of the building footprints. These were prepared by adding attributes holding height information and exported to SketchUp (cp. 3.2.1, 3D modelling method). Ground plans and façade drawings, which have to be prepared to request permission for constructions in Germany, were used as a guideline for 3D modelling. The façade drawings were further used as façade textures to increase visual detail without modelling windows, doors, and other details. As described earlier, the integration of the 3D models into the 3D city model is done via .3ds format and positioning vectors.

3.3 Integrating further Information through Attributes, Actions, and Legends

The methods presented so far focus on the integration of visual representations of geodata and plans into the 3D city model. In many cases, these visual representations can already be considered to be interfaces for further spatial information, e.g., the direct integration of vector features as terrain textures allows features to be selected and their attribute information to be queried. In other cases, the information is encoded in the visual representation, e.g., geo-coded raster maps as terrain textures or 3D models. Thus, it is necessary to decide if a chosen visual representation is suitable to communicate the intended information and, if not, how the representation can be enhanced or whether further methods can be used to achieve the aim of communicating specific information. Within the project, three methods were used: integration of information as attributes, integration through actions, and integration through legends.

3.3.1 Information as Object Attributes

In the case of 3D modelling, attribute information, which was added to the data earlier, gets lost during data export and import processes. It is possible to add this information manually in the authoring system after the import process. However, this method is time-consuming and error-prone. For this reason, externally modelled 3D models of buildings were converted to CityGML using built-in functions of the authoring system, and a function was developed to transfers attributes from the source building footprints to the CityGML data based on a spatial join (location-wise). The same function was used to transfer address information and building data from the cadastre map to the buildings.

3.3.2 Information Integration through Actions

Most spatial information integrated during the system development was related to other data or consisted of several documents. This additional data and information was made available by linking digital media and applications to 3D labels and 3D symbols. The method was applied, amongst others, to link 3D plan representations to a prototype web-based plan information system, to start-up GIS projects underlying the prototypic 3D Land Information System, and to link plans to additional data sets (e.g., text files, plan documents in PDF format, and images).

3.3.3 Information Integration through Legends

If thematic raster data sets are used as terrain textures, it is necessary to provide legends to translate the depicted signatures, symbols, and colours into information. This can either be done by using an action which relates to a legend file, or by integrating the legend as an image overlayed onto the 3D city model. Therefore, legends were prepared and stored as images with the terrain textures. The same method was used to prepare legends for 3D representations of plans.

4 Results

The primary result is a prototypic *3D Land Information System* of the city centre of Potsdam, which contains visual representations of three master plans (and plan versions), four development plans, two construction plans, cadastre data, environmental data, and public transportation network data. A screenshot depicting the integrated plans is shown in Figure 5, while

Figure 6 shows examples of integrated environmental data and the use of symbols.

Fig. 5. Screenshot of the prototypic 3D Land Information System with integrated planning data (from top left clock-wise): 1. Master plan 'Speicherstadt' as extruded buildings and terrain texture; 2. Master plan 'Alter Markt' and development plan 'Landtagsneubau'; 3. Development plan 'Babelsberger Strasse' and construction plans for the residential building plots; and 4. Master plan 'Reichsbahnausbesserungswerk'.

Fig. 6. Screenshot of the prototypic 3D Land Information System with integrated geodata including nature protection areas as raster-based terrain texture, water protection areas and polluted land cadastre as vector-based terrain textures.

4.1 Results from the Integration Methods Applied to Geodata

The integration of geodata as terrain texture is straightforward, and both methods – the direct integration of vector data and the integration of raster maps derived from vector data – resulted in an increase of information intensity of the prototypic 3D Land Information System. The results of these two methods must be differentiated, however. As can be observed from Figure 6, the integration of raster-based maps can be used to apply methods from cartography to visually communicate information included in the original source data. In contrast, the direct integration of vector data only allows us to apply colours and transparency. The vector features and attributes can be manually and rule-based selected, however, which increases the user's options to interact with the data. Moreover, the method can be combined with the use of 3D symbols to visually communicate the type of data represented by the vector features. Besides the integration of environmental data presented in this contribution (cp. Figure 6), further geodata such as land parcels, a topographic map, and public transportation network data have been integrated using the same methods.

4.2 Results from the Integration Methods Applied to Plans

In contrast to the integration of geodata, the integration of plans as 3D plan representations requires more data-processing effort. This is especially the case if only Adobe PDF documents or images are available as source data as was the case with the integrated master plans and development plans.
The first method applied to master plans, *extrusion-based modelling*, results in interactive block models, which were combined with terrain textures derived from the source plans as shown in Figure 7a. It is rated by stakeholders as being generally sufficient for communicating the basic idea of the planning proposal on the scale of master planning. Furthermore, it was rated as being especially useful for collaborative meetings because the building heights and appearance can be manipulated interactively.

The *3D modelling* approach, in contrast, results in a representation with more geometrical and appearance detail. In combination with additional 3D models (in the example a pier and two sailing boats) and the water shader applied to the water areas (cp. 4.1.2), a realistic representation is achieved, as shown in Figure 7b. Aside from the visual differences, the results of the two methods can be compared based on the included information. While the 3D modelling approach results in a gain of visual detail and a loss of attribute information, the extrusion functions for the generation of block buildings preserve attribute information assigned to the building data. The results of the 3D modelling approach were rated especially useful for presentations in meetings with decision-makers and for the promotion of projects.

a) b)

Fig. 7. The left image shows the result from using extrusion functions to create interactive block buildings and the right image shows a plan representation derived from the 3D modelling approach combined with a water shader.

The third method, *CityGML-based modelling*, results in 3D plan representations whose visual appearance is comparable to the results from the extrusion-based modelling method. However, the 3D representation contains much more semantic information. This can be attributed to the fact that the land-use concept is represented through objects, which are specified according to the CityGML-specification. Thus, it is generally possible to assess the fraction of vegetation areas, transportation areas, and building areas and whereby determine urban planning indicators. Furthermore, it is possible to define several appearance models for one data set as shown in Figure 8.

a) b)

Fig. 8. The images show the results from the CityGML-based modelling approach for master plans; the left image shows the plan representation as a CityGML data set without a defined appearance for the land-use objects, while in the right image the plan graphics was used to texture the objects.

In Figure 8a, no further textures or colours are applied to the land-use objects (the green colour is automatically assigned to CityGML vegetation objects by the system), while in Figure 8b the original raster plan was applied as a texture. Besides this flexibility with respect to the appearance, attributes assigned to the source plan features during the plan creation process are maintained and the data can be transferred to a CityGML-based database.

To include development plans within the 3D Land Information System, plans were integrated as terrain textures and transparent 3D building plots were created. The integration as terrain texture ensures that the plan graphic, which is specified by law, is maintained (cp. Figure 9a), while the transformation of the graphical elements into 3D representations of building plots visually communicates an idea of the spatial effect the plan might develop (Figure 9b). Additionally the 3D building plots can be queried to access further information assigned in the modelling process, such as maximum building height, maximum number of floors, or the maximum gross floor area. Thus the use of a 3D representation for building plots increases the interactivity and information intensity of the system.

Fig. 9. Representation of development plans as: a) terrain texture; and b) transparent 3D building plots

The integration results of the third plan category, construction plans, are depicted in Figure 10. Through the use of the documents needed to request construction permissions, the 3D models were modelled and textured (less than one hour per building). Their integration into the model, in combination with the corresponding development plan, efficiently communicates that a building permission has been submitted. Furthermore, it can be visually assessed, if the application fits into the building plots.

Fig. 10. Representation of construction plans through 3D models

4.3 Results from Integrating Information through Attributes, Actions, and Legends

The methods of integrating further information as attributes, through actions, and through legends as overlay images, result in a further increase of interactivity and information intensity of the system. The results of the integration as attribute information were briefly mentioned in the context of the conversion of master plans to a CityGML data set and the integration of 3D building plots. Both examples increase the information value of the model and enable querying information. Furthermore, the integration of address information into the system enables users to search for a location based on an address.

Actions assigned to 3D labels or 3D objects also raise the information value of the system as they can be used to access external applications and databases directly from the visual interface. This functionality was used to link the integrated plans to a prototype web-based planning information system, which includes further information and documents associated with the plans such as plan documents, press announcements, and architectural drawings.

The integration of legends as image overlay is useful and elementary. It enables users to decode the information contained in raster-based terrain textures. Figure 11 shows the use of overlay images to integrate legends, and illustrates the concept of using actions to link labels to external applications.

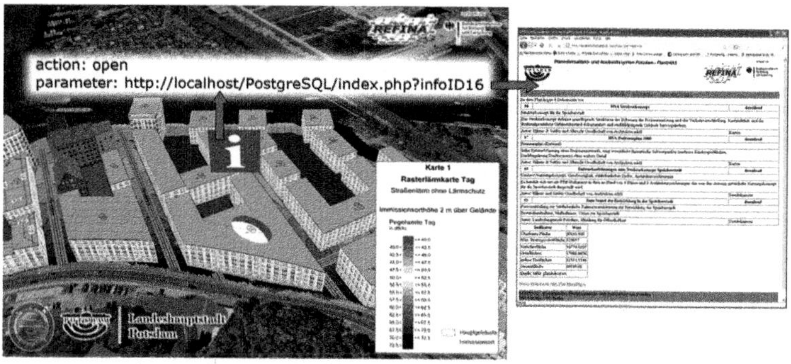

Fig. 11. Actions were used to relate 3D labels and 3D symbols to external applications such as web-based information portals and overlay images were used to integrate legend information into the 3D display.

4.4 Results from the Utilization of the System in Planning Processes

Although neither questionnaires nor extensive stakeholder analyses have been conducted so far, some preliminary results regarding usefulness and usability can presented. The system has proven to be very useful for visually assessing the height concepts of the master plan Speicherstadt in stakeholder meetings. The system was also used to present an agreed height concept to decision-makers, and in presentations and meetings with architects, investors, and authorities. In general, the responses from these stakeholder groups were positive. Several authorities within the administration are currently surveying the potential for future applications and discussing how the system can be permanently integrated into the information and communication infrastructure of the city administration. Moreover, several potential applications were formulated by different stakeholder groups, as shown in Table 2.

Table 2. Potential and implemented applications envisioned by different stakeholder groups

Stakeholder	Potential functionalities	Implemented
Authorities	- visual access to environmental data	+
	- presentation & visual assessment of planning proposals	+ (+)
	- city & business promotion	-
	- (public) participation meetings and web-service	-
	- process simulations and analyses (wind, shadows, etc.)	
Architects	- source of 3D data as basic planning information	-
	- environment to integrate planning proposals	+
Project developers	- promotion of projects	(+)
	- visual comparison of planning alternatives	+
	- visual interface to project data and database	-

+......implemented within the project
-.......not implemented so far
(+)...implemented through images and videos used by project partners

5 Discussion

This contribution shows how 3D city models can be applied and enhanced towards complex 3D Land Information Systems by integrating heterogeneous spatial information. The resulting system can be used to effectively support urban land management processes.

To implement such a system in a sustainable way, thorough modelling and integration strategies are needed. Within the administration and planning professionals, 3D city modelling expertise is still limited. Thus, system implementation requires a close cooperation and exchange of the involved stakeholders at the administrative, organizational and technical levels. Only if planning documents are made available by architects and engineers as geo-coded vector plans or geo-referenced 3D models does a continuous, systematic update of the underlying database become possible. In our case, a first step into this direction has been initiated by the city administration of Potsdam by formulating a directive, which requires planners to hand in geo-coded plans. Moreover, a GML-based standard for development plans in Germany (Benner and Krause 2007) is going to be adopted by the city of Potsdam. Since these development plans are GML-based, object-orientated, and geo-coded, they can be transformed to 3D representations and integrated into the system automatically. The utilization of an agreed data standard whereby ensures development security for the involved stakeholders. A comparable agreement for the digital exchange of 3D construction plans between administration and private companies and citizens respectively, could be used to automate the update of the 3D city model database.

To make full use of the system's potential, it will be advantageous to establish a direct connection to the spatial data infrastructure of the city, automating the integration of (geo-)data through services such as web map services or web feature services; such functionality has been demonstrated by Döllner and Hagedorn (2007). Transactional web feature services might also be used to provide access to the system for external users, such as architects and engineers, to enable collaborative use of the base 3D city model for planning issues. Of course, this would require secure connections, a user management system, and digital rights management to ensure the integrity of the system.

While most technology aspects could be identified and solved, organisational and human factors are still crucial. It would be necessary to adapt administrative processes and workflows and to train employees. Furthermore, acceptance of the system is not guaranteed. For example, the building conservation authority in Potsdam did not trust the height simulations, which were prepared for the master plan 'Speicherstadt' so an in situ simulation had to be conducted by the fire department. Only after this simulation drew the same results as the virtual simulation, the system's acceptance increased.

In summary, our thesis that 3D city models provide an innovative framework and medium for integrating and communicating heterogeneous spatial information in the context of urban land management is well supported. Nevertheless, many technological and organizational challenges, such as creating versions of models and their automatic, and systematic updating through communal business processes, remain unsolved. Moreover, further user and acceptance studies will be necessary.

Acknowledgements

The work was developed within the project "Land Information Systems based on Virtual 3D City Models" funded by the German Federal Ministry of Education and Research (BMBF, Project No. 0330782), and is part of the REFINA research program (www.refina-info.de). We also would like to thank Autodesk for providing the LandXplorer geovisualization system, which was used as technology platform.

References

Appleton, K. & A. Lovett (2003): GIS-based visualization of rural landscapes: defining ‚sufficient' realism for environmental decision-making. In: Landscape and Urban Planning, Vol. 65, pp. 117-131.

Benner J. and Krause K.U. (2007) Das GDI-DE Modellprojekt XPlanung. Erste Erfahrungen mit der Umsetzung des XPlanGML-Standards. In: Schrenk, M. (Ed.) REAL CORP 2007: To Plan is not Enough: Strategies, Concepts, Plans, Projects and their Successful Implementation in Urban, Regional and Real Estate Development; Proc. of the 12th Internat. Conf., Vienna, pp. 379-388.

Cartwright, W., Pettit, C., Nelson, A. & M. Berry (2005): Community Collaborative Decision-Making Tools: Determining the Extent of 'Geographical Dirtiness' for Effective Displays. Proceedings of the 21st International Cartographic Conference, 9-16th July, A Coruna, Spain.

Counsell J., Smith S. and Richman A. (2006): Overcoming some of the issues in maintaining large urban area 3D models via a web browser. In: Proceedings of the information visualization, pp 331-336, DOI 10.1109/IV.2006.82

Danahy J. W. (2005): Negotiating public view protection and high density in urban design. In: Bishop, I. & Lange, E. (eds.): Visualization in Landscape and environmental planning – Technology and Applications. Taylor & Francis, Oxon, UK.

Döllner J. (2005): Geovisualization and Real-Time 3D Computer Graphics. In: Dykes J., MacEachren A.M. and Kraak M.-J. (eds.): Exploring Geovisualization. Elsevier, Oxford, UK.

Döllner J. and Hagedorn B. (2007) Integrating Urban GIS, CAD, and BIM Data By Service-Based Virtual 3D City-Models. Online: http://cgs.hpi.uni-potsdam.de/publications/Public/2007/DH07/udms_2007_doha_draft.pdf, Last date accessed 12.11.2007.

Doyle S., Dodge M. and Smith A. (1998): The Potential of Web-Based Mapping and Virtual Reality Technologies for Modelling Urban Environments. Computers, Environment and Urban Systems, Vol. 22, No. 2, pp. 137-155.

Gröger G., Kolbe T.H., Czerwinski A. and Nagel C. (2008): OpenGIS City Geography Markup Language (CityGML) Encoding Standard, Version 1.0.0, International OGC Standard, Open Geospatial Consortium, Doc. No. 08-007r1, 2008.

Haala N. and Brenner C. (1999) Extraction of buildings and trees in urban environments. In: ISPRS J. Photogrammetry & Remote Sensing, Vol. 54, pp. 130-137, 1999.

Institute for Geodesy and Geoinformation Science (2008): citygml4j; java classes for handling CityGML data-sets, download and information: http://opportunity.bv.tu-berlin.de/software/projects/show/ citygml4j, last accessed 2008/12/10.

Kegel A. and Döllner J. (2007): Photorealistische Echtzeit-Visualisierung geovirtueller Umgebungen. In: Mitteilungen des Bundesamtes für Kartographie und Geodäsie 2007.

Kolbe T. H. (2009): Representing and Exchanging 3D City Models with CityGML. In: Lee J. and Zlatanova S. (Eds.): 3D Geoinformation Sciences. Springer Verlag, Berlin Heidelberg.

Lange E. and Hehl-Lange S. (2005): Future Scenarios of Peri-Urban Green Space. In: Bishop, I. & Lange, E. (eds.): Visualization in Landscape and environmental planning – Technology and Applications. Taylor & Francis, Oxon, UK

Ranzinger M. and Gleixner G. (1997): GIS Datasets for 3D Urban Planning. Computers, Environment and Urban Systems, Vol. 21, No. 2, pp 159-173.

Richman A., Hamilton A., Arayici Y., Counsell J. and Tkhelidze B. (2005) Remote Sensing, LIDAR, automated data capture and the VEPS project. In: Banissi et al. (eds): Proc. of the 9th Intl. Conference on Information Visualisation, London, pp 151-156, DOI 10.1109/IV.2005.106

Rottensteiner F., Trinder J. and Clode S. (2005) Data acquisition for 3D city models from LIDAR – Extracting buildings and roads. Geoscience and Remote Sensing Symposium, 2005, Proceedings. IEEE International Volume 1, DOI 10.1109/IGARSS.2005.1526226

Wang H., Song Y., Hamilton A. and Curwell S. (2007) Urban information integration for advanced e-Planning in Europe, Government Information Quarterly (2007), doi:10.1016/j.giq.2007.04.002

Wilson, T. (2008): OGC KML, Version 2.2.0, International OGC Standard, Open Geospatial Consortium, Doc. No. 07-147r2, 2008.

Matching River Datasets of Different Scales

Birgit Kieler[1], Wei Huang[2], Jan-Henrik Haunert[1], Jie Jiang[2]

[1] Institute of Cartography and Geoinformatics,
Leibniz Universität Hannover, Appelstraße 9A, 30167 Hannover,
{birgit.kieler, jan.haunert}@ikg.uni-hannover.de
[2] Department of Geo-spatial data product R & D, National Geomatics
Center of China, 1 Baishengcun, Zizhuyuan, Beijing, 100044,
{huangwei, jjie}@nsdi.gov.cn

Abstract. In order to ease the propagation of updates between geographic datasets of different scales and to support multi-scale analyses, different datasets need to be matched, that is, objects that represent the same entity in the physical world need to be identified. We propose a method for matching datasets of river systems that were acquired at different scales. This task is related to the problem of matching networks of lines, for example road networks. However, we also take into account that rivers may be represented by polygons. The geometric dimension of a river object may depend, for example, on the width of the river and the scale.

Our method comprises three steps. First, in order to cope with geometries of different dimensions, we collapse river polygons to centerlines by applying a skeletonization algorithm. We show how to preserve the topology of the river system in this step, which is an important requirement for the subsequent matching steps. Secondly, we perform a pre-matching of the arcs and nodes of the line network generated in the first step, that is, we detect candidate matches and define their quality. Thirdly, we perform the final matching by selecting a consistent set of good candidate matches.

We tested our method for two Chinese river datasets of the same areal extent, which were acquired at scales 1:50 000 and 1:250 000. The evaluation of our results allows us to conclude that our method seldom yields incorrect matches. The number of correct matches that are missed by our method is quite small.

Keywords: data matching, network, multi-scale representation, generalization, skeletonization

M. Sester et al. (eds.), *Advances in GIScience*, Lecture Notes in Geoinformation and Cartography,
DOI 10.1007/978-3-642-00318-9_7,

1 Introduction

In former times, the cartographer's job was to map the unexplored land. Today, however, we are rather faced with an excess than with a lack of data: for most parts of the Earth, digital geographic databases have been acquired multiple times, for multiple applications, and in multiple scales. This leads to two primary questions addressed by current cartographic research. First, how can we minimize the effort for keeping the databases up-to-date? Secondly, how can we combine the information given with different databases? An important prerequisite for answering these questions is to develop methods for database integration (Devogele et al. 1998). Subproblems of database integration are schema matching and data matching. Schema matching deals with the identification of corresponding concepts in data models (Volz 2005). Data matching aims to find corresponding objects in different datasets. In our paper we deal with data matching and focus on the matching of river datasets.

Data matching is useful for updating, since we can trigger an update from one object to a corresponding object in another dataset, once both datasets have been matched; for this purpose correspondences found by matching are stored as links in a database (Harrie and Hellström 1999; Dunkars 2004). Furthermore, we can combine the attribute sets given for both objects into a single detailed set, which, for example, allows users to perform complex analysis tasks. In this paper we assume that the schemas of both datasets were matched prior to the data matching process, for example, we can identify objects of river classes in both datasets and know that these classes represent similar concepts. Obviously, this knowledge is useful for data matching. Note, however, that data matching can also be applied to detect unknown correspondences between schemas (Kieler et al. 2007) and, when dealing with different scales, unknown generalization rules (Sester et al. 1998).

The identification of corresponding objects is often manually done or performed with semi-automatic procedures, which is expensive or even infeasible for large datasets. However, in recent years researchers have developed fully automatic methods for certain matching problems. Diez et al. (2008) consider the problem of matching road networks in datasets with different map projections; the transformation between the coordinate systems of both datasets is unknown. We, however, assume that the datasets are in the same coordinate system. The difficulty in our problem is not to find a global geometric map transformation but to deal with differences that are due to map generalization. Walter and Fritsch (1999) as well as Zhang and Meng (2007) developed methods for matching road datasets that have similar scales but were captured for different thematic domains, that is, a topographic dataset and a dataset for car navigation. The problem becomes more involved if the difference in scale increases, since small-scale datasets contain geometrically simplified shapes. Moreover, objects may be eliminated or aggregated through generalization (Timpf 1998). Therefore, matching algorithms that rely on comparisons of geometric features may fail. In order to match datasets of different scales, additional criteria need to be considered. Most existing methods for matching networks of lines exploit topological relations between map objects (Lüscher et al. 2007; Mustière and Devogele 2008; Zhang and Meng 2007). These relations do not so much depend on the scale. For example, the geometry of lines representing roads may be simplified to a high degree, but the topology of the road network is mainly preserved during generalization. Uitermark et al. (1999) developed a method for matching road datasets of different scales, where all roads are represented by area objects; in order to derive a network of lines, a skeletonization method is applied. To conclude, the matching problem is most explored for objects of the same geometric dimension, also considering different scales. However, there are still open problems, especially when of objects with different geometric dimensions are to be matched. For that reason we did not use a standard matching tool, like RoadMatcher (Vivid Solutions 2005). This open source software only handles line networks and only finds one to one matches.

Generalization often reduces the geometric dimension of objects, for example, a river may be represented by a polygon in a large-scale map or by a line in a small-scale map (Haunert and Sester 2008). Furthermore, there may be river objects of different geometric dimensions in a single dataset, for example, wide rivers are represented by polygons and narrow rivers are represented by lines. In this paper we address the matching of river datasets of different scales, also allowing for different geometric dimensions. However, our approach is not restricted to rivers. Roads in datasets of very large scale, for example, in cadastral maps, are represented by polygons. Our method may also be applied to match such a dataset with a topographic dataset of smaller scale, where roads are represented by lines.

The matching method that we propose comprises three steps. First, river polygons are collapsed to centerlines by applying a skeletonization algorithm. We show how to preserve the topology of the river system in this step, which is an important requirement for the subsequent matching steps. Secondly, a pre-matching of arcs and nodes is performed. In this step we detect candidate matches and define their quality. Thirdly, the final matching is performed by selecting a consistent set of good candidate matches.

The paper is structured as follows. We briefly sketch the context of our work, that is, we present the data we are dealing with and how they are captured and used in applications (Section 2). Then we present our matching method of three steps in Section 3, which is the main part of our paper. We evaluate and discuss the results of our experimental tests in Section 4 and conclude the paper in Section 5.

2 The Use Case: Chinese River Datasets

The national mapping agency of China manages topographic databases of four different scales, namely 1:50 000, 1:250 000, 1:1 000 000, and 1:4 000 000. Until now, these databases are collected and maintained independently, but for the future it is aimed to apply automated generalization and matching methods in order to ease the updating process. The databases are used to derive analog maps but also to directly support offices in their planning activities and decision-making procedures. Each database contains information on river systems; the geometric detail and the number of attributes reflect the particular scale. The lines representing rivers constitute so-called digital line graphs (DLGs). We use this term for the datasets we are dealing with, but we explicitly include polygons representing rivers. From now on, we refer to the river datasets of scales 1:50 000 and 1:250 000 as DLG 50 and DLG 250, respectively. We exclude the two datasets of smallest scales from our investigations. Figure 1 shows our test area, which has an extent of approximately 54 km^2 and is located in a rural area close to Shanghai. The Chinese datasets have attributes that allow for a distinction of natural and man-made waterways. There is also an attribute reflecting the name of the river. However, we do not consider these attributes in our approach, since this information has not been captured completely.

When we started to develop the matching strategy for rivers, we expected that the river network would have a structure similar to a tree, that is, we expected many confluences of rivers but only a few bifurcations. Our idea was to exploit this pattern for the matching procedure. However, we did not follow this idea, because the actual structure of the river network is not similar to a tree, see Fig. 1. The encountered structure with many bifurcations results from the extensive canal and dam system.

Finally it is remarkable that dataset DLG 250 is more up-to-date and contains a lot of new canals, which are not reflected in dataset DLG 50.

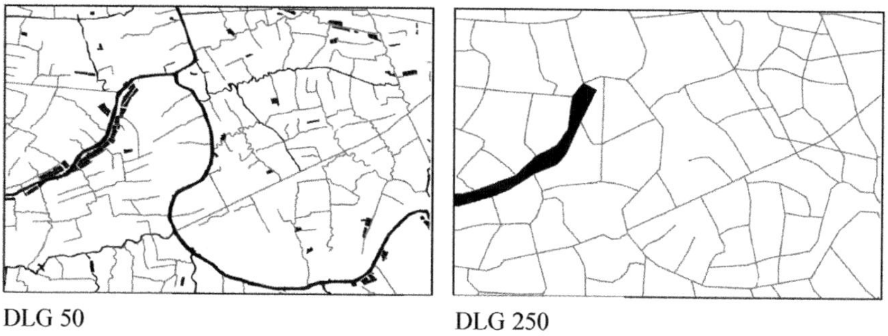

Fig. 1. Two river datasets of different scales in the test area: DLG 50 (large-scale dataset) and DLG 250 (small-scale dataset); lines are displayed in grey and polygons in black

3 Matching Procedure

Our matching method, which is illustrated as a process chart in Figure 2, is a three-steps procedure. First, in order to cope with geometries of different dimensions, we construct a network of lines, which is described in detail in Section 3.1. For this purpose, we collapse river polygons to centerlines by applying a skeletonization algorithm. The used basic method is presented in Section 3.1.1. Afterwards, in Section 3.1.2, we show how to preserve the topology of the river system in this step, which is an important requirement for the subsequent matching steps. The second step (Section 3.2) includes the pre-matching process, where we separately detect matching candidates from the arcs and nodes of the line network which we generate in the first step. Therefore we use distance criteria and angle difference criteria, in order to assess the quality of the matching candidates. In Section 3.3, we present the last step of our method. Based on the results of the pre-matching step of Section 3.2, we perform the final matching by selecting a consistent set of good candidate matches.

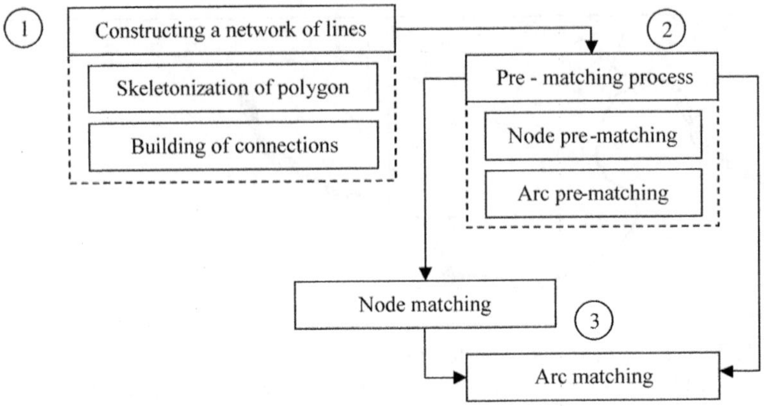

Fig.2. The process chart of our matching method deals with geometries of different dimensions and datasets of different scales.

3.1 Constructing a Network of Lines

In this section we present our method for automatically constructing a network of lines from the input data, which includes lines and polygons representing rivers. We first present a basic method for deriving centerlines for single polygons (Section 3.1.1) and then discuss how to preserve the topology of the river system (Section 3.1.2).

3.1.1 Creating Centerlines for a Single River Polygon

Haunert and Sester (2008) compare different types of skeletons that are commonly used in geographic information systems for deriving polygon centerlines. This includes the *medial axis*, which comprises straight lines and second-order lines. We do not select the medial axis, since handling second-order lines would cause computational overhead. An alternative skeleton is the *straight skeleton*, which only comprises straight lines. However, the existing algorithms for constructing the straight skeleton are too slow to handle large datasets. Therefore, we select a simple skeleton that is based on a constrained Delaunay triangulation of the polygon, see Figure 3. Penninga et al. (2005) discuss this method in detail. We give an outline of this method.

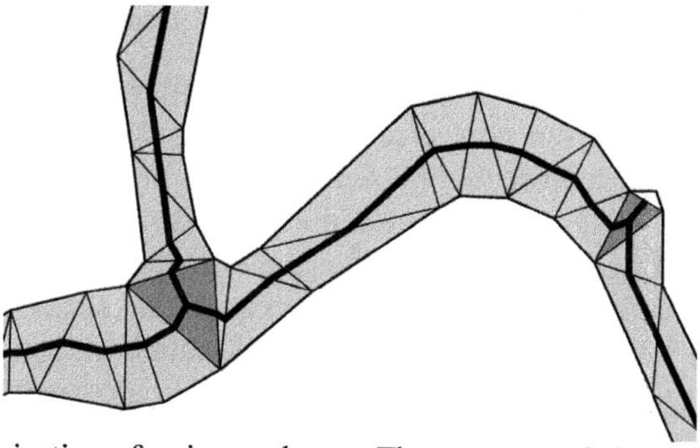

Fig. 3. Skeletonization of a river polygon. The constructed skeleton is bold black. There are two 0-triangles (dark shaded) and one 2-triangle (white); all other triangles are 1-triangles (light shaded).

The constrained Delaunay triangulation of a polygon is an exhaustive partition of the polygon into non-overlapping triangles. Most existing triangulation algorithms yield additional triangles in the exterior of the polygon; however, only the triangles in the interior of the polygon are used for constructing the skeleton. After constructing the triangulation, the triangles are handled independently. For each triangle, a piece of the skeleton is added. This procedure differs for different types of triangles:

- For each triangle that shares *two* edges with the polygon (that is, a *2-triangle*), no skeleton edge is added.
- For each triangle that shares *one* edge with the polygon (that is, a *1-triangle*), one skeleton edge is added. This edge is defined by connecting the midpoints of both other triangle edges.
- For each triangle that shares *no* edge with the polygon (that is, a *0-triangle*), three skeleton edges are added. Each such edge is defined by connecting the midpoint of a triangle edge with the triangle's centroid, that is, the point $((x_1+x_2+x_3)/3, (y_1+y_2+y_3)/3)$, where (x_1, y_1), (x_2, y_2), and (x_3, y_3) are the triangle vertices.

The main disadvantage of this skeleton compared to the medial axis is that it is not smooth. Often the centerline is zigzagging. However, we will define a matching method that is quite robust against these geometric distortions. We are mainly concerned with keeping the topology of the river network correct. We accept the disadvantage of a less smooth shape, since the described skeleton can be applied to construct a topologically correct river network, which we show in the next section.

3.1.2 Preserving the Topology of the River System

In the previous section we presented a method for constructing a skeleton of a single polygon. We now discuss how to preserve the topology of a river system that is represented by multiple lines and polygons. In order to accomplish this task, we perform a pre-processing prior to the skeleton construction and a post-processing after the skeleton construction. Both the pre-processing and the post-processing comprise two steps.

In the first pre-processing step, we amalgamate all mutually adjacent river polygons. Without this step we would obtain incorrect results at river junctions. Figure 4 shows the skeletonization result when calculating the triangulation for each polygon independently (Figure 4(a)) and when amalgamating the adjacent polygons before calculating the triangulation (Figure 4(b)). In the left figure there is a node on the boundary shared by both polygons; this causes the artifact during the skeleton construction. The latter result is better, since it represents the topology of the river network correctly.

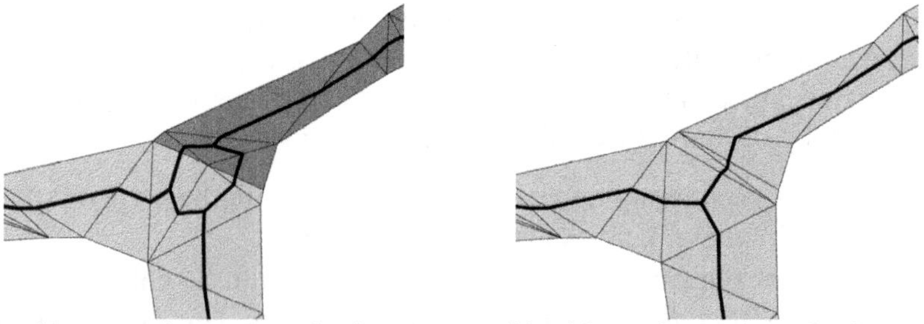

(a) without amalgamation of polygons (b) with amalgamation of polygons

Fig. 4. The skeleton for two adjacent river polygons (different shades).

In the second pre-processing step, we deal with the case that the endpoint of a line lies on the polygon boundary, for example, a narrow river (the line l) flows into a wide river (the polygon p). In this case we need to ensure that the shapes of the two rivers remain connected. We could try to solve this problem after the skeleton construction without any pre-processing, for example, by extending the line l in its original direction, until it touches the constructed skeleton. This approach, however, is too naive, since we would possibly create intersecting lines (see Figure 5). In order to avoid new intersections, we propose to define the connection of l and the skeleton of p based on the triangulation of p. Our approach requires that, if an endpoint v of l lies on the polygon boundary, the same point is a vertex of the triangulation. We can ensure this requirement simply by introducing v as an additional polygon vertex. This is done in the pre-processing, that is, before constructing the skeleton.

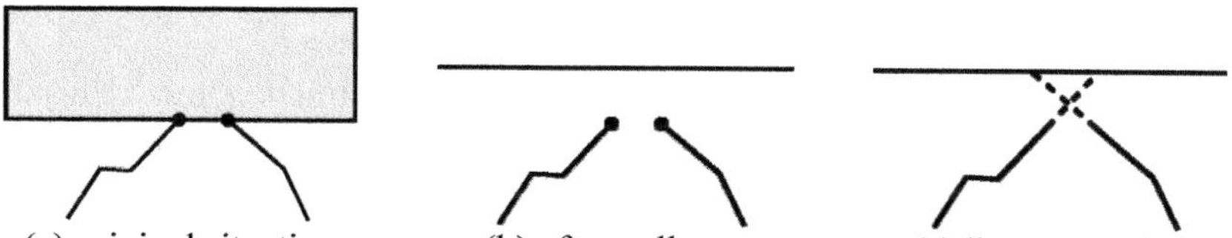

Fig. 5. When collapsing a river polygon, the connectivity of the river system can be affected ((a) and (b)). Extending lines to meet the skeleton line can result in unwanted intersections (c). Therefore, we suggest another approach, see Figure 6.

In the first post-processing step, that is, after applying the skeletonization method from Section 3.1.1, we construct the connections between the endpoints of the original river lines and the derived skeleton. Let v be a polygon vertex that is equal to the endpoint of a line. Since v is a vertex of the polygon, it is also a vertex of at least one triangle of the triangulation. We define T as the counter-clockwise ordered sequence of triangles that share the vertex v. We select a subsequence $T' = (t_1, t_2, ..., t_n)$ of T such that T' has the maximum number of elements among all subsequences of T that have the following property: each two subsequent triangles share a common edge. Note that the sequence T' is usually equal to T. However, the definition of T' is needed, since we can also construct special cases where two subsequent triangles in T do not share a common boundary. We now handle two different cases separately:

- If the number n of triangles in T' is *even*, we select the triangle edge e that separates $t_{n/2}$ and $t_{n/2+1}$. We insert a skeleton edge by connecting the vertex v and the midpoint of e, see Figure 6(a).
- If the number n of triangles is *odd*, we select the triangle $t = t_{(n+1)/2}$. Again, we consider different cases:
    - if t is a 2-triangle we insert a skeleton edge by connecting the vertex v and the midpoint of its opposite triangle edge as shown in Figure 6(b).
    - if t is a 1-triangle we insert a skeleton edge by connecting the vertex v and the midpoint of the skeleton edge that we added for t as shown in Figure 6(c).
    - if t is a 0-triangle we insert a skeleton edge by connecting the vertex v and the triangle's centroid as shown in Figure 6(d).

This approach indeed ensures that the topology of the river system is preserved. Intersecting connections as shown in Figure 5(c) are not possible. This is because, for each triangle, there is only a finite set of potential connections. These potential connections do not intersect.

In the second and final post-processing step we remove some 'dangling' arcs. More precisely, we discard any arc that terminates at a vertex of degree one and is shorter than the river width.

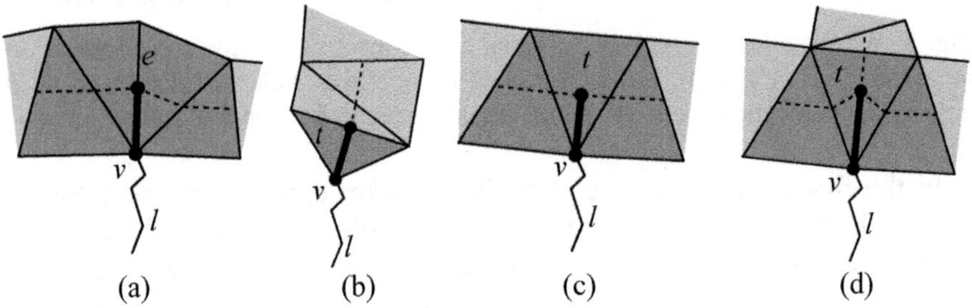

Fig. 6. A narrow river (line l) enters a wider river (dark and light shaded regions). To preserve the connectivity, we add a connection (bold black) between l and the constructed skeleton (dashed lines). The triangles in T' are shaded dark grey.

3.2 Pre-Matching Process

In this section we present our method for the selection of matching candidates, separately for nodes (Section 3.2.1) and arcs (Section 3.2.2) in order to reduce the input data for the final matching step. Furthermore, we define the quality of matching candidates.

3.2.1 Pre-Matching of Nodes

In the final matching step we will search, for each node n of the small-scale dataset, a corresponding node n' of the large-scale dataset. In this paper, we use the small-scale dataset as the reference dataset and the large-scale dataset as the target dataset. It is likely that a node corresponding to n exists, because the less detailed dataset usually does not contain more information than the detailed one. In the pre-matching step we search a set N'_n of nodes that *possibly* correspond to n. We should keep N'_n as small as possible while ensuring $n' \in N'_n$. The set N'_n may contain more than one node or could also be empty.

We perform the pre-matching of nodes based on a distance threshold δ. First, we calculate a buffer for each node n of the small-scale dataset, that is, a circle of radius δ with centre n. We define the candidate set N'_n as the set of nodes in the large-scale dataset that are contained in this buffer. In order to define an appropriate buffer size, we need to discuss two cases. First, if the chosen buffer size is too small, correct matches may be lost. Secondly, if the buffer size is too large, we need to resolve many ambiguous cases in the further process. Since lost matches cannot be recovered in the final matching step, an over-selection of candidates is better than an under-selection.

We store the results of the pre-matching of nodes in a table. Each row contains the identifiers of two nodes that form a potential match. The table also has a column that represents the distance of both nodes. This distance can be seen as a quality measure. Candidates with a small distance to the reference node have a high quality and candidates with a large distance have a low quality. All node candidates are analyzed concerning their suitability further in the node matching process of Section 3.3.1.

3.2.2 Pre-Matching of Arcs

Beside the detection of node candidates, we also perform a pre-matching of arcs based on another distance threshold ε. The pre-matching of arcs is similar to the pre-matching of nodes. We also store the results in a table. For each arc a of the small-scale dataset, the pre-matching yields a set A'_a of arcs of the large-scale dataset. Again, we need to ensure that A'_a is small and contains the actual match a'. To select an appropriate set of candidates for a, we calculate a buffer polygon β_a, which contains all points at distance ε from a or closer. We define the candidate set A'_a as the set of arcs of the large-scale dataset that are completely contained in this buffer polygon or cross its boundary.

We measure the quality of a potential arc match by comparing the direction of both arcs. The direction of an arc can be defined as the orientation angle of the straight line connecting both endpoints of the arc. However, an arc in the detailed dataset may correspond only to a part of an arc in the less detailed dataset. Therefore, the directions of a and a' can be very dissimilar when measured for the whole arcs. To define an appropriate quality measure, we perform a local comparison. For each arc b of the large-scale dataset, we define a buffer polygon β_b, which we also define based on the threshold ε. In order to assess the quality of the match (a,b), we calculate the intersection of both buffers, that is, we define
$\beta_{ab} = \beta_a \cup \beta_b$, see Fig. 7. The part of a that lies in β_{ab} may correspond to the part of b that lies in β_{ab}.. For both parts we can construct a straight line by connecting the start and endpoint. Comparing the orientation angles of these lines we can infer about the quality of the match (a,b).

If the difference of the angles is small, then the arcs have almost the same orientation. In this way the quality of the possible correspondence is high. However, if the angle difference is high, then the likelihood that the arcs have the similar orientation is quite low. Figure 8 displays all matching candidates that reach a certain quality; from left (Fig. 8 (1)) to right (Fig. 8 (3)) the quality increases. In Fig. 8 (1) all segments of the large-scale dataset are displayed in bold grey; these are possible matching candidates, because they are located in the buffer polygon of the investigated reference arc. In Fig. 8 (2) the segments with an angle difference of $\Delta\alpha \leq 45$ degrees are displayed in bold grey. Obviously the angle difference is sufficient, since all orthogonal river parts of the investigated reference part, which are apparent improper matching candidates, will get a low quality. In Fig. 8 (3) we show the matching candidates which fulfill the angle difference of $\Delta\alpha \leq 25$ degrees. In this case some correct matching candidates are missing, because the orientation of the arcs are too different. However, the remaining arcs get a high quality.

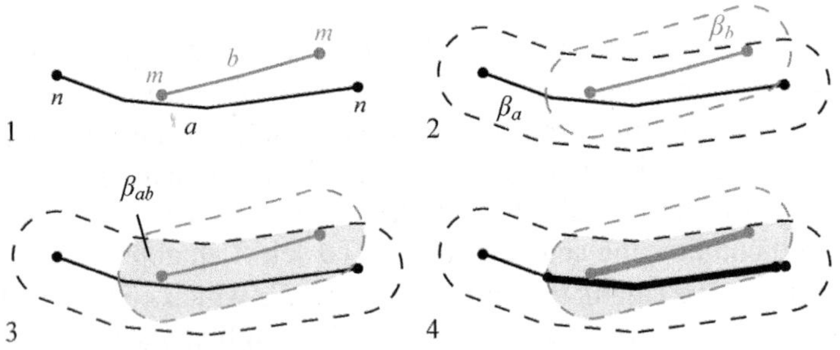

Fig. 7. Method for identification of arc segments for the comparison of orientation angles. (1) Arc a (black) of small-scale dataset and arc b (grey) of large-scale dataset; (2) Buffer polygons β_a and β_b; (3) Intersection area β_{ab} (light grey polygon); (4) Comparable parts of an arc: reference (bold black) and target (bold grey) dataset.

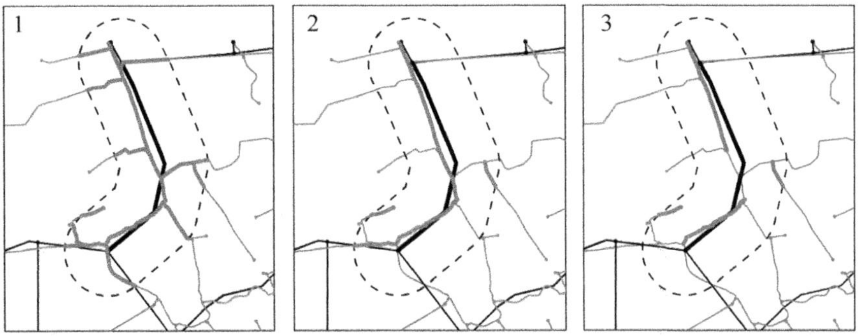

Fig. 8. (1) All matching candidates (bold grey) for an arc *a* (bold black) of the small-scale dataset; (2) all matching candidates whose orientation angle α is similar to the orientation angle of *a* ($\Delta\alpha \leq 45$ degrees); (3) all matching candidates whose orientation angle α is *very* similar to the orientation angle of *a* ($\Delta\alpha \leq 25$ degrees).

3.3 Final-Matching Process

The pre-matching results of the previous section are used as input for the final matching process, which comprises the matching process of nodes (Section 3.3.1) and, thereafter, the matching of arcs (Section 3.3.2).

3.3.1 Matching of Nodes

In this section we identify, for the nodes of the small-scale dataset, the nodes of the large-scale dataset that correspond best. For this we exploit the pre-matching results from Section 3.2.1 and analyze the node-arc topology. Therefore we have to extend the analysis to the arcs that are connected to the nodes.

First, we detect a set of node matches that are quite obvious – we call them *certain* matches. For this, we determine the node degree *deg(n)* for each node of the small-scale dataset and for all its node matching candidates. Normally, the less detailed dataset does not contain additional arcs. Therefore, we define that a node match *(n,m)* cannot be certain if the degree of the reference node *n* is greater than the degree of the candidate node *m*, that is, if $(deg(n) > deg(m))$. For example, the match in Fig. 9 (1) cannot be certain, but the matches in Fig. 9 (2) can be certain.

To decide whether a node match *(n,m)* with $deg(n) \leq deg(m)$ is certain, we analyze the set S_n of arcs that are incident to *n* and the set S_m of arcs that are incident to *m*. By querying the table that contains the pre-matching results for arcs, we select the set *P* containing all potential arc matches where one arc is in S_n and the other one is in S_m. Our approach is to define the node match *(n,m)* as certain only if there is a sufficiently good subset Q of P that has the following properties:

- Each arc in S_n belongs to exactly one match in Q.
- Each arc in S_m belongs to maximally one match in Q.
- The quality of each arc match *(a,b)* in Q is not worse than a certain threshold θ. As defined in Section 3.2.2, the quality is measured according to the angle difference of the arcs a and b.

In order to find such a subset, we propose a simple iterative approach. Initially, we set Q empty. We order the potential arc matches in P according to their quality, into a list. We iterate through this list. First, we select the arc match of highest quality and then we proceed with the next arc match. Let *(a,b)* be the arc match selected in a certain iteration. We add the match *(a,b)* to Q if its quality is high enough and if Q does not contain another match with a or b. After we reach the end of the list, we test whether each arc in S_n belongs to exactly one match in Q. In this case we define the node match *(n,m)* as certain.

After selecting the certain matches, we select additional node matches. We iteratively assess the potential node matches, ordered by decreasing quality. We call a node *free* if it has not yet been matched to another node. In each iteration we apply the following rule:

- we accept a potential node match *(n,m)* if both n and m are free and if there is no other potential match *(n,o)* such that o is free.

After assessing this first rule for all potential node matches, we apply a second rule:

- we accept a potential node match *(n,m)* if both n and m are free and if m is the free node closest to n.

Note that there may still be reference nodes that are unmatched.

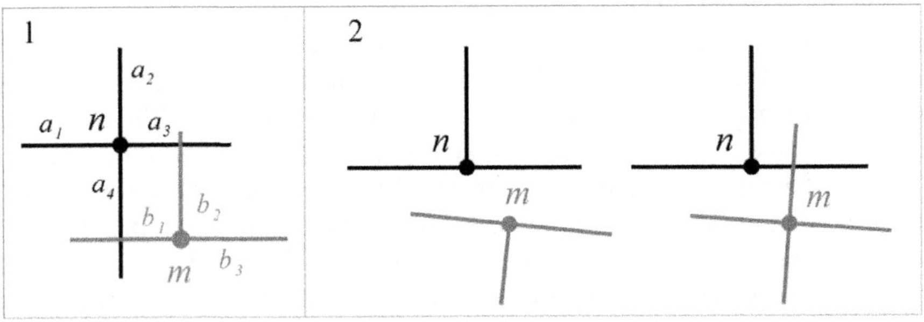

Fig. 9. (1) Exclusion of such candidate nodes m of the large-scale dataset, because *deg(n)>deg(m)*; (2) Further investigation of such kind of matches, because the defined conditions *deg(n)=deg(m)* or *deg(n)<deg(m)* are fulfilled.

3.3.2 Matching of Arcs

We perform the final matching of arcs based on the pre-matching of arcs and the matching of nodes. We select the arcs one by one from the small-scale dataset and search for the corresponding set of arcs in the large-scale dataset. In doing so, three cases need to be considered: (1) both endpoints of the arc of the small-scale dataset are matched with nodes in the large-scale dataset, as shown in Fig. 12a; (2) only one endpoint is matched with a node in the large-scale dataset; (3) none of the endpoints is matched with a node in the large-scale dataset, as shown in Fig. 12b. Different methods will be applied to the three cases.

Case (1): In this case we search a path that comprises pre-matched arcs of the large-scale dataset and connects the nodes that are matched to the endpoints of the selected small-scale arc. In order to find a good path, we minimise a cost function. This approach for matching arcs has been proposed by Mustière & Devogele (2008). For each arc a included in the path we charge a cost equal to the product of its length and the angle difference to the small-scale arc as defined in Sect. 3.2.2. This minimisation problem can be solved with a shortest-path algorithm, for example, the algorithm of Dijkstra (1959).

Case (2): In this case we first select all pre-matched arcs that have an endpoint that is matched with an endpoint of the small-scale arc and whose angle difference with the small-scale arc is smaller than a certain threshold μ. We match the small-scale arc with one arc of this selection, more precisely, with the arc that is closest to the unmatched endpoint of the small-scale arc.

Case (3): The small-scale arc remains unmatched.

4 Experimental Results

Our proposed matching process was tested in the test area (see Fig. 1) for the DLG 250 as the small-scale dataset and the DLG 50 as the large-scale dataset. In the first step we constructed a network of lines and ensured that the topology of the river network is complete. The result is shown in Fig. 10. For the distance criterion in the pre-matching of nodes we applied a threshold of $\delta = 440$ m, which was empirically determined. For the following pre-matching of arcs we applied the buffer size of $\varepsilon = 330$ m in order to get, for each arc a of the small-scale dataset, the set A'_a of arcs of the large-scale dataset.

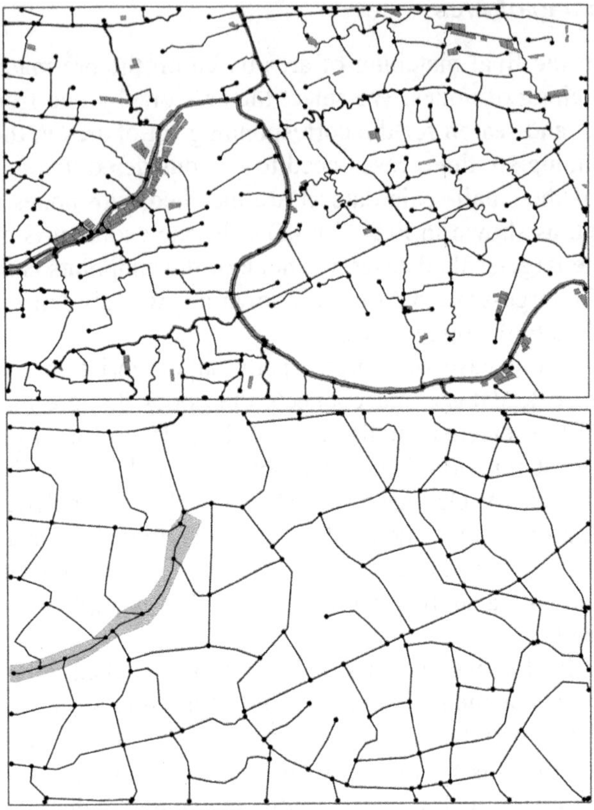

Fig. 10. The original datasets (grey) overlaid by the result of the construction of the topological correct line networks (black): (top) large-scale dataset DLG 50 and (bottom) small-scale dataset DLG 250.

Based on the results of the pre-matching step, we perform the final matching step. First, in the node matching step, we define the threshold $\theta = 60$ degrees in order to decide whether a node match is certain. Secondly, we define $\mu = 12$ degrees in the arc matching step in order to compare the direction of arcs. Figures 11 and 12 show our matching results.

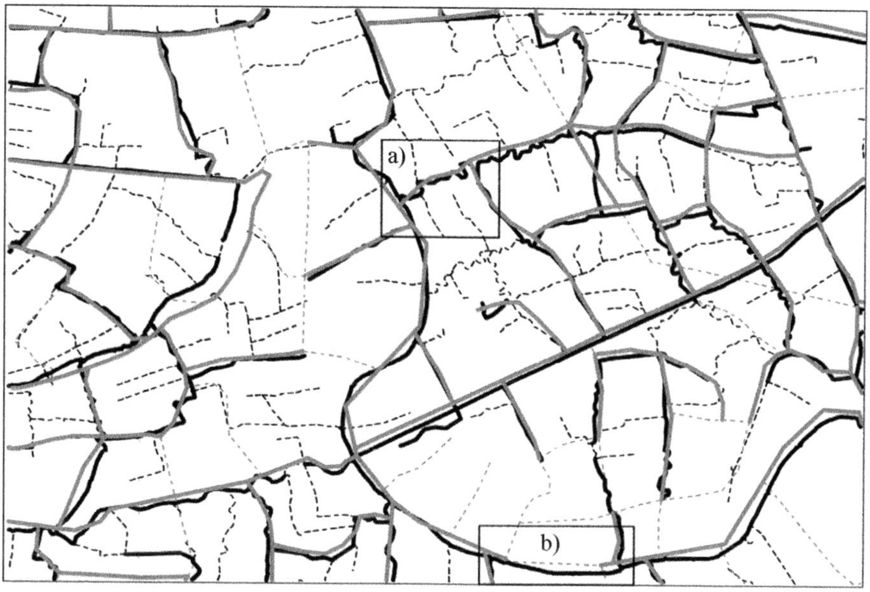

Fig. 11. Matches found by our method for the sample in Figure 10; line networks of the small-scale dataset are displayed dashed in grey and of the large-scale dataset dashed in black; matched arcs are displayed bold.

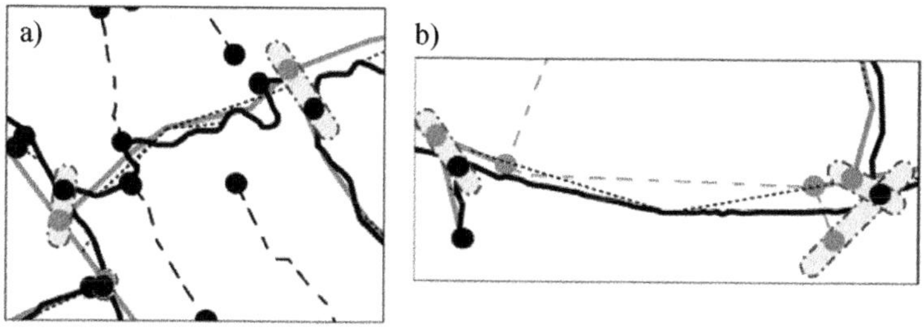

Fig. 12. Two magnified parts of figure 11; a) Relation one to many: the small-scale arc (grey) that was matched to four arcs of the large-scale dataset (black) marked by dotted black lines; b) the small-scale arc displayed in dashed grey is not matched, because none of the endpoints is matched.

We evaluated our matching results by comparing them with results that were obtained manually. In order to decide whether two objects should be manually matched, we compared their location, shape and orientation. Furthermore we took the network topology into account. Table 1 summarizes our results. For each arc match found by our method we assessed whether it is correct or not. We compare the total length of all correctly matched arcs and the total length of all arcs that were incorrectly matched (first line of Table 1). We see that only a small part of the found matches is wrong (2.3%). Furthermore, for each small-scale arc that was not matched by our method, we assess, whether there is indeed no corresponding large-scale object or whether an existing correspondence was missed. Again, we aggregate our results by summing up the lengths of the involved arcs (second line of Table 3). The human expert was able to match arcs of a total length of 85057.13 m + 8983.76 m = 94040.89 m. Compared to this value, our method failed to find 9.5% of the matches.

Table 1. Statistics of our matching results

	Total Length (m)	Correct (m)	Wrong (m)	Precision
Matched	87020.19	85057.13	963.06	97.7%
Not-matched	33573.85	24590.09	8983.76	73.2%
Sum	120594.04	109647.22	9946.82	90.9%

When both endpoints of an arc were matched, also the arc match is usually correct. However, if there is only one or even no matched endpoint, our method usually matches the arc only with some corresponding parts, but not with all.

We also compared our test results for parts of the river system that were originally represented by lines and parts that were originally represented by polygons. In this case the matching was done based on the derived skeleton lines. In conclusion, we did not observe a difference in the performance for both parts of the river system. Therefore, we assume that our skeletonization method is suited to support the matching method.

5 Conclusion and Outlook to Future Work

We have presented a new method for matching river datasets of different scales. In particular, we have shown how to cope with different geometric dimensions. For this purpose we have proposed a skeletonization method that constructs a topologically correct network of lines. The actual matching of the networks is based on a pre-matching of arcs and nodes and the final matching step. Our method was tested for different Chinese datasets. The results were compared with those obtained by a human expert.

We conclude that our method finds 90.5% of the actual correspondences. Only 2.3% of the matches found are wrong. These numbers are satisfactory, in particular, since large parts of the river system were represented by polygons in the large-scale dataset and by lines in the small-scale dataset. For these parts of the river system, we observe, on the whole, a similar performance compared to parts that are represented by lines in both scales. Therefore, we assume that our skeletonization method is suitable for matching tasks.

A possibility to improve our method is to also consider semantic attributes that are given for the rivers objects, for example, attributes that reflect the name or expressing whether a river is natural or man-made. Future research should also consider that generalization operators like aggregation and typification influence how objects are represented in different scales. For example, two rivers running parallel to each other may be represented by a single river line.

Acknowledgements

This research is funded by the German Science Foundation (DFG) and the National Nature Science Foundation of China (NSFC): 40620130438. The support is gratefully acknowledged.

References

Devogele, T., Parent, C., and Spaccapietra, S. (1998): On spatial database integration. International Journal of Geographical Information Science 12(4), pp. 335-352.

Diez, Y., Lopez, M.A., and Sellarès, J.A. (2008): Noisy Road Network Matching. T.J. Cova et al. (Eds.): GIScience 2008, LNCS 5266, pp. 38–54, 2008.

Dijkstra, E. W. (1959): A note on two problems in connexion with graphs. Numerische Mathematik, 1, 269-271.

Dunkars, M. (2004): Multiple Representation Databases for Topographic Information. Ph.D. thesis, Royal Institute of Technology (KTH), Stockholm, Sweden.

Harrie, L., and Hellström, A.-K. (1999): A prototype system for propagating updates between cartographic data sets. The Cartographic Journal 36(2), pp. 133-140.

Haunert, J-H., and Sester, M. (2008): Area Collapse and Road Centerlines based on Straight Skeletons. GeoInformatica 12(2), pp. 169-191.

Kieler, B., Sester, M., Wang, H., and Jiang, J. (2007): Semantic Data Integration: Data of Similar and Different Scales. Photogrammetrie Fernerkundung Geoinformation (PFG), vol. 6, pp. 447-457.

Lüscher, P., Burghardt, D., and Weibel, R. (2007): Matching road data of scales with an order of magnitude difference. Proc. XXIII International Cartographic Conference, Moscow, Russia, August 3–10, 2007.

Mustière, S. and Devogele, T. (2008): Matching Networks with Different Levels of Detail. GeoInformatica 12(4), pp. 435-453.

Penninga, F., Verbree, E., Quak, W., and van Oesterom, P. (2005): Construction of the planar partition postal code map based on cadastral registration. GeoInformatica 9(2), pp. 181-204.

Sester, M., Anders, K.-H., and Walter, V. (1998): Linking Objects of Different Spatial Data Sets by Integration and Aggregation. GeoInformatica 2(4), pp. 335-358.

Timpf, S. (1998): Hierarchical Structures in Map Series. Ph.D. thesis, Technical University Vienna, Austria.

Uitermark, H., Vogels, A., and van Oosterom, P. (1999): Semantic and Geometric Aspects of Integrating Road Networks. A. Vckovski, K.E. Brassel, and H.-J. Schek (Eds.): INTEROP'99, LNCS 1580, pp. 177-188, 1999.

Vivid Solutions (2005): RoadMatcher User Guide - RoadMatcher Version 1.4. http://www.vividsolutions.com/products.asp?catg=spaapp&code=roadmatcher (accessed 2009/02/10)

Volz, S. (2005): Data-Driven Matching of Geospatial Schemas. A.G. Cohn and D.M. Mark (Eds.): COSIT 2005, LNCS 3693, pp. 115-132, 2005.

Walter, V. and Fritsch, D. (1999): Matching spatial data sets: a statistical approach. International Journal of Geographical Information Science 13(5), pp. 445-473.

Zhang, M., and Meng, L. (2007): An iterative road-matching approach for the integration of postal data. Computers, Environment and Urban Systems 31(5), pp. 597- 615.

An Approach to Facilitate the Integration of Hydrological Data by means of Ontologies and Multilingual Thesauri

Miguel Ángel Latre, Javier Lacasta, Eddy Mojica, Javier Nogueras-Iso, Francisco Javier Zarazaga-Soria

Computer Science and Systems Engineering Department
University of Zaragoza, María de Luna 1, E-50018 Zaragoza, Spain
{latre, jlacasta, eddyma, jnog, javy}@unizar.es
http://iaaa.cps.unizar.es/

Abstract. The general concern about environmental issues has involved the creation of national and international policies that require, at a technical level, the analysis, merging and processing of data obtained from very different sources. This paper proposes an approach for the integration of hydrological data that is based on the use of a multilingual ontology to facilitate the mapping across the local data models in the different sources. The novelty of the proposal is that the multilingual domain ontology is generated automatically by the merging and pruning of existing lexical ontologies. This approach has been tested in the context of the European Water Framework directive for the development of reporting applications in cross-border scenarios. Nevertheless, this approach could be easily extended to other domains.

1 Introduction

The general concern about environmental issues in recent years has involved the creation of national and international policies encouraging the development of information infrastructures to facilitate the cooperative access and exploitation of data coming from different sources from public and private institutions.

An example of this general concern about environmental issues can be found in the European context. The environmental protection is one of the

M. Sester et al. (eds.), *Advances in GIScience*, Lecture Notes in Geoinformation and Cartography,
DOI 10.1007/978-3-642-00318-9_8,

interests of the European Union and different initiatives and policies in this field are taking place, such as the Water Framework (European Commission, 2000) and INSPIRE (European Commission, 2007) directives.

INSPIRE (*INfrastructure for SPatial InfoRmation in Europe*) aims at the creation of a European spatial information infrastructure that delivers integrated spatial information services, being environmental information the first application domain tackled by this directive. It is also interesting to take into account that a considerable amount of these environmental initiatives are related to the hydrology domain. The European Water Framework Directive (WFD) is considered to be the most important piece of legislation in this aspect (Usländer, 2005). Its main objective is to achieve an accurate management of all water bodies and reach a "good status" for them by 2015.

This paper proposes an approach for the integration of hydrologic data that aims at discovering implicit relations between hydrologic features that are not usually made explicit in database models. In this particular domain, hydrologists must monitor a great variety of features and phenomena that, although initially disconnected, may affect the status of water bodies. An information retrieval system for this kind of data is presented in this paper.

The approach proposed here is based on the use a multilingual lexical ontology or thesaurus. An ontology is usually defined as "an explicit formal specification of a shared conceptualization" (Gruber, 1993). It is considered as a means for the integration of data because it enables the establishment of a common reference model that facilitates the mapping across the local data models in different sources. Additionally, a domain ontology may help to infer relations that are not usually explicit in the local models and facilitate the combination of different feature types. Although the use of ontologies and thesauri for data integration is not new, the novelty of the proposal is that the multilingual thesaurus focused on the hydrologic domain is automatically generated by the merging and pruning of existing thesauri. The applicability of thesauri for searching and retrieval in digital libraries has promoted the creation and diffusion of well-established thesauri in many different domains. Thus, thesauri can facilitate an important source of information for the development of ontologies focused on specific domains. This automatically generated thesaurus is the main element that allows the information retrieval system presented in this paper to work.

The rest of the paper is organized as follows. Section 2 summarizes the state of the art in ontology based discovery and retrieval. Section 3 describes the information retrieval system this paper is based on, including the methodology for the multilingual thesaurus generation and the results obtained in the hydrologic domain (3.3). The last section concludes and introduces some ideas on future work.

2 State of the Art in Ontology Based Discovery and Retrieval

In a Spatial Data Infrastructure context, discovering and accessing suitable geographical information is a crucial task. However, semantic heterogeneity caused by natural language ambiguity (e.g., synonymy, homonymy) makes it difficult to interpret feature property names and user queries.

Some research works have been focused on advancing in the solution for overcoming this heterogeneity. Bernard et al. (2004) describe the architecture of an ontology based discovery and retrieval system of geographical information. In this system, different Web Feature Services are described with metadata which includes a reference to an application ontology that describes the feature types in terms of a shared domain ontology. User queries are processed as follows: users state their queries in terms of the shared domain ontology; then the system expands the user query restrictions with the names of the stored features. Lutz and Klien (2006) work shows the evolution of the previous system. This latter version defines a query language and provides a user interface that helps users to formulate queries.

Other works in this line are the ones proposed by Hübner et al. (2004) and Navarrete (2006). The first one describes an ontology based reasoning system that allows integrating heterogeneous geographical information by resolving structural, syntactic and semantic heterogeneities. The query system supports the specification of queries of the type *concept@location in time*. The user selects a set of registered domain-specific application ontologies (in the thematic, spatial, and temporal domains) based on a common vocabulary and use them to select search terms that are expanded by selecting all equivalences and subconcepts (for the thematic search term), spatially related place names (for the spatial search term), and relevant time periods (for temporal ones). The second one provides a framework to represent semantic relations among the concepts from different datasets of a repository. The system is based on a high level ontology constructed by merging the knowledge provided by the datasets of the repository that describe in a precise and formal way the content of the repository. This ontology is then used to define semantic services or queries that enable agents to find and integrate thematic information. It specifically focuses on finding datasets containing information on a particular theme (including theme subclasses if they are considered of interest); translating the content of a dataset to another compatible vocabulary; and integrating heterogeneous content from different datasets.

Not related to the geo-spatial area but also focused on improving the discovery and retrieval of information using ontologies is the work of Tudhope et al. (2006). It describes a system that uses terminological ontologies (faceted thesauri) to perform query expansion in indexed collections. It describes a semantic closeness algorithm that creates a neighbourhood of semantically related concepts (for retrieval purposes) from a selected one and gives them a weight according to their closeness. Somewhat in the same line is the work of Miles (2006). It analyzes the information retrieval issues caused by the language heterogeneity and proposes a formal theory to describe the ways in which a structured vocabulary may be used to construct and index over a collection of objects. Finally, it compares different expansion techniques in user queries to improve recall and precision.

The information retrieval system presented in this paper goes along the lines of these works, although it presents some differences. We want to combine powerfulness of allowing expert users to make their queries by means of an ontology with the simplicity of offering novice users a search mechanism based on terms from a thesaurus as base for the queries. Additionally, we aim at a system where the data resources are just standard OGC Web Feature Services created and maintained by the appropriate organizations, but where the addition of new information resources to the information retrieval system can be done in a easy and effortlessness way.

3 Design and Architecture of the System

This section is devoted to present the information retrieval system and the mechanism to generate the multilingual thesaurus. An overview of the system, its functionality, including an example of use, and its architecture are presented below. The generation of the multilingual thesaurus is described next in 3.3, followed by a description of the different components composing the information retrieval system (3.4).

3.1 System Overview

The system presented here aims at searching and automatically integrating hydrologic features from different sources on demand, just by providing a hydrology related term or concept. The functionality of the system has been tested in the context of the European Water Framework Directive

(WFD) (European Commission, 2000) for the development of reporting applications in cross-border scenarios. This is a similar scenario to the one of the SDIGER (Zarazaga-Soria et al., 2007), an INSPIRE pilot project whose aim was to test the feasibility of developing a cross-border inter-administration SDI to support WFD information access. A web application was developed in order to provide on-line the different map reports about the WFD implementation. In that project, the integration of data coming from French and Spanish data repositories was facilitated by means of ad-hoc software applications applying crosswalks between local data models and common reference models.

The original SDIGER web application aims at using INSPIRE principles for fulfilling the WFD reporting requirements established by the European Commission (European Commission, 2004). The application generates automatically the established reports from data and services belonging to the different WFD Competent Authorities involved in the SDIGER project: The Ebro River Basin Authority *(Confederación Hidrográfica del Ebro,* CHE) and the Adour-Garonne Water Agency *(Agence de l'Eau Adour-Garonne,* AEAG).

This time we want to go a step forward and enable users to search and automatically integrate hydrologic features from different sources on demand. What it is intended is to expand the SDIGER application to be able to work not only with a set of fixed feature types (the SDIGER application worked only with surface water and groundwater bodies) and a fixed set of data sources (the web feature servers of the water agencies involved in the project), but also with other kinds of features, through the use of a hydrology domain ontology and with the help of a multilingual hydrologic thesaurus. The domain ontology we have chosen is built upon the data model proposed by the Common Implementation Strategy of the WFD in the "Guidance document on implementing the GIS elements" (Vogt, 2002). However, an ontology based on the data model proposed by the "Data Specifications" Drafting Team or the Hydrography Thematic Working Group of the INSPIRE directive (INSPIRE Drafting Teams, 2008b, 2008a), once it has been adopted, would be not only equally appropriate in the field of the WFD, but even more generic when applied to the hydrography field.

By making use of the domain ontology, linked with the multilingual thesaurus, the user may request the combination and merging of hydrologic features not necessarily connected in the local data sources due to the use of different modelling approaches, as it is explained in the following sections.

3.2 Architecture

Figure 1 shows the overall architecture of the application. The main component is the *Ontology based IR system*, which takes a term as input and returns a set of features as output. It orchestrates the interactions with the other components of the system: a *Web Ontology Service* (WOS), a *Services Catalog*, a *WFS Query Resolver* and a *WFS Broker*.

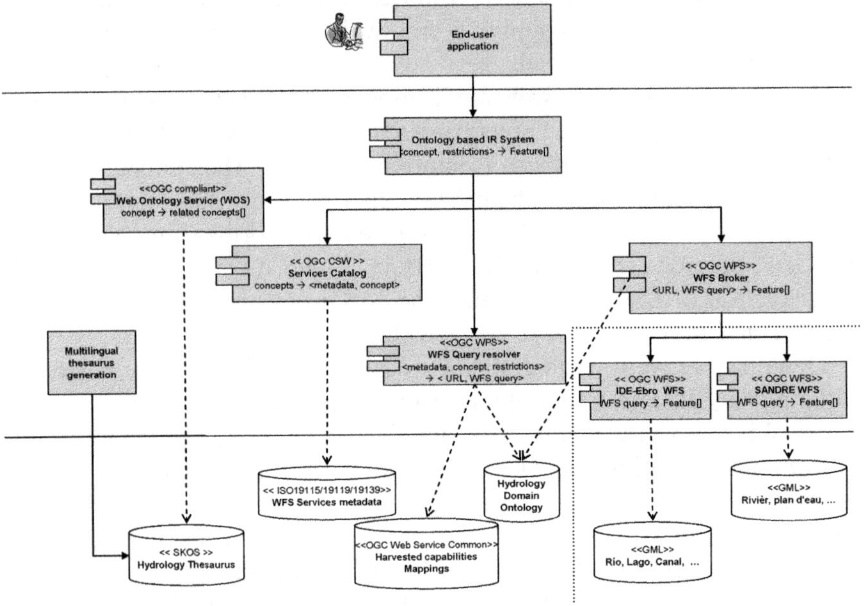

Fig. 1. Architecture of the application for the integration of data

The function of WOS is to expand the search term with related ones, according to the hydrology thesaurus obtained as it is explained in section 3.3. The component is used in order to increase the number of Web Feature Services that are going to be queried for features, and, thus, improve the recall of the information retrieval system. The objective of the query expansion is to solve some of the problems derived in terms of user queries: synonymy, multilinguality and lack of explicit knowledge of the domain (like hierarchical relationships, for instance).

The *Services Catalog* is used to provide WFS instances and feature types that are linked to the term the user is searching for.

The *WFS Query Resolver* purpose is to build the queries that are going to be requested to the services found by the *Services Catalog* according to their local models. It is in charge of selecting the appropriate feature types and translating the restrictions the user may have imposed into a filter encoding query.

Once the appropriate WFS, feature types and, if applicable, the filter encoding query in the local feature models have been obtained, the *WFS Broker* actually performs the queries to the different WFS. It is also in charge of combining the results the system must return.

The relation between the searched concept and the concepts that are related to local repositories is found via the thesaurus.

Figure 2 shows a schema that describes how the information retrieval system works. The example shows the particular case of a report about the term "water bodies", where two different web feature services are part of the system: the WFS of *IDE-Ebro*, the SDI of the Ebro River Basin Authority, and the *SANDRE WFS* of the French Ministry for the Environment. The hydrologic data needed can be accessed through these two servers, located in different data repositories according to the category of the water body (river, lake, transitional or coastal), where different modelling approaches have been used for each server.

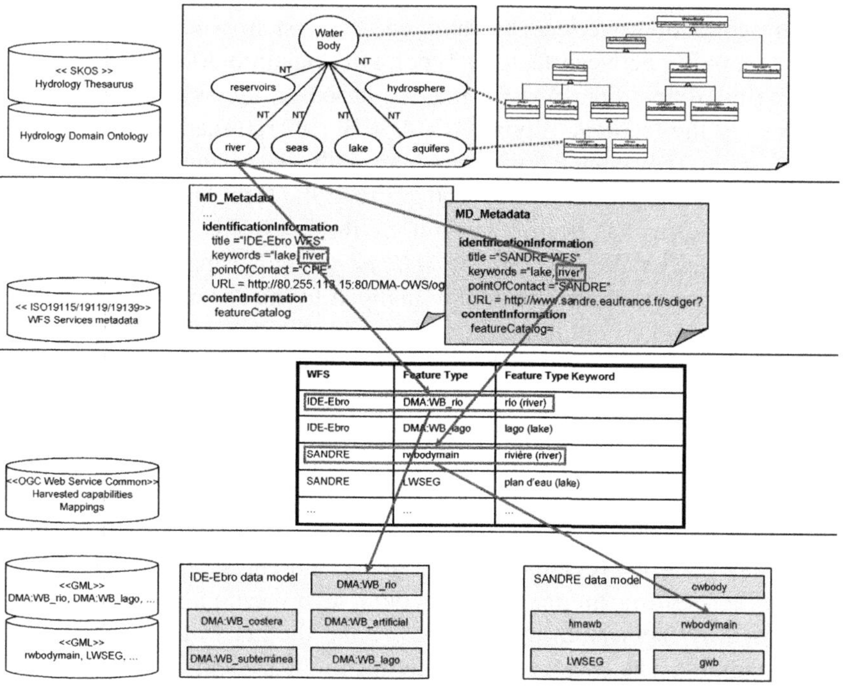

Fig. 2. Example of use

The end-user application makes a request for “water bodies”, which triggers a search of this concept in the hydrologic thesaurus, linked through “narrow-term” relations to the terms “river”, “lake”, “reservoirs”, “seas”, “aquifers” and “hydrosphere” (and their translations to other languages present in the source thesauri). All these concepts but “hydrosphere” appear in the metadata records of the system *Services Catalog*. In the case of “river”, we found that the “DMA:WB_rio” feature type of the *IDE-Ebro WFS* and the “rwbodymain” feature type of the *SANDRE WFS* have a translation of that term as keyword, and the same happens for the rest of the expanded terms (“lake”, “reservoirs”, “seas” and “aquifers”). Thus, the *WFS Broker* performs ten *getFeature* requests: five to the *IDE-Ebro WFS* to get features of types “DMA:WB_rio”, “DMA:WB_lago”, “DMA:WB_artificial”, “DMA:WB_costera” and “DMA:WB_subterranea”, and another five to the *SANDRE WFS* to get features of types “rwbodymain”, “LWSEG”, “hmawb”, “cwbody” and “gwb”.

If the results are going to be used for portrayal, the merging component renders them and returns an image to the user application. In the case that a feature collection needs to be returned, the system would propose to the user a mapping between features types and domain ontology concepts. Assuming the domain ontology is the one based on the WFD data model, in the case of the features of type “DMA:WB_rio” returned by the *IDE-Ebro WFS*, the system would propose a map with the concept “riverWaterBody”, which the system has deduced because the term “river” of the multilingual thesaurus has been found among the keywords describing the feature type “DMA:WB_rio” in the *IDE-Ebro WFS* capabilities, and since this same term has been previously mapped to the “riverWaterBody” concept of the domain ontology. In addition to this, for each attribute within the features of type “DMA:WB_rio”, the system would provide a list of possible matches for mapping with attributes of the concept “riverWaterBody” based on the coincidence of the data type used for its representation, in order to allow the user to map it to the correct one. For instance, the “DMA:WB_rio” attribute “nombre” of type string could be mapped to every string attribute of the concept “riverWaterBody”. The user, if no other has done it previously, could select, among all these, the attribute “name” to perform the map. Users could save these mappings in order to be reused by themselves or others, avoiding the need of perform this manual mapping. The user can make queries to the system by either selecting a concept form the multilingual thesaurus, or by choosing a concept of the domain ontology. In the first case, the users are just interested in obtaining features related to a certain term. In the second, they can also provide a set of restrictions (selection of features within a feature type and/or projection of attributes) on the features related to the concept they are interested in, in order to find features satisfying certain conditions.

3.3 Generation of a Multilingual Thesaurus in the Hydrology Domain

As base for the expansion of queries performed by the system described in this paper, it has been necessary to create a multilingual thesaurus describing the terminology in the hydrology domain. The created terminological model has been based on the list of hydrology related concepts referenced within the European Water Framework Directive. There were two main goals for construction of this thesaurus. On the one hand, we wanted a multilingual resource, while on the other one, we wanted to enrich it with more concepts related to hydrology than the ones described within the Directive text.

The creation process is based on the merging of a set of multilingual terminological ontologies following the thesaurus structure that contain hydrology related concepts with the selected set of hydrologic terms. The output obtained is a multilingual thesaurus specialized in the desired area of knowledge.

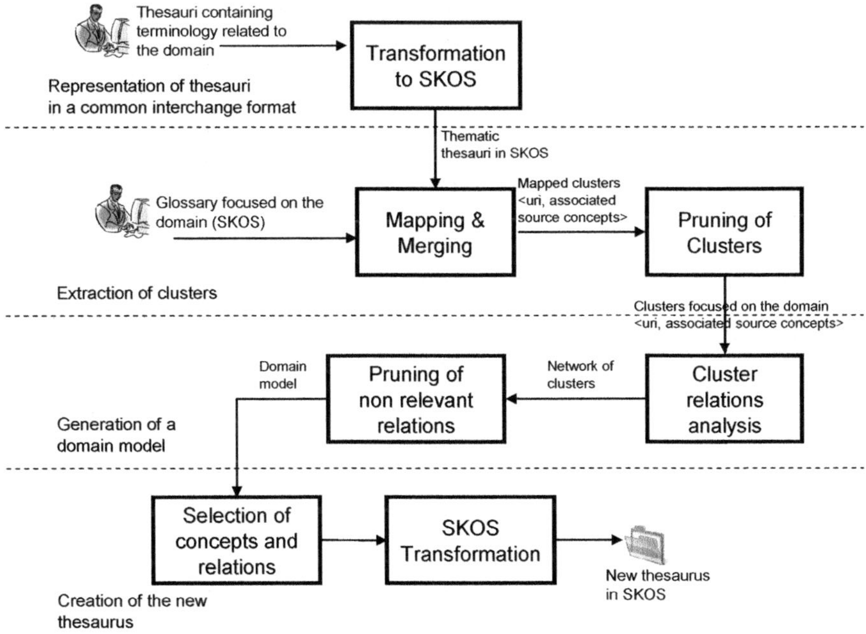

Fig. 3. Work-flow for the generation of a domain specific thesaurus

Figure 3 depicts the different steps of the process, showing the inputs and the produced results. Four different tasks can be highlighted (A detailed description of the three first ones is shown in Lacasta et al. (2007c) work):

- Representation of input thesauri in a common format. This task is devoted to the transformation of the input thesauri into SKOS (Miles et al., 2005) (a W3C initiative for the representation of Knowledge Organization Systems) with the objective of having a homogeneous input to the generation system. Apart from the list of hydrologic terms, the terminological models used as input for the method are the following thesauri: GEMET (the GEneral Multilingual Environmental Thesaurus of the European Environment Agency)[1], AGROVOC (the FAO Agricultural Vocabulary)[2], EUROVOC (the European Vocabulary of the European Communities)[3] and the UNESCO thesaurus[4]. They provide a shared conceptualization in the areas of economics, politics, culture and environment.
- Extraction of clusters. This is the main step and it consists in the detection of intersections between concepts in the different input thesauri, and the set of terms selected in the area of hydrology through the analysis of their lexical similarities. Each set of mapped concepts is grouped into a cluster, which is the name given to a concept in the output thesaurus. A cluster represents a group of equivalent concepts and it is identified with one of the URIs of the original concepts. With the objective of focusing on the hydrology theme, only those clusters that are found to be related to theme are stored. The selection of the clusters is performed with the following criteria:
  - If the cluster contains a concept from the selected European Water Framework Directive list of terms, the cluster is stored.
  - If a concept in the cluster is related to another one in other cluster that fits in the previous case, the cluster is preserved. This is done to add additional terms related to the original concepts of hydrology selected from the Water Framework Directive text.
  - In other case, the cluster is deleted.
- Generation of a domain model. This step consists in connecting the clusters previously extracted. The relations between the concepts assigned to the different clusters are converted into relations between the clusters

---

[1] http://www.eionet.europa.eu/gemet
[2] http://www.fao.org/aims/ag intro.htm
[3] http://europa.eu/eurovoc/
[4] http://www.ulcc.ac.uk/unesco/

that contain them. The relations between clusters are labelled with: the types of relations, which are derived from the original types of relations between concepts; and a weight that represents the number of occurrences for each original relation type between the concepts of the interrelated clusters. Additionally, it is possible to reduce the size of the resulting network by selecting only those concepts related between them by relations of at least a minimum weight. For example, it is possible to obtain a reduced network with only those clusters that have relations of at least weight 3 (that is, they have been found in three of the original thesauri).

- Generation of a new thematic thesaurus. The last step of the defined process is to transform the network of clusters into a thesaurus. The generation of the thesaurus consists in taking the clusters of the network and organizing them into a hierarchical model. The clusters are transformed into concepts of the new thesaurus; one of the labels of the original concepts within the cluster is selected as preferred label. With respect to the thesaurus structure, each relation is marked with the type that has more occurrences. Additionally, those concepts that do not have broader relationship are marked as top terms. Finally, the generated structure is reviewed to verify that the BT/NT relationships structure does not contains cycles. If a cycle is found, it is removed by replacing the NT/BT relationship that generates the cycle by a *related* relationship.

The result thesaurus has been generated using the complete network of concepts as base (all the generated concepts and relations). It contains 322 concepts with 966 preferred labels and 2424 alternative labels. With respect to the relations, the number of *broader/narrower* pairs is 239 and the number of *related* relationships is 203. Figure 4 shows a subset of the generated thesaurus using the ThManager tool[5] (Lacasta et al., 2007b). The figure shows a branch of the thesaurus starting from the "land cover" concept. It shows the generated hierarchy containing the different types of water bodies.

---

[5] ThManager is an OpenSource tool for the creation and visualization of terminological ontologies stored in SKOS format (see http://thmanager.sourceforge.net/).

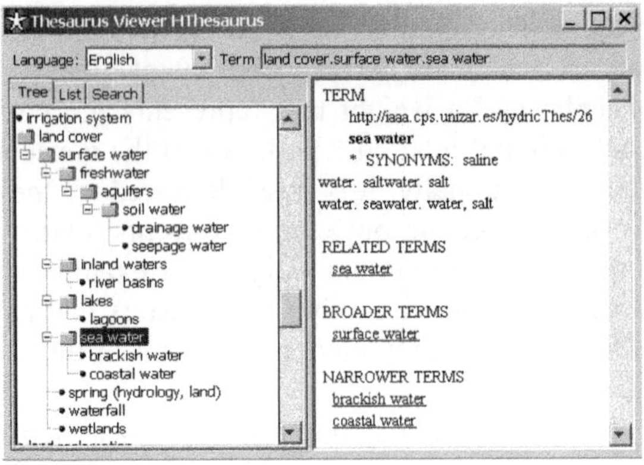

Fig. 4. Visualization of a part of the generated hydrologic thesaurus

3.4 Components of the System

3.4.1 *Web Ontology Service*

The *Web Ontology Service* is an OGC Web Services Architecture specification compliant component whose purpose is to facilitate the management and use ontologies and thesauri to other web components requiring them. Designed as a centralized service, the architecture of this service aims at reducing the cost of creation of a new ontology or thesaurus, improving reusability and avoiding duplicities and inconsistencies. The architecture and a detailed revision of the functionality of these services are described in Lacasta et al. (2007a). In the context of this paper, this service is used to solve some of the problems derived from the ambiguity of user queries (e.g., synonymy, multilinguality, or lack of explicit knowledge of the domain). This is done by expanding the user search terms with other related ones, according to the multilingual thesaurus obtained in the previous section. This reformulation of the queries has as final objective to increase the number of WFS to be queried for features, and, thus, improve the overall recall of the information retrieval system.

3.4.2 *Services Catalog*

The function of the *Services Catalog* is to provide to the system with WFS that are linked to the term the user is searching for. It is a standard OGC Services Catalog that access ISO19115/19119/19139 metadata (Nogueras-Iso et al., 2009), which have been created automatically through a crosswalk from the different capabilities XML files the Web Feature Services included in the system return to the *getCapabilities* request. This approach has two main advantages. On the one hand, in order to include a new WFS into the system, just its URL and its capabilities files are needed. On the other hand, the capabilities XML files include a list of the feature types, together with a list of keywords for each feature type served by a particular WFS. If the capabilities file is thorough enough, the transformed metadata record can be used perfectly to find appropriate WFS related to a hydrologic concept, avoiding the need of manually editing and completing them. Since we are working with an thesaurus built from a set of multilingual thesauri, the language in which the keywords are written in the capabilities is irrelevant (provided that the language is supported by the original thesauri).

3.4.3 WFS Query Resolver

The *WFS Query Resolver* purpose is to make up the actual queries that are going to be requested to the Feature Services found by the *Services Catalog*. It is in charge of selecting the appropriate feature type to query the WFS. In order to do so, it requests to the WFS its capabilities and finds out which feature type is linked with the keyword or concept the user is interested in searching.

Additionally, it is also in charge of translating the restrictions users may have imposed. This set of restrictions on the features they are interested in must have been established by using the domain ontology, and the *WFS Query Resolver* translates them into the appropriate filter encoding queries. In order to do this, the user restrictions terms (attributes and, possibly, values) are translated into the appropriate ones for a particular WFS. In this case, the data model of any WFS that can be accessed by our information retrieval system must be carefully analyzed, through the metadata provided by *getCapabilities* and *describeFeatureType* requests (and it is even probable that an additional detail of information about the data model would be needed). Then, a manual mapping between elements of the WFS feature model and the domain ontology must be done. Not only feature types must be mapped to the domain ontology concepts, but also their attributes and their domain values must be mapped (Fallahi et al., 2008). In the case of quantitative attributes, it could be necessary to identify measurement units, since conversion of units could be necessary. And, in the case of qualitative attributes, a mapping between the possible values the attributes can have in the feature model and in the domain ontology must be also performed if these values are a controlled list or a thesaurus. An OGC Feature Catalog would have been used if it has been possible, but there is no possibility of storing keywords of each feature type and attribute, and it does not allow establishing mappings between feature types of different collections.

The approach where users provide just a hydrologic term to perform the search is less powerful, but it is definitely much more simple from the use and maintenance points of view: users do not have to make a query using a domain ontology (they can freely use any term to perform the search), and the mapping between the user request and the WFS feature models can be done automatically and on-the-fly (just WFS capabilities information is needed to perform the mapping). Thus, the scalability of this approach is enormous. In the second approach, where users must express their searches in terms of a domain ontology, a previous work of mapping feature types, attributes and domain values must have been done. Anyway, this work of maintenance in order to add new WFS can be alleviated by getting the users to do it, as it is explained in the next section.

3.4.4 WFS Broker

Once the appropriate Web Feature Services, feature types and, if applicable, the filter encoding query of the local feature models have been obtained, the *WFS Broker* actually performs the actual queries to the different services. The *WFS Broker* is also responsible of combining the results the system must return:

- The simplest way of combining the results consists in generating a map with the spatial data of the returned features and, then, returning an image instead of a feature collection. Obviously, since the returned features are provided in the form of a map, the only use of this returning mechanism is the portrayal of them. In this case, it can be considered that the *WFS Broker* is acting like a Web Map Service with Style Layer Descriptor capabilities.
- Combining the results into a single GML file and into a unique feature type. In order to do that, the system would provide the user with a mapping to the domain ontology. Since there is a map between the term in the multilingual thesaurus and one of the keywords that appear describing the feature type in each of the capabilities of the WFS that have been queried, and that the terms of the thesaurus and the domain ontology concepts have already mapped, the system can automatically propose mappings between the different feature types and the domain ontology. Furthermore, the system can also propose syntactically correct mappings between the different attributes of the feature type and the ones of the domain ontology concept, based on equivalence of data types of the attributes. Users could select the correct mappings between attributes,

and, this way, make mappings between the local models of the WFS and the domain ontology. These mappings can be reused, by the same user in posterior queries, and by the system itself, to be used by the *WFS Query Resolver* when transforming complex user queries into the appropriate filter encoding queries, as it has been mentioned previously. In this last case, the mappings should be validated by the system administrator, since the may be used by queries of other users of the system.

At this moment, we have implemented the first approach, and we are working on the second one.

4 Conclusions

This paper has presented an information retrieval system that facilitates the integration of hydrologic data and the discovery of implicit relations between features, not usually found directly in local data repositories. The relation between the searched concept and the concepts that are related to local repositories is found via a multilingual thesaurus generated starting from a set of thesauri from different knowledge areas and a selected list of terms focused on the domain. The system takes as input a search term or query, uses the multilingual thesaurus to expand it, locates and queries appropriate Web Feature Services, and returns the results as a map or as a feature collection.

Further work will be devoted to the study of some inferred relations and how they can contribute to the improvement of hydrologic models, to the returning mechanism of data as a feature collection and how to orchestrate the process as a service chain using formal languages for service composition such as BPEL (Business Process Execution Language).

Acknowledgments

This work has been partially supported by the Spanish Ministry of Science and Innovation through the projects TIN2007-65341 and PET2006_0760 from the National Plan for Scientific Research, Development and Technology Innovation and Zeta Amaltea, S.L. The work of Eddy Mojica has been partially supported by the University of Zaragoza, Banco Santander and Funcación Carolina through the program "Ayudas para realizar estudios de doctorado dirigidas a estudiantes latinoamericanos. Curso 2008–2009."

References

Bernard, L., Einspanier, U., Haubrock, S., Hübner, S., Klien, E., Kuhn, W., Lessing, R., Lutz, M., and Visser, U. (2004). Ontology-based discovery and retrieval of geographical information in spatial data. Geotechnology Science Report 4, Institute of Geoinformatics, Münster.

European Commission (2000). Directive 2000/60/EC of the European Parliament and of the Council establishing a framework for Community action in the field of water policy. Official Journal of the European Union.

European Commission (2004). Common Implementation Strategy for the WFD. Reporting sheets for 2005 reporting. Version 5.0. Technical report, European Commission - DG Environment D.2.

European Commission (2007). Directive 2007/2/EC of the European Parliament and of the Council of 14 March 2007 establishing an Infrastructure for Spatial Information in the European Community (INSPIRE). Official Journal of the European Union.

Fallahi, G. R., Frank, A. U., Mesgari, M. S., and Rajabifard, A. (2008). An ontological structure for semantic interoperability of GIS and environmental modeling. *International Journal of Applied Earth Observation and Geoinformation*, 10(3):342–357.

Gruber, T. (1993). A translation approach to portable ontology specifications. *ACM Knowledge Acquisition, Special issue: Current issues in knowledge modeling*, 5, Issue 2(KSL 92-71):199–220.

Hübner, S., Spittel, R., Visser, U., and Vogele, T. (2004). Ontology-based search for interactive digital maps. *IEEE Intelligent Systems*, 19(3):80 – 86.

INSPIRE Drafting Teams (2008a). D2.8.I.8 INSPIRE Data Specification on Hydrography – Draft Guidelines. Technical report, European Commission.

INSPIRE Drafting Teams (2008b). Drafting Team "Data Specifications" — deliverable D2.3: Definition of Annex Themes and Scope. Technical report, European Commission.

Lacasta, J., Nogueras-Iso, J., Béjar, R., Muro-Medrano, P., and Zarazaga-Soria, F. (2007a). A web ontology service to facilitate interoperability within a spatial data infrastructure: applicability to discovery. *Data & Knowledge Engineering*, 63(3):947–971.

Lacasta, J., Nogueras-Iso, J., López-Pellicer, F. J., Medrano, P. M., and Zarazaga-Soria, F. J. (2007b). ThManager: An open source tool for creating and visualizing SKOS. *Information Technology and Libraries (ITAL)*, 26(3):39–51.

Lacasta, J., Nogueras-Iso, J., Zarazaga-Soria, F., and Muro-Medrano, P. (2007c). Generating an urban domain ontology through the merging of a cross-domain lexical ontologies. In *Proceedings of 2nd Workshop of COST Action C21 - Towntology: Ontologies for urban development: conceptual models for practitioners*, pages 60–74.

Lutz, M. and Klien, E. (2006). Ontology-Based Retrieval of Geographic Information. *Journal of Geographical Information Science*, 20(3):233–260.

Miles, A. (2006). Retrieval and the semantic web. A theory of retrieval using structured vocabularies. Master's thesis, Oxford Brookes University.

Miles, A., Matthews, B., and Wilson, M. (2005). SKOS Core: Simple Knowledge organization for the WEB. In *Proc of Int Conf on Dublin Core and Metadata Applications*, pages 5–13, Madrid, Spain.

Navarrete, A. (2006). Semantic integration of thematic geographic information in a multimedia context. PhD thesis, Pompeu Fabra University.

Nogueras-Iso, J., Barrera, J., Rodríguez, A. F., Recio, R., Laborda, C., and Zarazaga-Soria, F. (2009). *Spatial Data Infrastructure Convergence: Research, Emerging Trends, and Critical Assessment*, chapter Development and deployment of a services catalog in compliance with the INSPIRE metadata implementing rules. The Netherlands Geodetic Commission (NGC), The Netherlands. accepted for publication.

Tudhope, D., Binding, C., Blocks, D., and Cunliffe, D. (2006). Query expansion via conceptual distance in thesaurus indexed collections. *Journal of Documentation*, 62(4):509–533.

Usländer, T. (2005). Trends of environmental information systems in the context of the European Water Framework directive. *Environmental Modelling & Software*, 20(12):1532–1542.

Vogt, J. (2002). Guidance document on implementing the GIS elements of the Water Framework Directive. Technical report, Commission of the European Communities.

Zarazaga-Soria, F. J., Nogueras-Iso, J., Latre, M. A., Rodríguez, A., López, E., Vivas, P., and Muro-Medrano, P. (2007). *Research and Theory in Advancing Spatial Data Infrastructure Concepts*, chapter Providing SDI Services in a Cross-Border Scenario: the SDIGER Project Use Case, pages 107–119. ESRI Press.

Applying Instance Visualisation and Conceptual Schema Mapping for Geodata Harmonisation

Thorsten Reitz, Arjan Kuijper

Fraunhofer Institute for Computer Graphics Research (IGD)
Fraunhoferstr.5, 64283 Darmstadt, Germany
{thorsten.reitz, arjan.kuijper}@igd.fraunhofer.de

Abstract. This paper gives an introduction to concepts for a better expert knowledge extraction for geodata harmonisation. Geodata harmonisation involves overcoming heterogeneities on the syntactic, schematic and se-mantic levels, but user interaction is mostly required for the semantic interoperability level. Consequently, the goal is to provide tools to geo-domain experts which allow them to use their knowledge to describe the conceptual schemas of their domain as well as alignments to other domain's conceptual schemas in a sufficiently expressive way without be-coming experts in ontology engineering. We describe an approach for this that includes both visual analysis on the basis of geospatial instances and specific interaction guidance processes. Based on this approach, a prototypical implementation of a tool called the HUMBOLDT Alignment Editor (HALE) is introduced.

1 Introduction and Motivation

There are two main directions that are apparent in today's spatial data infrastructure implementations. One that can be seen is a data-provider driven approach focused on the needs of provision and catalog-oriented discovery, as indicated by the relatively high availability of metadata clearinghouses (Janssen and Vandenbroucke, 2006; Nowak and Craglia, 2006). Another one is a silo-like approach such as in the oceanography or meteorology communities, where data is mostly provided for one application domain in that domain's well-known formats and classification systems. Both approaches provide the basis for further development by acting as

M. Sester et al. (eds.), *Advances in GIScience*, Lecture Notes in Geoinformation and Cartography,
DOI 10.1007/978-3-642-00318-9_9, © Springer-Verlag Berlin Heidelberg 2009

seedlings for others SDIs and applications to grow. However, when aiming towards the goal of establishing a spatial data infrastructure that is useful to a wide range of users across domains, sufficient cross-domain, cross-border and cross-language interoperability of the data sets and services offered remains an unresolved requirement and therefore a major topic of research.

1.1 Problem

To reach such a comprehensive interoperability level, an understanding and description of the data sets from different application domains is required on multiple dimensions. This understanding includes syntactic and schematic aspects, and is especially difficult for semantic aspects, since it can only be provided by persons very well versed in working with the terms of an application domain. However, extracting the knowledge of these experts for (automated) harmonization processing can be quite difficult – an ontology engineering expert often needs to invest a high amount of effort to precisely formalize concepts from the application domain, whereas the application domain expert often finds working with tools for ontology experts counter-intuitive and not adopted to the specifics of his domain.

This is in part due to the nature of the process – creating a description of the meaning of information, given in such a way that the information is reusable in different contexts afterwards – is a very complex task. The complexity tends to increase even more when describing relations between different application domains, such as by creating a mapping between two formalized application schemas. Such a mapping of conceptual application schemas requires very good understanding of two application domains and can contain a lot of mismatches and false assumptions.

As an example, consider the case of a protected areas management application that also needs to use hydrographical data. For hydrography and protected areas management, the properties of a water body that are important are rather different. Because of that, even aspects that superficially have the same meaning can actually have quite a different meaning. Figure 1 shows this exemplarily, because parts of the geographic data set expressed in the two different schemas are directly visible.

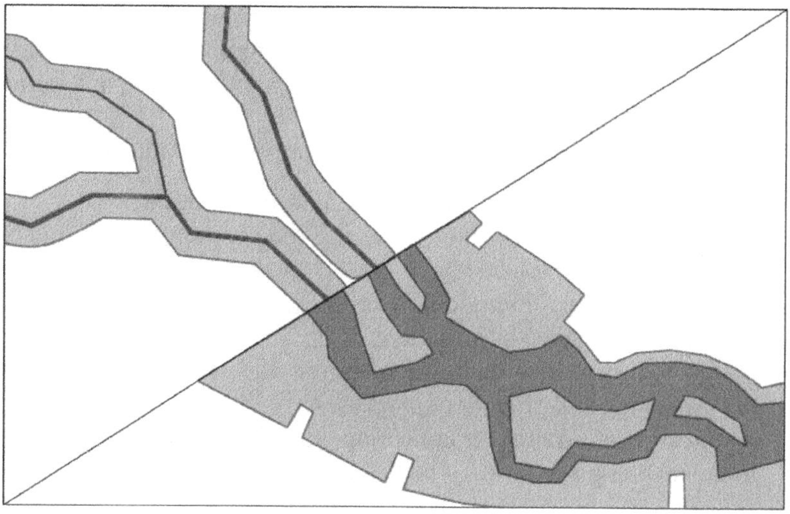

Fig. 1. A direct comparison of two geographical datasets with different geometric attributes for the same Feature Type (*River*) and different classification rules applied (*Floodplain*)

From the visual analysis that this image allows, a domain user can easily infer that the two domains do not use the same definition for flood plain (the green areas). In the schema displayed in the lower right part, flood plains are those areas designated to be flood plains by the zoning/development plan. In the other case, flood plains refer more to specific types of vegetation, which appears because of high ground water levels and of regular flooding. Inferring this just from the schemas, where in both cases the geometric attributes are of the same type (such as a GML *Multi-Polygon*), would not have been possible.

1.2 Approach

This paper describes a new approach for supporting domain expert users in both describing the conceptual schemas of their application domains and also in relating those schemas to those of other domains by creating mappings of application schemas. It exploits both geographic instances and their conceptual-level descriptions to make an alignment an interactive, deterministic and understandable process, which so far was problematic (Jantke and Dötsch, 1997).

This work contributes to the goal to provide an interaction concept specifically adapted to geodata harmonisation by adopting and extending several methods from general schema engineering, GI Science and cartography.

The following aspects of this concept are presented in this paper:

1. Definition of a general geodata harmonisation process;
2. Definition of an interaction process of the creating of alignments between application schemas;
3. Definition of a quality model to assess the quality of a mapping using schema and instance information;
4. Usage of a map as a central interaction and feedback component in the modelling and alignment processes;
5. Application of ontology engineering, specifically alignment, methods, to the mapping process;

Furthermore, a maximum-impact model is introduced which is used to value individual tasks and to allow users to quickly identify which mappings between parts of a schema are crucial in terms of attached reference data sets.

This paper is organized as follows: The next section provides an overview of related work and also presents an analysis of the applicability of existing alignment software to the issue of geodata harmonization. The third section presents the conceptual framework developed for schema creation and alignment, while the fourth section focuses on the prototypical implementation of the alignment tool. The final section concludes with a summary and an outlook.

2 Related Work

The harmonisation of geodata as a means of making an SDI effective is a complex task that involves a large set of aspects. There are several different approaches to this. Some are based on database integration techniques and employ information on the logical schema as well transformations on the physical level. For such approaches, there is robust commercial software available such as Safe's Feature Manipulation Engine that contains a wealth of file and database importers and exporters, as well as high numbers of transformation functions (Kaufman, 2004). Working at the physical and logical level however means that no assertions on a conceptual level can be used, which can be very useful in assessing the accuracy of a trans-

formation. Also, there are several more aspects to geodata harmonisation that need to be taken into account.
Out of these aspects, which have been summarized by Hall (2007) and are shown in Fig. 2, the method presented here addresses the aspect of different *Application Schemas, Terminology, Consistency* and *Quality*.

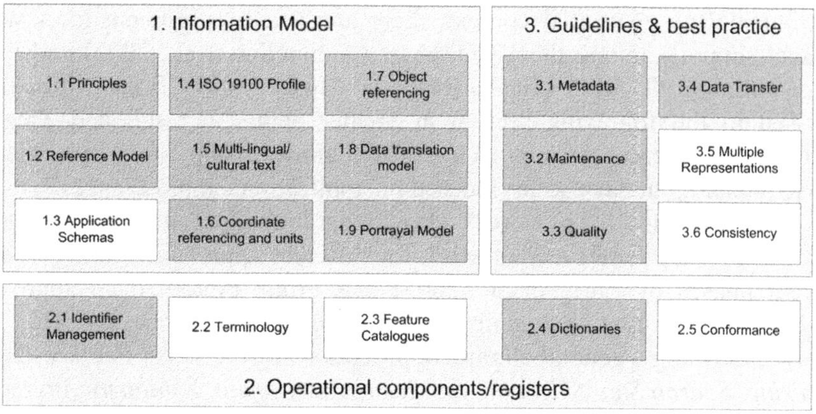

Fig. 2. Aspects of geodata harmonization (Adopted from Hall (2007), pg. 8). Aspects highlighted in white are of importance to the work presented in this paper

Researching the specifics of alignment of geospatial conceptual application schemas and the development of tools for that is a rather new field. However, the alignment of conceptual schemas and specifically of ontologies has been investigated by some research groups.

Using ontologies as a way of expressing formalized, computer-readable application schemas and as a means to facilitate interoperability and as a means for semantic harmonization in the geospatial world has become quite a common, if not undisputed methodology, within the GIScience community in the last years (Agarwal, 2005). This ontological view focuses, as suggested by Fonseca and Martin, on the *"explanation and information integration grounded in assumptions about invariant conditions that define the domain of interest"* (Fonseca and Martin, 2007). A conceptual schema focuses *"on enabling the measurement and classification of the observed facts"* (Fonseca and Martin, 2007) and can be understood as a link between ontologies and data. Another way to summarize this distinction would be to think of the conceptual schema providing all typing and right-hand-side definitions.

Agarwal et al. (2005) provide a formal framework for the description both of mappings between concepts of different ontologies and also formal description of possible mismatches. This formal description gives definitions for four different types of mismatches, categorized into mismatches based on equivalence (*Term Mismatch, Attribute Mismatch, Abstraction Mismatch*) and mismatches based on subsumption (*Structure Mismatch*).

Cruz et al. (2007) have presented a tool for visual alignment of geospatial ontologies and shows its application on three wetland ontologies. The ontologies used are, however, purely hierarchical classification schemes.

Outside the geospatial context, there are several alignment tools with visual support, among them COMA++ (Aumueller et al., 2005) and Protégé with the PROMPT plug-in (Noy and Musen, 2003). Both will be discussed in the following section in greater detail. Jerroudi and Ziegler (2007) also present an approach to do alignments supported by a visual environment, but have so far focused on applications in biosciences.

There has also been a work classifying individual step in alignment processes and to offer a framework for different approaches fitting in different phases by Ehrig et al. (2005). His group presented an approach called APFEL used to organize ontology alignment processes. This involved defining a general alignment process with five steps: *Feature Engineering, Search Step Selection, Similarity Assessment, Similarity Aggregation* and *Interpretation*. This general ontology and alignment engineering process has also been taken up and extended for the process of data harmonization presented in section 3.

For a more extensive description of tools and methodologies generally applied in ontology alignment, please refer to Ehrig (2006).

2.1 Applicability of Ontology Alignment Tools for Geospatial Application Schema Mapping

Based on this survey of related research results and software, an analysis of the alignment approaches and tools was conducted. The goal was to see how they can be used for conceptual schema mapping geodata harmonization and whether they would be accepted by geodomain experts.

This analysis was done in two phases, first comparing current software based on an internal assessment. In this phase, several requirements were identified on the base of the requirements given by Kaltz et al. (2006) for interactive Ontology Engineering processes, such as a graphical user interface supporting a semi-automatic alignment process. In the second phase, the assessment was complemented by conducting a formal usability experiment with the two programs that best fulfilled the requirements.

These were PROMPT for Protégé and COMA++, which both use subsumption reasoning and several forms of (literal) string matching of concept titles to do a semi-automatic alignment. However, the two differ in their approach to user guidance, with COMA++ establishing mappings first and then allowing a user to modify them, whereas PROMPT for Protégé provides suggestion sets out of which the user has to select actual mappings.

The experiment was conducted as an on-line test, with the participants logging into a pre-configured computer via remote desktop software. For the two alignment programs, each participant had to work on two tasks with the goal of creating alignments between parts of wetland ontologies with a total duration of about 60 minutes. These wetland ontologies were also used for the evaluation done by Cruz on her visual alignment tool evaluation (Cruz et al., 2007). All interaction was screen-captured for later analysis.

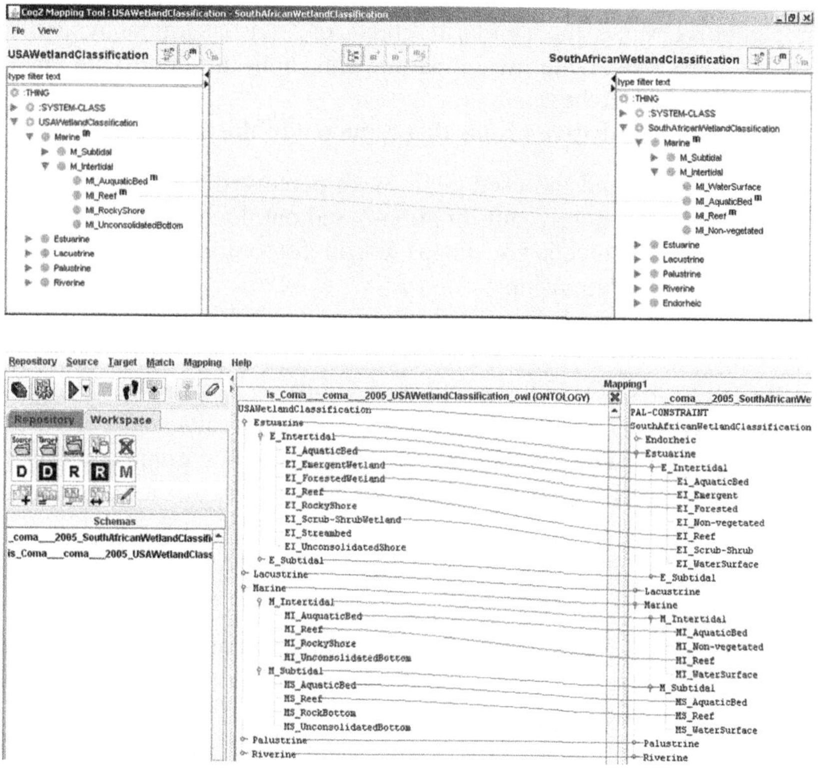

Fig. 3. The COMA++ alignment visualization (top) and the CogZ mapping window in Protégé PROMPT (bottom)

The ordering of the tasks (and consequently of the alignment programs) was reversed after the first half of participants to see whether there would be a transfer of learned knowledge from one program to the other. Every participant had to fill out an ISO 9241-based questionnaire so that we would be able to compare subjective and objective views.

This round of experiments was only conducted with a rather small group of eight participants, all of whom are domain experts from different fields of geoinformation application, such as protected areas management, analysis of time series of remote sensing data for hazard analysis and urban planning. Most participants have had experience in using classification schemas, in modelling data, and most of these have at least had some exposure to UML. However, when it came to ontology engineering, only two out of the eight participants indicated that they had some basic understanding of the concepts and tasks involved.

The quality of the alignments created was about equal for both programs, with PROMPT on average requiring about 30% more time for task completion. Despite the similarity, the CogZ visualization available in PROMPT was rated better by the participants, both for identifying matches and possible mismatches.

Some of the qualitative results that came out of this experiment are:

- Both programs and the used GUIs were perceived as being well-suited for alignment of purely subsumption-based ontologies, but the tree-style visualization elements are not as useful for ontologies also allowing other types of relationships;
- Users are not sure they can trust the results of the automatic matching parts of the software;
- Especially in PROMPT, unified user guidance is missing, as there are many interaction elements whose interrelations are not easily learned;
- Supporting visualizations such as the neighbourhood graph employed by PROMPT are not adopted to learned styles of visualization in the geospatial area;
- Visualizations are not expressive enough for identifying possible mismatches.

However, even within the small group of participants, the acceptance of individual elements of the tools varied quite a bit and especially in the self-assessment, variation in responses was quite high.

The qualitative findings of this small study have been used as starting points for deriving requirements on the user interaction processes for developing a data harmonization toolkit, as described in the following sections.

3 A Framework for the Schema Mapping Process

As explained in the beginning, it is important to make geospatial conceptual schema mapping a repeatable, deterministic process. The method employed towards this was to formalize the process of geodata harmonisation, as also done by Ehrig et al. (2005), and of the schema mapping itself, and then to formalize individual steps in this process as far as useful.

Basically, a task-based approach was followed, starting with a user and the tasks he can perform on a given set of conceptual schema elements and ending with concrete supporting workflows, services and algorithms.
When referring to a user, the concept always assumes a person knowledgeable about concepts used in at least one geospatial application domain, with basic experience in describing those (such as in UML class diagrams or using text templates) and with no specific experience in ontology engineering and alignment.
For these users, there are two main goals towards supporting harmonization; i.e. the integration of geodata available in a local schema into a shared schema. One is the description of non-harmonized (source) resources, the other a description of how that description of non-harmonized resource relates to a harmonized (target) description.

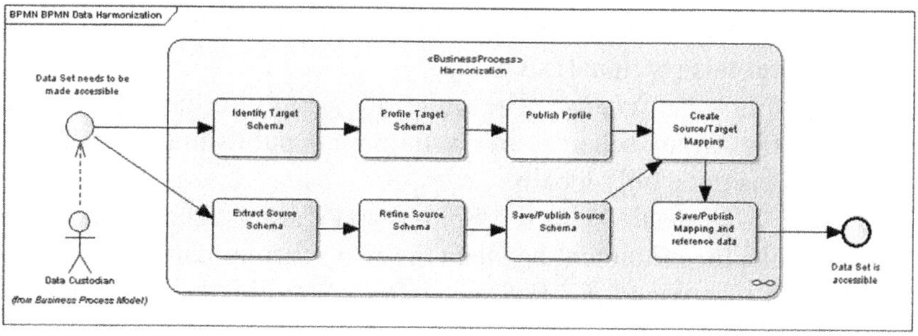

Fig. 4. The individual steps of the preparation process for conceptual schema geodata harmonisation

The presented concept generally assumes that a target schema is available, such as an INSPIRE schema. In addition, a reference data set for the target schema, and access to a source data set optionally also with a schema, have to be available for maximum quality. This reference data set can be a data set that has any spatial relation to the source data set. In the example from the introduction, where a detailed floodplain analysis is requested, the reference data set would be the one with the fully-modelled floodplains and rivers.

The overall process of extracting a schema and of creating an alignment for integrating a non-harmonized data set with a Spatial Data Infrastructure is outlined in Figure 4 which can be seen as the frame for the task-based processes detailed in the next sections.

1. *Identify Target (Harmonized or Local) Schema:* As explained above, it is assumed that a target schema is available. This can be one of the schemas created by the data specification teams responsible in an SDI, or one used to define the concepts used within an organization. The target schema can be selected from a catalogue of known schemas or can be identified via keyword or query-by-example mechanisms.
2. *Profiling Target Schema:* In cases where the shared schema used by the SDI is not sufficient, the schema can be profiled to add elements specific to the (sub-)domain.
3. *Publish Profile:* Only profiles that are "published", which can happen in different scopes, can be used for an alignment.
4. *Extracting Source Schema:* To be able to perform a data transformation of the source data set to the Target schema, the source schema needs to be described. This can be done partly automatic by applying schema extraction techniques if no schema is available.
5. *Refining Source Schema:* If the Source Schema has been extracted automatically, some refinements might be necessary, which are conducted in this (optional) step.
6. *Publish Source Schema:* The same rules apply to the source schema as to the target schema, but usually the publishing of the source schema is done only locally.
7. *Create Alignment:* With the Source and Target Schemas now being available, the alignment between the two schemas can be created.
8. *Publish Alignment and Reference Data Set:* Finally, all created artefacts are published.

With this process completed, a data set can be considered as conceptually integrated into a SDI or into a local system environment. This process provides the context for the activity that is being investigated in-depth in the next section – the creation of the alignment.

3.1 Elements of Work

To develop a process, the artefacts being defined and modified in it have also to be defined. In the current concept and implementation, Feature Catalogues as defined in ISO 19110, with a few added elements (*Context* and *Rule*), are used as Conceptual Schema Language. The metamodel for these extended Feature Catalogues has been formally described in Notation3 (N3) and can consequently be transformed into RDF(-schema) and OWL structures easily when needed or into a GML application schema (Reitz et al., 2005). In any such Feature Catalogue, the following elements can be found:

- *Context (CO):* The Context of a Conceptual Schema or a Class, defined as a set of Axioms (see Rule). These have to be fulfilled for the usage of that schema. In this way, the specific purpose that a schema usually has been created for is expressed.
- *Conceptual Schema (= Feature Catalogue, CS):* Formalized model of a universe of discourse, such as a schema of the concepts used in a certain application domain. Consists of a set of Classes with Relations.
- *Class (= FeatureType, CL):* A concept summarizing common characteristics of a set of individuals and thus describing this set.
- *Property (PR):* A (non-complex) property, attribute or characteristic of a class. Can be used as a predicate in a Rule.
- *Relation (RE):* Links concepts within a single Conceptual Schema. Allowed types of relations are generalization, aggregation and equality.
- *Rule (RU):* A Rule about what the Conceptual Schema (Axiom) or the instances created from it's classes (Constraint) has to fulfill, such as an attribute only taking certain values so that the instance can belong to a class.

In addition to the elements contained by the schema itself, there are instances of the classes of that schema that are abbreviated as i and are always assumed to be part of a data set DS. Instances belong to at least one class.

We use an abstract mapping model that can used both for making assertions and for defining attributive transformations:

- *Alignment (AL):* A Set of *Mappings* between the *Classes* of two Conceptual Schemas.
- *Mapping (MA):* An association of Classes from different Conceptual Schemas. It is modelled as a subtype of *Relation*.
- *Function (FU):* An association of Properties of one Class with Properties of another Class.

In the first prototype, a very simple grounding was used for this abstract mapping model, which is currently being replaced by a more complex and powerful one (see section 5).

3.2 Defining Tasks and Goals

The next step, after the coarse-grained process definition and the analysis of the artefacts used for the schema mapping process, is ensuring the repeatability and deterministic nature of the process. This includes the further discretisation of the alignment process into individual tasks and the definition of goals for each task.

For each of the elements in a conceptual schema and an alignment, there are basically three possible tasks classes: *Create*, *Refine* and *Remove*. All tasks are assumed to have one goal: Increasing the coverage of the conceptual schema or the mapping over a given set of instances, both in terms of completeness of individuals and in terms of completeness of attributes.

To describe this goal in a way suitable for task success evaluation within this approach, a quality model is defined as follows based on ISO 19113 definitions, to maintain consistency of terms within data sets, alignments and schemas. It should be noted that quality as fitness-for-purpose can, according to this model, only be fully determined in concrete usage based on availability of concrete data sets to harmonize, and that the model is therefore not suitable for usage in "isolated" alignment without instances.

The quality model for a schema has these elements:

- *Completeness of classes:* This element can be defined in relation to one or many given data sets and is defined via two sub-parameters:
  - *Excess (1):* Which portion of classes CL of the application schema CS has no instances i in a given data set DS?

$$Ex_{CL} = \frac{|\neg\{cl \in CS.CL \mid \exists i \in DS : f(cl) = i\}|}{|CS.CL|} \quad (1)$$

  - *Omission (2):* What portion of instances i in a data set does not belong to at least one of the classes CL of a given schema CS?

$$Om_{CL} = \frac{|\neg\{i \in DS \mid \exists cl \in CS.CL : f(i) = cl\}|}{|DS|} \quad (2)$$

- *Completeness of properties:* This can also only be defined on the basis of a given data set and is also defined via two sub parameters:
  - *Excess (3):* What portion of instances in a data set does not have a property defined via the class of the instance, i.e. which properties are not represented in instance in the data set? "null"-values are not counted, only those where the property is not even defined, since the schemas' completeness is analysed.

$$Ex_{PR} = \frac{|\neg\{i \in DS \mid \forall\, pr \in i.CL.PR : \exists\, pr \in i.PR\}|}{|DS|} \quad (3)$$

- *Omission (4):* What portion of properties in a data set does not even have a property defined via the class of the instance? As with *Excess*, "null"-values are not counted, only those where the property is not even defined.

$$Om_{CL} = \frac{|\neg\{pr \in DS.IN.PR \mid \exists\, pr \in IN.CL.PR\}|}{|pr \in DS.IN.PR|} \quad (4)$$

- *Correctness:* This is defined as the correctness of the schema, in respect to it's axioms, the logical consistency of the schema and the correctness of the schema with respect to concrete values in a given data set.
  - *Axioms (5):* Which portion of classes in the schema fails to fulfil one or multiple of the axioms postulated on the schema?

$$Cor_{AX} = \frac{|\neg\{cl \in CS.CL \mid \exists\, ax \in CO : f(cl, ax) = false\}|}{|CS.CL|} \quad (5)$$

  - *Schema (6):* Which portion of classes has relations to other classes that lead to non-deterministic behaviour or to outright logical inconsistency, e.g. cyclic inheritance?

$$Cor_{CS} = \frac{|\{cl \in CS.CL \mid \exists\, rel \in CS.CL.RE : f(cl, rel) = false\}|}{|CS.CL|} \quad (6)$$

  - *Constraints (7):* Which portion of property values of the instances in a given data set conforms to the Constraints imposed on values of attributes in the schema?

$$Cor_{CON} = \frac{|pr \in IN.PR \mid \exists\, co \in CS.CO : f(pr, co) = false|}{|IN.PR|} \quad (7)$$

The quality model of an alignment is quite similar in terms of the elements used, but has lesser dependency on a set of available data sets. Besides representing the quality elements known from ISO 19113, they also represent formalized mismatches (Agarwal et al., 2005) described in Section 2.

- *Completeness of mappings (8):* Which portion of classes from the source schema CSs has been directly mapped to a class in the target schema?

$$Com_{AL} = \frac{|cl_s \in CS_s.CL \mid \exists\, ma \in AL : ma.cl_s = cl_s|}{|CS_s.CL|} \quad (8)$$

- *Completeness of mappings over instances (9):* Which portion of instances of a data set has already been mapped via their class having been mapped?

$$Com_{INMA} = \frac{|i \in DS \mid \exists ma \in AL : ma(i.CL) \neq \emptyset|}{|DS|} \quad (9)$$

- *Completeness of Functions (10):* Which portion of properties of the source schema is mapped via a function to an attribute in the target schema?

$$Com_{FU} = \frac{|pr \in CS_s.CL.PR \mid \exists fu \in pr.FU : fu(pr) \neq \emptyset|}{|CS_s.CL.PR|} \quad (10)$$

- *Correctness of mappings:* This is defined as the correctness of the alignment as a whole, meaning the logical consistency of the mappings in respect to the two application schemas now aligned and the correctness of the alignment with respect to concrete values in a given data set.
  - *Consistency of Alignment over Mappings (11):* Which portion of mappings defines relations that lead to logical inconsistency, e.g. equality defined on classes that via a different relation have been declared non-equal?

$$Con_{AL} = \frac{|ma \in AL.MA \mid ma(CS_s, CS_t) = false|}{|AL.MA|} \quad (11)$$

  - *Consistency of Functions over Instances (12):* Which portion of property values of the instances in a given source data set is transformed to values allowable by the type and value constraints on properties in the target schema?

$$Con_{INFU} = \frac{|pr \in DS_s.IN.PR \mid fu \in pr.FU : \forall PR_t.FU = true|}{|DS_s.IN.PR|} \quad (12)$$

These individual quality elements are – depending on the task – used to define what value a task will have, together with an analysis of the data set used as reference. This allows an ordering of tasks. Summarized, each task now has the following attributes:

1. A type that indicates what kind of artefact is involved (*Class, Attribute, ...*)
2. A source that identifies what implementation of a task class has created that task. This source implementation is a reasoner or rule-based engine, which is also responsible for describing the task and it's reason in a user-understandable form;
3. A value that identifies the impact the solving of a task will have in terms of the metrics used in the quality model;

4. A severity level that identifies whether the task is required to clear up a logical error in the mapping or schema, whether it is a logical warning that indicates a possible mismatch or erroneous modelling, and a normal task indicates a simple open point that will improve the quality of the schema or mapping. As an example for a warning, take the case that two classes are declared equal via an equality relation, but an algorithm finds they share no substructures like attribute names and types.

Based on the objects being handled and the basic operations (*create, refine, remove*), there is a total of 24 possible task types. Each can be conducted manually, semi-automatically or automatically, and there are actually implementations available for all of them on one of these three levels. As an example, the PROMPT plug-in for ontology alignment that is available in Protégé: It first creates a list of possible matches (*Alignment.addMapping*), effectively creating a task for each possible match, and then asks the user to manually confirm this possible match (*Alignment.refineMapping*). For simplicity, we actually consider all task implementation approaches that require user interaction to be manual, even if they are supported in some way and might actually be only half-manual. From these task types a process of steps can now be built that, at the end, has an extracted schema or a created alignment as a result which is of sufficient quality.

3.3 The Iterative Conceptual Schema Mapping Process

With tasks and goals having been defined, they are applied to form a complete process which at the end has an alignment of sufficient quality. This alignment process consists of the following steps:

1. When a source and target schema have been selected and adopted as required, a precomputing step is executed which creates a list of suggestions for mappings on the class level, which are added as *Alignment.RefineMapping* tasks to the task list. This can be the output of any of the known alignment libraries.
2. The user can now go through the list of tasks, select one of them and open that task's working context. This will present him with all information typically required for the task, such as the classes affected by the mapping, allowing him to modify it or just confirm it. After doing so, a post computing phase is initialized, with the following possible outcomes:

1. For any confirmed or modified mapping, tasks are created to map mandatory attributes;
2. For a modified mapping, the logical consistency of the mappings as a whole is checked, and if necessary, additional tasks are added, i.e. in the case of a likely logical error or mismatch;
3. Both for confirmations and modifications, the list of tasks is updated, removing tasks that are not necessary any more (because they have been indirectly confirmed) and adding new ones, such as confirmation tasks for subclass mappings or warning based on different attribute constraints in the schemas.
4. This continues until the alignment quality reaches all the thresholds for the quality model indicated by the user.

As can be seen, there is a cycle of selecting a task, working on its resolution, computing the outcomes of the task and then again selecting a new task. This loop is ended by fulfilling the quality constraints required for the alignment being created.

To summarize, an overall conceptual harmonisation process has been described, it has been broken down into the objects required and into individual tasks, and a model how to work with individual tasks has been presented. The following section shows how this concept is implemented.

4 Implementation of a Task-Based Conceptual Schema Alignment Tool

Since the goal of this work is the efficient extraction of expert knowledge for geographic application schema mapping, implementation started with developing user interaction concepts, before defining back-end components and technology that have been used in an ongoing reference implementation.

4.1 User Interaction

The interaction concept is focused on three requirements:

1. Direct feedback should be given to a user's alignments in terms of changes on a map view.
2. The user should always get the information required for the task type addressed, so the views actually change depending on the task class. This concept has worked well in other areas that have complex tasks and processes as well as rather intangible results, such as software

production (Kersten and Murphy, 2006) and knowledge extraction (Jantke and Dötsch, 1997).

3. The user can choose different schema organization modes that make navigation in complex schemas and alignments easier.

A core aspect of the user interaction is the visualization of the source data set to be harmonized in a classical map view. This map view furthermore allows a reference data set using the target schema to be displayed in several splitting modes, so that one can get a direct impression how far the two data sets have been aligned via the schema alignment process.

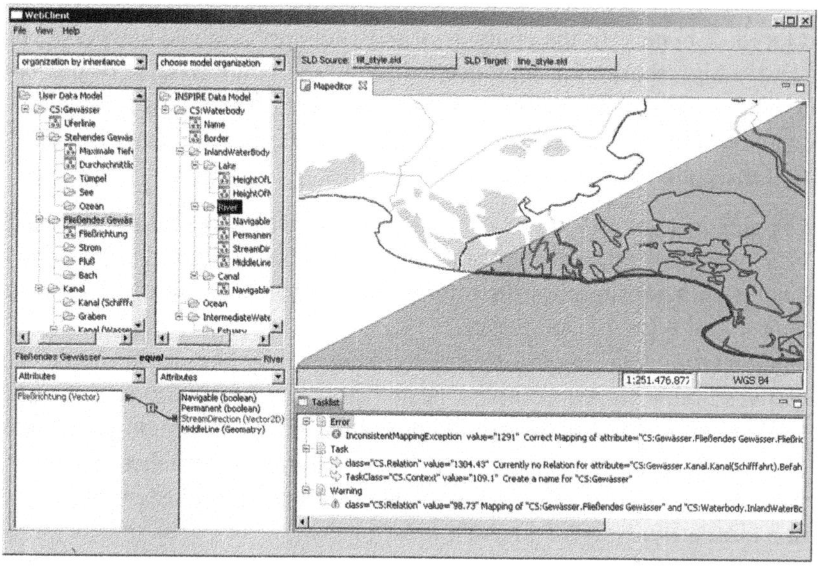

Fig. 5. A screen capture of the HALE application with some of it's key views, including source and target schema browsers, attribute browsers, the source/target splitted map view and the task list.

The splitting modes that can be selected are a diagonal split mode, as shown in Figure 5, a horizontal and vertical split mode, as well as an overlay mode which is particularly useful when two data sets that share a common border are supposed to be seamless. In this way, a user can visually judge how far he has already come in aligning the two schemas.

Since the map view is central to geospatial users, the possibility of applying different styled layer descriptors (SLDs) to both reference data sets has also been added. For each class mapped from source to target, the instances in the source data set are displayed using the target SLD, for those yet unmapped, the source SLD is used. In this way, it is easy to differentiate between features already mapped and those which still have to be mapped.

Another core aspect is to show users which tasks are important and which ones might have less impact. For this purpose, all tasks are ordered by their value, which is calculated according to the elements of the quality models for schemas and alignments presented in section 3.2. Consequently, tasks with low effort and high effect (on schema and instances) are prioritized and are given a higher value. This ordering is done within the three severity levels, but errors should be handled before warnings and normal tasks.

Additional efforts have been invested into the presentation of the source and target schemas. In the analysis of existing applications, we have seen a relatively high score of acceptance for the CogZ window-style of presentation, so the approach of having dual hierarchical tree views seemed to be acceptable. However, we have added several different means of organization of the content of these trees, specifically:

- *Organization by Inheritance:* This is the default way and most classical way of organization for an ontology, but it has the disadvantage of focusing on a subsumption hierarchy and might consequently not be suitable for all types of ontologies.
- *Organization by Aggregation:* Organization by Aggregation is especially suitable to see complex nested property associations and can be applied well to schemas originating from XML or relational data bases.
- *Organization by Similarity:* For this, similarity approaches are used to provide an ordering of the classes in a schema into clusters with a certain similarity.

For both the source and the target schema, organization can be selected individually. Furthermore, all schema views can be filtered by the task context, by a constraint on the structure (such as maximum depth) or properties (such as abstract or concrete) or by entering a keyword.

When a mapping between two concepts is established or being created, the Qualification area is used. This view has different contents varying by the type of the task. In Figure 5, the Function Panel is shown which allows adding functions to transform attributes of the source and target class. Other panels allow to qualify a mapping or to view information on a schema element.

4.2 Architecture and Technology

The presented approach is implemented as a part of a framework and toolset for data harmonisation and service integration. As shown in Figure 6, there are several components that are used in the implementation of the approach. While there is a wide range of additional (service) components that are implemented in HUMBOLDT, including transformation web services, again the focus lies on the implementation of the HUMBOLDT Alignment Editor and on the purpose of the components that it directly interacts with.

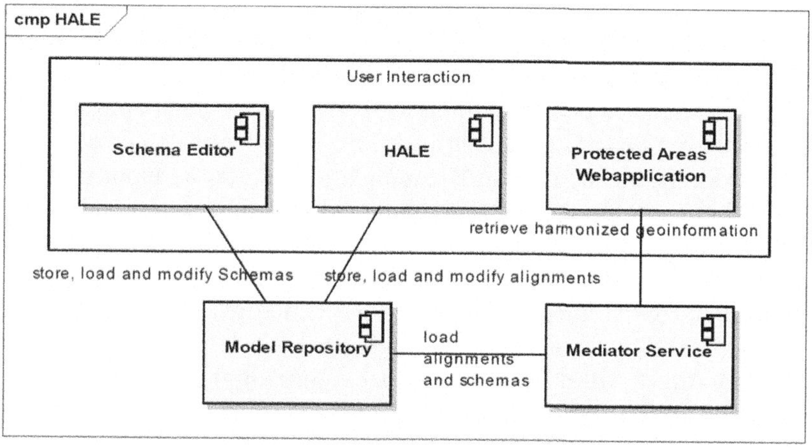

Fig. 6. The components of the implementation of the protected areas scenario

HALE itself is a Java SE 1.6 desktop application based on the Eclipse RCP. GeoTools components are used for rendering the map perspective and also for handling the import of various geodata file formats, such as Shapefiles and GML. Java Rules (Drools) are used to validate the constraints placed on properties as well as to add tasks based on properties. For this, rules are created and executed at run-time based on rule templates and information gathered from constraints in the schemas.

Conceptual Schemas and alignments are persisted in a Web Service called the Model Repository and offers operations for storage, loading and manipulation of alignments and schemas.

5 Conclusions and Outlook

There are three main contributions this paper offers: an analysis of the applicability and usefulness of general-purpose alignment and schema editors for geo-experts, a concept for a more targeted knowledge extraction process based on that analysis, and a task and quality model supporting this.

The HALE application that evolved as an implementation of the approach described within this paper is being developed in close collaboration with users from one scenario, but has not yet been tested and evaluated in a formal way. A formal evaluation with a second round of usability tests will be started soon, based on the tests conducted on Protégé PROMPT and COMA++. The results will be fed back into the on-going development process of HALE and the surrounding toolkit.

HALE and its supporting framework of service components will be made available under LGPL. Furthermore, it is planned to contribute the conceptual schemas and possibly exemplary data sets as input to the Ontology Alignment Evaluation Initiative (OAEI), so that a test bed for comparisons based on geospatial data can be established. Also, we have already started to evaluate mapping formalisms for their suitability, and have conducted an experiment with OMWG's Ontology Mapping Language (Scharffe, 2008) that was very promising.

One point that will have to be evaluated thoroughly is the suitability of the metrics used to value tasks. These might in turn be dependent of the context, both of a user and of the conceptual schema.

Acknowledgements

This work was supported by the HUMBOLDT project, which is co-funded by the European Commission under the sixth framework programme within contract SIP5-CT-2006-030962. The usability evaluation of the ontology alignment tools was executed by Katharina Reitz, and the development of HALE was supported by Christian Jauss.

References

Agarwal, P. (2005). Ontological considerations in GIScience. International Journal of Geographical Information Science, Vol. 19 (5): pp 501-536.

Agarwal, P., Huang, Y., and Dimitrova, V. (2005). Formal Approach to Reconciliation of Individual Ontologies for Personalisation of Geospatial Semantic Web. GeoSpatial Semantics, Proceedings of the First International Conference, Mexico City, Mexico, pp 195-210.

Aumueller, D, Do, H.-H., Massmann, S., and Rahm, E. (2005). Schema and ontology matching with COMA++. Proceedings of the 2005 ACM SIGMOD international conference on Management of data, Baltimore, Maryland, pp. 906-908.

Cruz, I., Bathala, S., Makar, N., and Sunna, W. (2007). A visual tool for ontology alignment to enable geospatial interoperability. Journal of Visual Languages and Computing, Vol. 18 (3): pp 230-254.

Ehrig, M. (2006) Ontology Alignment: Bridging the Semantic Web Gap. Springer, Berlin.

Ehrig, M., Staab, S., and Sure, Y. (2005). Bootstrapping ontology alignment methods with APFEL. Special interest tracks and posters of the 14th international conference on World Wide Web, Chiba, Japan, pp 1148-1149.

El Jerroudi, Z. and Ziegler, J. (2007). Interaktives Vergleichen und Zusammenfuhrungen von Ontologien (Interactive Ontology-Mapping and Merging). i-com: Zeitschrift für interaktive und kooperative Medien, vol. 6 (2007) ; no. 3, pp 44-49

Fonseca, F. and Martin, J. (2007). Learning the Differences Between Ontologies and Conceptual Schemas Through Ontology-Driven Information Systems. Journal of the Association for Information Systems, Vol. 8 (2): pp 129-142.

Hall, M. (2007). Automatisierte Semantische Harmonisierung von Landnutzungsdaten. Proceedings of the 2007 AGIT Conference. Salzburg, Austria.

Janssen, K. and Vandenbroucke, D. (2006). Spatial Data Infrastructures in Europe: State of play 2006. Online Document: http://www.ec-gis.org/inspire/reports/stateofplay2006/INSPIRE-SoP-2006%20v4.2.pdf. Accessed: 1. Jul. 2008.

Jantke, K.P. and Dötsch, V. (1997). The Necessity of User Guidance in Case-Based Knowledge Acquisition. Proceedings of the 10th International Florida Research Symposium, Daytona Beach,Florida, USA, May 10-14, Douglas D. Dankel (ed.), pp 312 – 336.

Kaltz, J.W., Lohmann, S., Hussein, T., Lang, E., and Ziegler, J. (2008). Ontologiebasiertes Engineering kontextadaptiver Webanwendungen (Ontology-based Engineering of Context-adaptive Web Applications).
Online Document: http://www.atypon-link.com/OLD/doi/abs/10.1524/icom.2005.4.3.22. Accessed: 4. Jun. 2008.

Kaufman, B. (2004): GIS inside an independent Oil and Gas company. ESRI 2004 User Conference Paper #1558, Aug 2004, San Diego, CA

Kersten, M. and Murphy, G.C. (2006). Using task context to improve programmer productivity. Proceedings of the 14th ACM SIGSOFT international symposium on Foundations of software engineering, Portland, Oregon, USA.
pp 1-11.

Nowak, J. and Craglia, M. (2006). INSPIRE Metadata Survey Results. Online Document: http://www.ec-gis.org/inspire/reports/INSPIRE_Metadata_Survey_2006_final.pdf. Accessed: 1. Jun. 2008.

Noy, N.F. and Musen, M.A. (2003). The PROMPT suite: interactive tools for ontology merging and mapping. International Journal of Human-Computer Studies, Vol. 59 (6): 983-1024.

Reitz, T., Herter, H., and Haist, J. (2005) Integrating Semantics into the interoperable 3D-GIS CS3D. Proceedings of the 25th Conference on Urban Data Management (UDMS), Aalborg, Denmark, April, 2005.

Scharffe, F.s (2008). Correspondence Patterns Representation. Dissertation submitted to the Faculty of Mathematics, Computer Science and Physics of the University of Innsbruck, Austria.

Transferring Segmented Properties in the Conflation of Transportation Networks

Alberto Belussi[1], Federica Liguori[2], Jody Marca[2], Sara Migliorini[1], Mauro Negri[2], Giuseppe Pelagatti[2], Paolo Visentini[2]

[1] Dipartimento di Informatica, Università di Verona, Verona, Italy, alberto.belussi@univr.it
[2] Dipartimento di Elettronica e Informazione, Politecnico di Milano, Milan, Italy, giuseppe.pelagatti@polimi.it

Abstract. In Spatial Data Infrastructures (SDIs) one of the layers that require particular attention in the integration process is the one describing the transport network. Different models can be adopted to represent and manage transport networks data, however, in most cases we have to handle properties that change their value along the road paths, for instance the road width or the number of lanes. This kind of data, usually called "segmented properties" can be represented in different ways, although at conceptual level they can be described as a function from the geometry to a specific domain of values.
In this paper we define a conceptual data model for representing "segmented properties" and map it into two implementation approaches: the *structural implementation*, with explicit geometry representation for each homogeneous segment, and the *dynamic implementation*, with the implicit representation of the segments by means of a measure (abscissa) on the route path. Then, we propose a method for transferring data from one implementation to the other one and to manage the problems arising from the integration of networks with different geometry and different implementations of segmented properties. Finally, we illustrate the application of the proposed approach to a practical case involving data of the Lombardy Region (Italy).

M. Sester et al. (eds.), *Advances in GIScience*, Lecture Notes in Geoinformation and Cartography, DOI 10.1007/978-3-642-00318-9_10,

1 Introduction

Data describing transport networks are often involved in practical applications, for example in location based services, traffic management, and many others. As a consequence different data models have been adopted for describing it. These models must include constructs to represent the properties whose value changes along the network routes, called here "segmented properties", like, for example, the number of lanes or the speed limits.

Moreover, many different implementations of a segmented property can be adopted; however, they can be grouped into the following two classes:

- **The structural implementation**: it requires to store explicitly the geometry of each homogeneous segment, that represents a portion of a given route having the same value of the segmented property.
- **The dynamic implementation**: the geomery of each segment is represented by a start and an end measure (called abscissa) along the route path, this measures delimit the portion of the curve in which the segmented property has a particular uniform value.

The work presented in this paper is the result of a practical project funded by Regione Lombardia in Italy, that aims to investigate the problem of integrating the road network of two spatial databases (shown in Fig. 1): (i) the database of the regional administration (**Rdb**), containing less accurate geometry and many segmented properties represented with a dynamic implementation, and (ii) the database of a local administration (**Ldb**), containing more accurate and up to date geometries but using a structural implementation for the segmented properties.

The goal of the project was to design and implement a methodology and some tools for supporting the integration process. The main issues that had to be faced are the following:

1. to perform a *schema integration* between **Rdb** and **Ldb,** thus determining the mapping between feature classes and properties of the two databases;
2. to apply a *conflation* between the road geometries of the two databases, with the aim to obtain a unique representation of the road network (harmonized geometry);
3. to transfer the segmented properties of both databases on the harmonized geometry in order to obtain the complete integration between **Ldb** and **Rdb**.

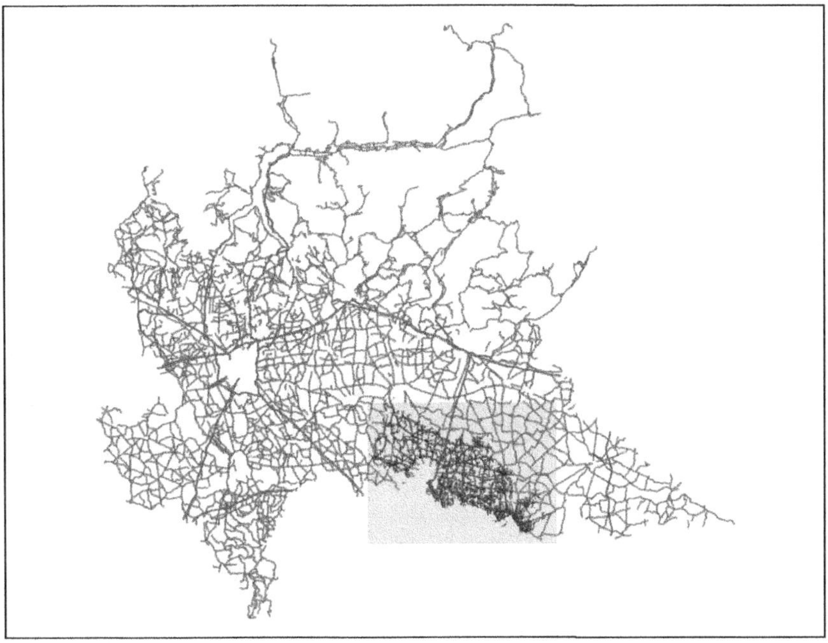

Fig. 1. Road network of the Lombardy Region: the Rdb covers all the region territory, while the Ldb covers only the highlighted rectangle.

In Fig. 2 we show an example of a situation that had to be faced during the project. In the figure, the red (solid) lines belong to the **Rdb**, while the blue (dashed) lines belong to the **Ldb**. Notice that:

- Only in **Rdb** the instances of the class Road (this class represents entire Roads at the application level) are represented, while in **Ldb** only the class Road Elements (road elements are only portions of roads delimited mainly by topological rules) are represented, but in **Ldb** the geometry is up to date and has higher accuracy.
- In **Rdb** there are some segmented properties like: the number of lanes, the type of roadway, the surface type and others.

In this paper we focus on the third problem listed above, namely the transfer of segmented properties on the harmonized geometry; however, the different implementations of segmented properties in the databases has impacted also the schema integration process, since segmented properties having a structural representation appear in the implementation schema as features, not as properties, making more difficult the discovery of feature classes that have common meaning in the two schemata. Although this paper does not deal with the problem of schema integration, it is important to state that in the integration methodology there has been a "reverse engineering" step in order to produce from the two original implementation schemata two corresponding conceptual schemata; during this step also segmented properties are recognized and defined at conceptual level, independently from their implementations.

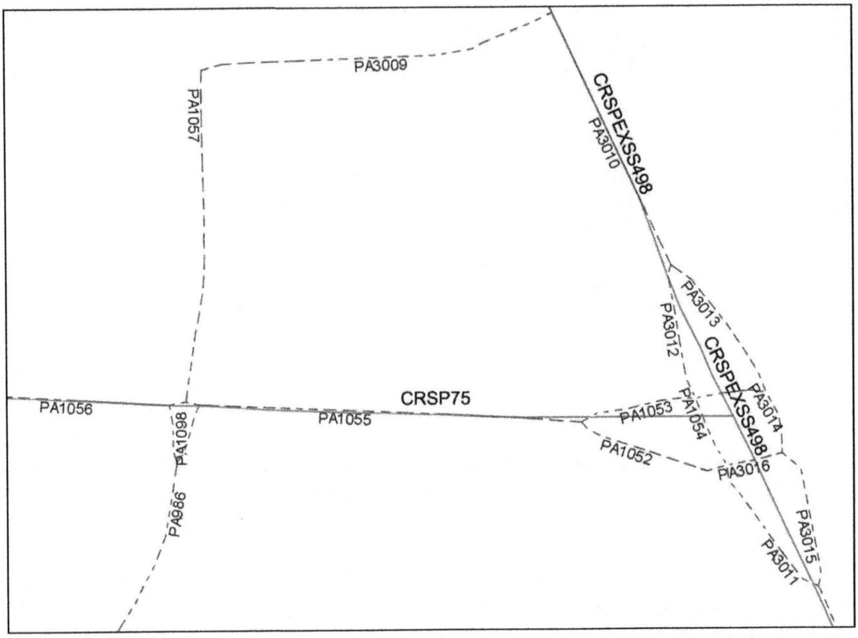

Fig. 2. Road network of the Lombardy Region: the Rdb roads are represented in red (solid) and have road-ID starting with "CRSP", while the Ldb roads are blue (dashed) and have ID starting with "PA".

Therefore, the paper is organized as follows: in section 2 it presents the definition of segmented properties at conceptual level and the corresponding two implementation approaches mentioned above: the *structural implementation* and the *dynamic implementation*. In section 3 it shows how data represented in one implementation can be converted into the other one, assuming to have one common geometry. Finally, it shows how to transfer the segmented properties defined on one geometry on a harmonized geometry derived from a conflation process and applies this general approach to the specific case of Lombardy Region.

1.1 Related Works

Several works in literature have proposed solutions for the integration of geographical data, focusing on general aspects concerning spatial data conflation (Lynch and Saalfeld 1985; Walter and Fritsch 1999; Cobb et al. 1998) or, more specifically, on the integration of spatial database contents (Duckham and Worboys 2007; Hariharan et al. 2005). Some works are tightly related to this paper since they present solutions regarding the representation and integration of data describing road networks (Dueker and Butler 1998; Hage et al. 2003; Scarponcini 2002).

Hage et al. (2003) describe the data model deployed by a Danish mobile content integrator and service delivery system. This model captures multiple, integrated representations of the transportation infrastructure, in particular it considers the following representations: (1) *kilometer-post representation*, in which a location is expressed in terms of one road, a given distance marker on that road (e.g. kilometer post), and an offset from the distance maker. (2) *Link-node representation*, which is based on the concepts of undirected and directed mathematical graphs, where a node is a place with a significant change of traffic properties (e.g. road intersection) and a link is a route that connects two nodes, this representation describes the topology of the transportation infrastructure. (3) *Geographical representation* which captures the geographical coordinates of the transportation infrastructure and an internal representation, named (4) *segmented representation,* which models an infrastructure as a collection of segments that intersect at connections, the position of any content references directly segments. All these infrastructure representations must be interrelated and in Hage et al. (2003) the authors explain how it is possible to translate one representation into another. They also consider the problem of query and update data modelled with this representation. The model we are proposing in this paper for the dynamic implementation is similar to the ideas presented in Hage et al. (2003). In addition we consider also a structural implementation of the segmented properties, which has the advantage to preserve the internal subdivision of the road in segments expressing each event in terms of a segment and a value.

Scarponcini (2002) presents a generalized model for linear referencing in transportation network; in the paper the author claims that his goal is to define a unified theoretical framework for representing and translating linear locations. In particular, he separates the concept of the linear element, which is the geometry where the measure is performed, from the method of measurement, and he formalizes the concept of location expression as the combination of a linear element, the distance of the location from the beginning of this element and a measurement method. We define our dynamic representation of segmented properties according to this theoretical definition which is also adopted by the ISO Standard for linear referencing (ISO 19148).

2 Conceptual Model of Segmented Properties and Implementations

The following conceptual model and implementations of segmented properties are defined using standard terms of the spatial database community. In particular, for the spatial attribute we adopt the types proposed by ISO in the "Spatial Schema" (ISO 19107).

2.1 Conceptual Definition of a Segmented Property

Given a class of objects *C* with a spatial attribute *g* of type *GM_MultiCurve* (namely an aggregation of one or more curves without any constraint among them), a segmented property *sp* with domain *D* defined on *g* can be represented at conceptual level with the following function f_{sp}:

$$f_{sp}: PSet(g) \rightarrow D$$

where *PSet(g)* is the pointset representation of g. The function f_{sp} returns for each point of *PSet(g)* the value of the property *sp*. Thus, the value of *sp* can theoretically change in each point of *g* but, in practical cases, it has a constant value along a portion of *g*, so that we can divide *g* into several parts, each one having the same value of *sp*. We call these parts **homogeneous segments**.

Notice that the fact that *g* is a multicurve allows one to deal with the most general case and is required by most real-life applications, but makes the implementation of segmented properties more difficult, as we will see. In simple cases *g* can also be of types *GM_Curve* (a primitive curve).

Usually the domain *D* is a finite domain, also called *enumeration* in UML. In this case it can be useful to introduce also the following function for property *sp*:

$$Segments_{sp}: D \rightarrow Set(GM\_Curve)$$

This function returns, for each value *v* in *D*, the set of curves where the property *sp* has value *v*.

Finally we suppose that the function f_{sp} is a total function; if f_{sp} is a partial function, since it is not defined everywhere on *g*, then a total function can be obtained by supposing to extend the domain D with a null value ($D \cup \{NULL\} = D_{NULL}$):

$$f_{sp}: PSet(g) \rightarrow D_{NULL}$$

Now we present the implementations of the conceptual definition.

2.2 Structural Implementation of a Segmented Property

The structural implementation of a segmented property is obtained by explicitly representing the geometry of each homogeneous segment. Thus, given an instance f of the class C with m alphanumeric attributes $Attr_1, \ldots, Attr_m$ and a spatial attribute g:

$$f = (a_1, \ldots, a_m, \gamma)$$

(where $a_1, \ldots, a_m$, are the values of the alphanumeric attributes and γ the geometry of g), the segments describing the segmented property sp on g are represented as a set of objects having two attributes, called *Value* and *Geometry*, as follows:

$$f.sp = \{(v_1, \gamma_1), \ldots, (v_n, \gamma_n)\}$$

where v_i is the value of the *value* attribute, γ_i is the spatial value of the *geometry* attribute and the following condition holds:

$$PSet(\gamma) = PSet(\gamma_1) \cup \ldots \cup PSet(\gamma_n)$$

Example 1

The situation of Fig. 3 will be used as an example for showing the different implementations; it represents the speed limits in Km/h along a road as a segmented property.

The structural implementation of the segmented property "*speed limits*" for the road is represented in Fig. 4. The segments describing the property are:

$$f.speed\_lim = \{(110, \gamma_0), (110, \gamma_1), (40, \gamma_2), (70, \gamma_3), (40, \gamma_4), (40, \gamma_5), (70, \gamma_6), (40, \gamma_7), (110, \gamma_8)\}.$$

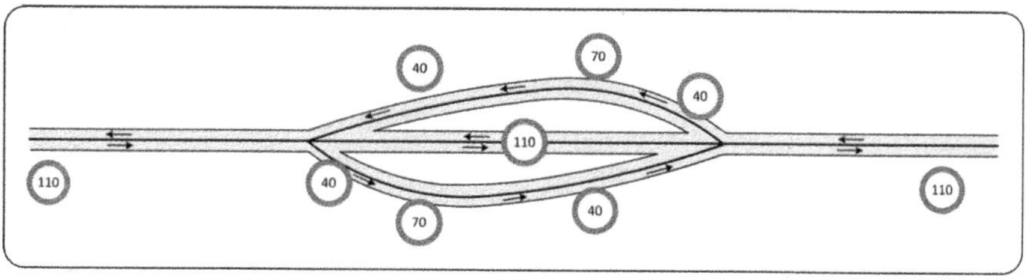

Fig. 3. Case study for the representation of segmented properties

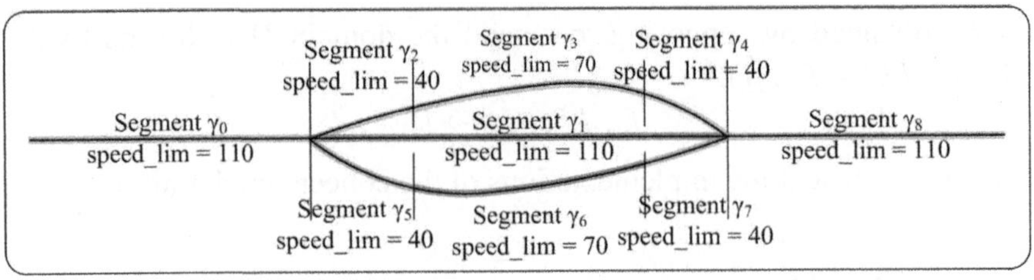

Fig. 4. Structural implementation of the segmented property *speed_lim*.

2.3 Dynamic Implementation of a Segmented Property

The dynamic implementation of a segmented property is obtained by representing each homogeneous segment indirectly by means of an *abscissa* defined on the spatial attribute g. Thus, given a class instance $f = (a_1, ..., a_m, \gamma)$, first it is necessary to define the abscissa on γ.
Regarding the representation of an abscissa on a linear geometry (also called Linear Referencing System), the following aspects must be considered:

- since γ is a multicurve, it might be not connected and it might have bifurcations; thus, in order to define an abscissa on γ, we have to divide γ into connected and simple curves, called *topological arcs*;
- the value of the abscissa can be *calculated*, i.e. derived from the geometry (length), or be specified independently from the geometry;
- the value of the abscissa can be specified from the beginning of the multicurve, thus requiring an ordering of the topological arcs, or be specified from a reference point (see below) contained in the same topological arc.

A *reference point (RP, α)* indicates the value that the abscissa assumes in that point; each arc must have a reference point at the start and another one at the end, optionally other reference points can be located along the arc, but they must always be inside the arc geometry (in real cases reference points can also be located near and not exactly on the arc geometry; however, in order to use them in the abscissa management, these reference points must be projected onto the arc. In this paper, we suppose that this projection has already been performed by a pre-processing phase).

By combining the above choices we obtain the following 4 basic implementations:

1. **Absolute abscissa**: the abscissa is measured from the start point of the linear geometry and is independent from the geometry length.

2. **Absolute calculated abscissa**: the abscissa is measured from the start point of the linear geometry, but it is calculated from the length of the geometry.
3. **Local abscissa**: the abscissa is defined and used locally on each topological arc of the linear geometry. It is zero at the starting point of each arc, but at the end point its value could be different with respect to the geometry length.
4. **Local calculated abscissa**: in this case the abscissa at the end point is equal to the length of the topological arc.

The structure for storing topological arcs and reference points is the following one:

$$f.arcs_{\gamma} = \{(arc_1, L_1), \ldots, (arc_n, L_n)\}$$

where:

- $PSet(\gamma) = PSet(arc_1) \cup \ldots \cup PSet(arc_n)$.
- $L_i = \{(RP_1, \alpha_1), \ldots, (RP_m, \alpha_m)\}$ where RP_j is a reference point and α_j is the measure of the abscissa on arc_i at the location RP_j. Each RP_j is contained in the arc_i, that is: $RP_j \in PSet(arc_i)$.
- $cardinality(L_i) \geq 2$.
- The start point of the arc $arc_i \in f.arcs_{\gamma}$ is called $RP_{i,start}$, the abscissa of $RP_{i,start}$ is called $\alpha_{i,start}$. In the same way, the end point of the arc arc_i is called $RP_{i,end}$, the abscissa of $RP_{i,end}$ is called $\alpha_{i,end}$.
- If an absolute or absolute calculated abscissa is represented, then $\alpha_{i,start}$ ($\alpha_{i,end}$) is the distance (measured or calculated) of the point $RP_{i,start}$ ($RP_{i,end}$) from the start of γ.
- If the abscissa is local or local calculated, then $\alpha_{i,start}$ is zero and $\alpha_{i,end}$ is equal to the length (measured or calculated) of arc_i.
- For absolute or absolute calculated abscissa, we call α_{start} the minimum $\alpha_{i,start}$ among all the arcs of $f.arcs_{\gamma}$ and α_{end} the maximum $\alpha_{i,end}$.

Let us call *event* the dynamic implementation of a homogeneous segment; given the representation of topological arcs, there are two possible structures for representing events:

- One based on the absolute abscissa, where the events are geo-referenced by means of two abscissa values (β) as follows:

  $$f.event_{sp\ on\ \gamma} = \{(v_k, arc_i, \beta_{k,start}, \beta_{k,end}), \ldots\}$$

  where the following condition holds:

- Consistency: $\forall(v_k, arc_i, \beta_{k,start}, \beta_{k,end}) \in f.event_{sp\ on\ \gamma}$:
  $\beta_{k,start} \leq \beta_{k,end}$ and $\exists(arc_j, L_j) \in f.arcs_\gamma$: $arc_i = arc_j$
  $\alpha_{j,start} \leq \beta_{k,start} \leq \alpha_{j,end}$ and $\alpha_{j,start} \leq \beta_{k,end} \leq \alpha_{j,end}$
- Completeness: $\forall(arc_i, L_i) \in f.arcs_\gamma \forall\alpha$: $\alpha_{i,start} \leq \alpha \leq \alpha_{i,end}$:
  $\exists(v_k, arc_j, \beta_{k,start}, \beta_{k,end}) \in f.event_{sp\ on\ \gamma}$: $arc_i = arc_j$ and $\beta_{k,start} \leq \alpha \leq \beta_{k,end}$

- One based on the local abscissa, where the events are geo-referenced by means of a distance measure from a reference point, as follows:

$$f.event_{sp\ on\ \gamma} = \{(v_k, arc_i, RP_{k,start}, \Delta_{k,start}, RP_{k,end}, \Delta_{k,end}), \ldots\}$$

where $\Delta_{k,start}$, ($\Delta_{k,end}$) represents the distance, measured or calculated along arc_i, of the event start (event end) from $RP_{k,start}$ ($RP_{k,end}$) and the following conditions hold:

- Consistency: $\forall(v_k, arc_i, RP_{k,start}, \Delta_{k,start}, RP_{k,end}, \Delta_{k,end}) \in f.event_{sp\ on\ \gamma}$:
  $\exists(arc_j, L_j) \in f.arcs_\gamma$: $arc_i = arc_j$ and
  $\exists(RP_l, \alpha_l), (RP_h, \alpha_h) \in L_j$: $RP_l = RP_{k,start}$ and $RP_h = RP_{k,end}$ and
  $(\alpha_l < \alpha_h$ or $(\alpha_l = \alpha_h$ and $\Delta_{k,start} \leq \Delta_{k,end}))$
- Completeness: $\forall(arc_i, L_i) \in f.\ arcs_\gamma \forall\alpha$: $\alpha_{i,start} \leq \alpha \leq \alpha_{i,end}$:
  $\exists(v_k, arc_j, RP_{k,start}, \Delta_{k,start}, RP_{k,end}, \Delta_{k,end}) \in f.event_{sp\ on\ \gamma}$:
  $arc_i, = arc_j$ and $(\alpha_{k,start}+\Delta_{k,start}) \leq \alpha \leq (\alpha_{k,end}+\Delta_{k,end})$

Example 2

In order to represent the segmented property *"speed limits"* for the road of Fig. 3 using the dynamic representation, we have to define five topological arcs, because we need to break the main road into the minimum number of simple curves covering the same geometry. We start with the dynamic implementation based on *absolute calculated abscissa*, as represented in Fig. 5 and we suppose to have two reference points on each arc, that represent the distance of the arc start point and of the arc end point from the start of the road calculated on the geometry. The set of topological arcs is the following one:

$$f.arcs_\gamma = \{(arc_1, L_1), (arc_2, L_2), (arc_3, L_3)\ , (arc_4, L_4), (arc_5, L_5)\}$$

where:

$L_1=\{(RP_{1,start}, 1000), (RP_{1,end}, 2200)\}$, $L_2 = \{(RP_{2,start}, 2200), (RP_{2,end}, 2340)\}$
$L_3=\{(RP_{3,start}, 2200), (RP_{3,end}, 2350)\}$, $L_4 = \{(RP_{4,start}, 2200), (RP_{4,end}, 2300)\}$
$L_5=\{(RP_{5,start}, 2300), (RP_{5,end}, 8000)\}$

Given this set of topological arcs, the event *speed limits* can be represented as follows:

$$f.event_{speed\_lim\ on\ \gamma} = \{(110, arc_1, 1000, 2200), (40, arc_2, 2200, 2250), (70, arc_2, 2250, 2300), (40, arc_2, 2300, 2340), (40, arc_3, 2200, 2260), (70, arc_3, 2260, 2300), (40, arc_3, 2300, 2350), (110, arc_4, 2200, 2300), (110, arc_5, 2300, 8000)\}$$

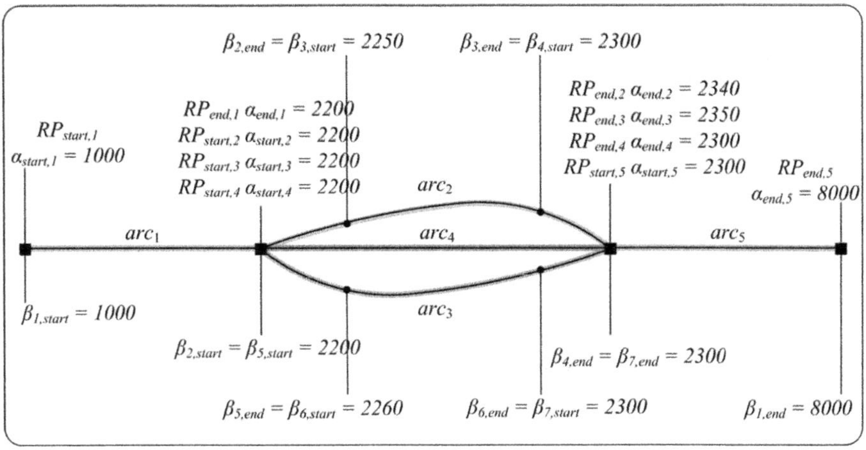

Fig. 5. Dynamic implementation of the segmented property *speed limits* with absolute calculated abscissa.

Let us now consider the same example using a dynamic implementation based on *absolute abscissa*. Usually the absolute abscissa (not calculated) represents a measure along the road path and can be used to locate on the ground the point where an event starts (or ends); therefore, it is correlated but often not identical, with the length of the arc. In particular, if we compute the difference $(\alpha_{i,end} - \alpha_{i,start})$, we expect that this value is equal to the length of the arc arc_i, however for several reasons this is not always true. This discrepancy occurs for example:

- when the geometry is 2D, the road has a relevant slope and the abscissa comes from a survey on the ground, then a large discrepancy could result;
- when the geometry is very simplified with respect to the real shape of the road, again the discrepancy could be significant.

In the first case the abscissa has the goal to store the 3D length of the road and thus the difference is an additional information, in the second case it is only an accuracy problem. In any case this discrepancy can be calculated and stored together with the arc, thus each arc can be represented as follows:

$$(arc_i, L_i, \mathbf{err}_i)$$

where err_i is the ratio between the $(arc_i.length)/(\alpha_{i,end} - \alpha_{i,start})$.

In Fig. 6 we present the same example using a dynamic implementation based on *absolute abscissa*. In this example we assume that, beyond the reference points at the ends of each topological arc, we have two other reference points, that "stretch" the abscissa along that arc independently from the geometry length. The set of topological arcs is:

$$f.arcs_\gamma = \{(arc_1, L_1), (arc_2, L_2), (arc_3, L_3)\ , (arc_4, L_4), (arc_5, L_5)\}$$

where:

$L_1 = \{(RP_{1,start}, 1000), (RP_A, 2000), (RP_{1,end}, 2250)\}$
$L_2 = \{(RP_{2,start}, 2250), (RP_{2,end}, 2500)\}$
$L_3 = \{(RP_{3,start}, 2250), (RP_{3,end}, 2500)\}$
$L_4 = \{(RP_{4,start}, 2250), (RP_{4,end}, 2500)\}$
$L_5 = \{(RP_{5,start}, 2500), (RP_B, 5000), (RP_{5,end}, 8000)\}$

Given this set of topological arcs, the event *speed limits* can be represented as:

$$f.event_{speed\_lim\ on\ \gamma} = \{(110, arc_1, 1000, 2250), (40, arc_2, 2250, 2300), (70, arc_2, 2300, 2450), (40, arc_2, 2450, 2500), (40, arc_3, 2250, 2300), (70, arc_3, 2300, 2450), (40, arc_3, 2450, 2500), (110, arc_4, 2250, 2500), (110, arc_5, 2500, 8000)\}$$

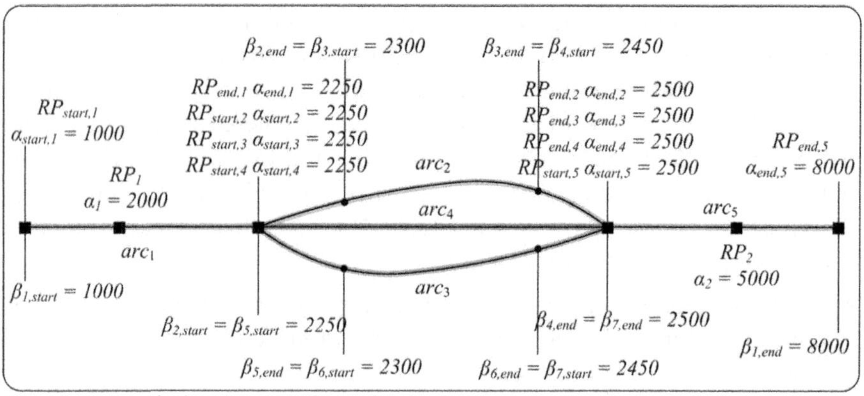

Fig. 6. Dynamic implementation of the segmented property *speed limits* with absolute abscissa.

In Fig. 7 we present the same example of Fig. 5 with *local calculated abscissa*, namely with abscissa values calculated from the start of the corresponding topological arc, instead of from the start of the road. The set of topological arcs is:

$$f.arcs_\gamma = \{(arc_1, L_1), (arc_2, L_2), (arc_3, L_3)\ , (arc_4, L_4), (arc_5, L_5)\}$$

where:

$L_1 = \{(RP_{1,start}, 0), (RP_{1,end}, 1200)\}$, $L_2 = \{(RP_{2,start}, 0), (RP_{2,end}, 140)\}$
$L_3 = \{(RP_{3,start}, 0), (RP_{3,end}, 150)\}$, $L_4 = \{(RP_{4,start}, 0), (RP_{4,end}, 100)\}$
$L_5 = \{(RP_{5,start}, 0), (RP_{5,end}, 5700)\}$

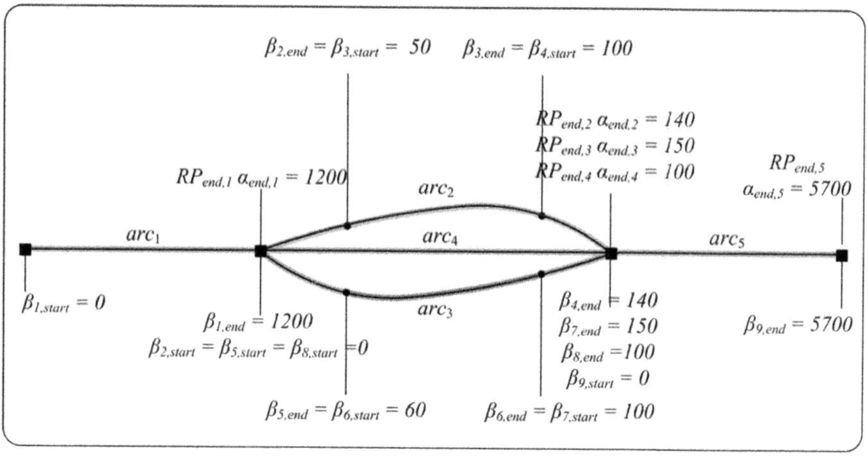

Fig. 7. Dynamic implementation of the segmented property *speed limits* with local calculated abscissa.

Notice that the reference points in this case can also be discarded completely since the arc length can be used instead. Given this set of topological arcs, the event *speed limits* can be represented as:

$$f.event_{speed\_lim\ on\ \gamma} = \{(110, arc_1, 0, 1200), (40, arc_2, 0, 50), (70, arc_2, 50, 100), (40, arc_2, 100, 140), (40, arc_3, 0, 60), (70, arc_3, 60, 100), (40, arc_3, 100, 150), (110, arc_4, 0, 100), (110, arc_5, 0, 5700)\}$$

As shown for the absolute abscissa, we can extend the example of local calculated abscissa to the case of local abscissa, by adding new reference points and assigning abscissa values that are not consistent with the arcs length.

3 Transferring Segmented Properties Among Different Implementation Structures

In this section we consider the problem of transferring segmented properties from one implementation to the other one.

3.1 Dynamic to Structural Implementation

Given a segmented property in the dynamic implementation, it is necessary to convert each event it is composed of into a geometry. Consider first the absolute (calculated or not) abscissa: each event is defined by a pair of abscissa values; for example, if we consider an object *f* with a dynamic property *sp*, we start from the following data structure:

$$f.event_{sp\ on\ \gamma}=\{(v_k, arc_i, \beta_{k,start}, \beta_{k,end}), \ldots\}$$

each event:

$$e=(v_k, arc_i, \beta_{k,start}, \beta_{k,end})$$

can be converted into the following two points on the arc arc_i:

$$P_1=arc_i.locate(\beta_{k,start}),\ P_2=arc_i.locate(\beta_{k,end})$$

where $arc_i.locate(\beta)$ returns the point P on arc_i such that: $arc_i.distance(RP_{i,start},P)=(\beta - \alpha_{i,start})$ and $arc_i.distance(RP_{i,start},P)$ measures the distance, along the arc_i geometry, that exists between $RP_{i,start}$ and P, taking into account the stretching due to the additional reference points.

In the case of local (calculated or not) abscissa, we start from the following data structure:

$$f.event_{sp\ on\ \gamma} = \{(v_k, arc_i, RP_{k,start}, \Delta_{k,start}, RP_{k,end}, \Delta_{k,end}), \ldots\}$$

each event:

$$e = (v_k, arc_i, RP_{k,start}, \Delta_{k,start}, RP_{k,end}, \Delta_{k,end})$$

can be converted into the following two points on the arc arc_i:

$$P_1=arc_i.locate\_relative(RP_{k,start}, \Delta_{k,start}),$$
$$P_2=arc_i.locate\_relative(RP_{k,end}, \Delta_{k,end})$$

where $arc_i.locate\_relative(RP,\Delta)$ returns the point P on arc_i such that: $arc_i.distance(RP,P) =\Delta$.

Once we have located the points P_1, P_2, either with absolute or local abscissa, each event generates the corresponding homogeneous segment as follows:

$$f.sp = \{(v_k, arc_i.geometry(P_1,P_2)), \ldots\}$$

where $arc_i.geometry(P_1,P_2)$ returns the curve obtained by splitting in P_1 and P_2 arc_i into three curves and choosing the one with the abscissa between P_1 and P_2.

Moreover, in order to maintain the spatial integrity constraint that requires the exact containment of each homogeneous segment into the geometry of the spatial attribute g, the points P_1 and P_2 are added as vertices also to the geometry γ.

Fig. 8 shows an example of transformation from a dynamic to a structural implementation, starting from the case in Fig. 5 (absolute calculated abscissa). We calculate the points P_1, P_2, P_3, P_4, P_5, P_6, P_7 and P_8 ,which represent the locations of the start and end abscissa of the events on the corresponding arc, then we split the road using these points and find the nine segments $\gamma_0,\ldots,\gamma_8$ which are shown in Fig. 8.

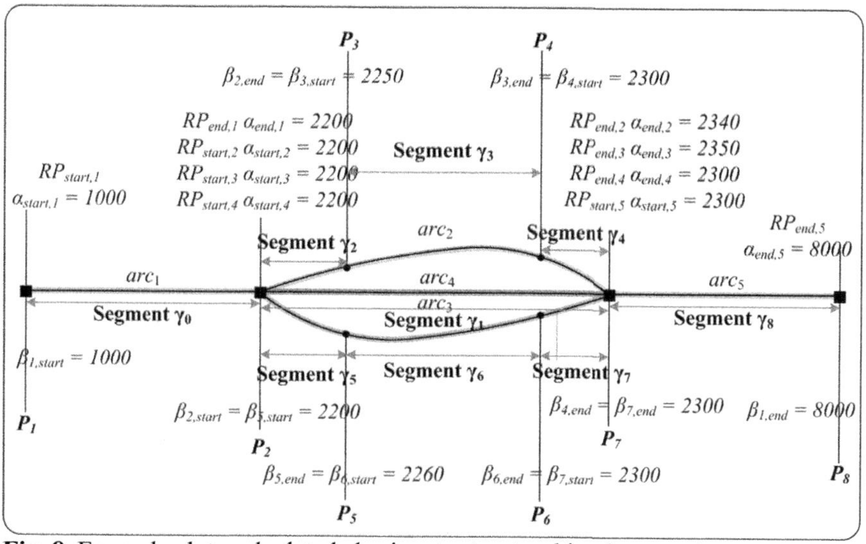

Fig. 8. From absolute calculated abscissa to structural implementation.

3.2 Structural to Dynamic Implementation

Given a segmented property in the structural implementation, it is necessary to convert each homogeneous segment, it is composed of, into an event. This requires first to define the topological arcs on the multicurve representing the geometry of g; for example, if we consider an object f with a dynamic property sp, we start from the following data structure:

$$f.sp = \{(v_1, \gamma_1), \ldots, (v_n, \gamma_n)\}.$$

The definition of the topological arcs on γ requires an algorithm that starting from the multicurve γ produces the structure:

$$f.arcs_\gamma = \{(arc_1, L_1), \ldots, (arc_n, L_n)\}$$

This can be done in different ways according to the availability or absence of additional information about the definition of an abscissa along the multicurve; in other words, the following cases can occur:

- If no additional information is available, we can split γ into a set of simple curves $\{arc_1, \dots, arc_n\}$, each one having exactly two end points and we order them starting from one of the curves having an end point not coincident with the end point of any other curve. Then, for each arc arc_i, we compute the reference points $RP_{i,start}$ and $RP_{i,end}$ and the corresponding abscissa $\alpha_{i,start}$ and $\alpha_{i,end}$. If we choose an absolute abscissa we calcutate $\alpha_{i,start}$ and $\alpha_{i,end}$ as follows:

$$\alpha_{1,start} = 0;\ \alpha_{1,end} = arc_1.lenght()$$

$$\dots$$

$$\alpha_{i,start} = \alpha_{i-1,end};\ \alpha_{i,end} = \alpha_{i,start} + arc_i.lenght()$$

  while if we choose a local abscissa we calcutate $\alpha_{i,start}$ and $\alpha_{i,end}$ for each arc arc_i as follows:

$$\alpha_{i,start} = 0;\ \alpha_{i,end} = arc_i.lenght()$$

- If additional information is available, usually it consists of a set of reference points $RP=\{(RP_1, \alpha_1), \dots\}$ (at least one point for each end point of the multicurve and one for each internal node) defining the abscissa along the multicurve γ. Thus, in this case we split γ into a set of simple curves $\{arc_1, \dots, arc_n\}$, having exactly two end points and then, according to the assigned reference points, we can order the arcs. Finally, for each arc arc_i, we compute the reference points $RP_{i,start}$ and $RP_{i,end}$ and the corresponding abscissa $\alpha_{i,start}$ and $\alpha_{i,end}$ as follows:

$$\alpha_{i,start} = \alpha_1 \text{ where } (RP_1, \alpha_1) \in RP \wedge RP_1 = RP_{i,start}$$

$$\alpha_{i,end} = \alpha_2 \text{ where } (RP_2, \alpha_2) \in RP \wedge RP_2 = RP_{i,end}$$

  If some of the end points have no corresponding reference points, then its abscissa can be computed as follows: let's call this orphan reference point RP_x, we compute the two nearest reference points of RP then we assign to RP_x an abscissa proportional to the distance on the arc between RP_x and the reference points, whose abscissas are known. This approach is also proposed in Scarponcini (2002).

At this point it is necessary to convert each homogeneous segment in one or more events as follows: given $(v_i, \gamma_i) \in f.sp$ the end points of the geometry γ_i, called EP_1 and EP_2, are intersected with the arcs of $f.arcs_{\gamma}$, that were generated as described above. Since γ_i is a simple curve, EP_1 and EP_2 will intersect only one arc arc_i that contains them. At this point the abscissa $\alpha_{i,start}$, $\alpha_{i,end}$ of the event v_i can be computed considering the distance along arc_i between the reference points of arc_i and EP_1, EP_2 respectively. For example, we want to translate the structural implementation of the *speed limits* property presented in Fig. 4 into a dynamic one using an absolute calculated abscissa. We have first to split the road obtaining a set of simple curves (topological arcs) $f.arcs_\gamma = \{arc_1, arc_2, arc_3, arc_4, arc_5\}$, as illustrated above. Then we transfer on the topological arcs the available reference points and, if necessary (if they are missing), we create additional pseudo reference points, one for each arc termination. The abscissa associated to each reference point can be calculated using the geometry distance (calculated abscissa) or the real distance, if the information is available. Now we intersect the end points of each segment γ_i of $f.speed\_lim$ with the arcs in $f.arcs_\gamma$ and obtain the points EP_1, EP_2, EP_3, EP_4, EP_5, EP_6, EP_7 and EP_8. The set of events is:

$$f.event_{speed\_lim\ on\ \gamma} = \{(110, arc_1, \beta_{EP1}, \beta_{EP8}), \\ (40, arc_2, \beta_{EP2}, \beta_{EP3}), (70, arc_2, \beta_{EP3}, \beta_{EP4}), (40, arc_2, \beta_{EP4}, \beta_{EP7}), \\ (40, arc_3, \beta_{EP2}, \beta_{EP5}), (70, arc_3, \beta_{EP5}, \beta_{EP6}), (40, arc_3, \beta_{EP6}, \beta_{EP7}), \\ (110, arc_4, \beta_{EP2}, \beta_{EP7}), (110, arc_5, \beta_{EP7}, \beta_{EP8})\}$$

where each α_{EPi} is calculated as explained before depending on the particular dynamic implementation we have choose.

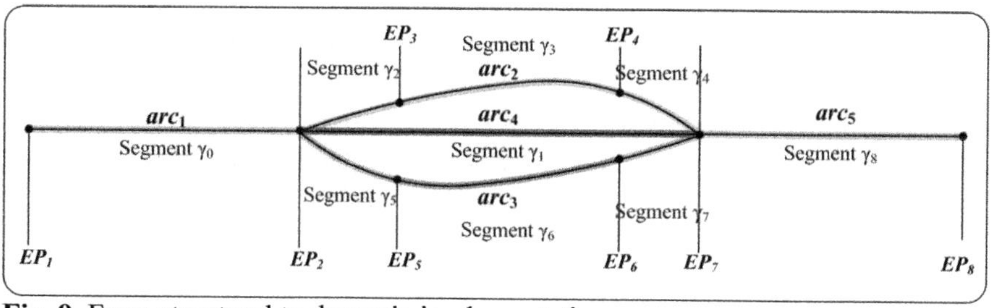

Fig. 9. From structural to dynamic implementation.

4 Managing Integration of Segmented Properties after Geometry Conflation

Considering the integration of segmented properties, the more interesting situation occurs when the integration has to be performed between networks having different geometries. Indeed, in this case a conflation phase is necessary. Conflation has been extensively studied in several research

areas (Lynch and Saalfeld 1985; Walter and Fritsch 1999; Cobb et al. 1998; Duckham and Worboys 2007; Blasby et al. 2003), however, the proposed methods focus in many cases only on the geometry harmonization without taking into account the instances and instance properties integration. In particular, as we will show in this section, a considerable amount of work is required for the integration of "segmented properties".
During the project in Lombardy Region we faced such situation. In particular, considering the formal definitions shown in the previous section, the starting point of the project can be described as follows:

- The **Rdb** contains *Main Roads,* with segmented properties implemented using an absolute abscissa, and *Road Elements*. Topological arcs of main roads are explicitly represented and many reference points are stored for each topological arc. The geometry of the network is 2D. Finally, the geometry of each main road is composed of the geometries of a set of road elements.
- The **Ldb** contains *Road Elements* with a structural implementation of segmented properties. Neither road instances nor any kind of abscissa are represented. The geometry of the network is 3D.

In this case we cannot just transform the segmented properties of **Rdb** into segmented properties of **Ldb** or vice versa: a more complex procedure is necessary; in particular, in the project we have applied a methodology based on the following steps:

1. Data selection: it consists of the selection from **Rdb** only of data that overlap **Ldb**.
2. Conflation: it consists of running a conflation tool for performing the feature matching between the networks, considering **Ldb** as reference network. In particular, we use the *RoadMatcher Tool* (open source) available in the Jump platform (Vivid solution Inc. (Blasby et al. 2003)).
3. Detect and adjust missing matches: it consists of identifying the cases where the RoadMatcher has not found a match (usually this is due to an update that has been registered in the new network and that is not present in the old one). For managing these cases we have implemented a post-processing tool that is able to guide the user in identifying the missing matches and allows her to specify manually the matching; such matching can also be one to many, i.e. one road element of **Rdb** can match with one or more road elements of **Ldb**, but also many to one, since a road element of **Ldb** could possible match with a portion of a road element of **Rdb**. These situations are not handled by the *RoadMatcher Tool.* In Fig. 10 a case of missing

matches is shown. The adjustment of missing matches was necessary in all the situations where the Ldb provides updates of the Rdb. Using the proposed tool the experts were able to solve the 99% of these cases.

4. Creating the road instances: after completing the network matching, the road instances must be reconstructed on the new network (i.e. on the geometry of **Ldb**).
5. Transferring the dynamic segmented properties of **Rdb** onto **Ldb** by one of the following approaches:
   A - Maintain the absolute abscissa of **Rdb**. This requires: (i) to build the topological arcs on the path of the created road instances; (ii) to project the reference points of **Rdb** onto the new network and create, if necessary, the reference points at the start and end of each arc; (iii) copy the events from **Rdb** to **Ldb** without changing their abscissa values. If one event is defined on an **Rdb** topological arc that is matched with *n* **Ldb** topological arcs, it will be split in *n* events and the abscissa values will be adjusted accordingly.
   B – Convert the events of **Rdb** *into homogeneous segments of* **Ldb**. This requires: (i) to convert each event of **Rdb** into the corresponding homogeneous segment following the approach presented in Section 3.1 (ii) to project the homogeneous segment onto the **Ldb** road elements that have been matched (in step 3) with the **Rdb** road elements composing this segment; multiple matches can produce more homogeneous segments on the **Ldb** network.
   C – Convert the absolute abscissa of **Rdb** *into a local abscissa on* **Ldb**. This requires the same operations shown by the previous two cases, in addition the homogeneous segments have to be converted into events with a local abscissa, by computing the distance between the event start (event end) and the nearest reference point.

Regarding the case shown in Fig. 10, the match process (step 3) will be completed manually by assigning the **Ldb** road elements to **Rdb** road elements as follows: 12 and 13 to A; 16, 17, 52 and 53 to B and 11, 14, 15, 54 to C.

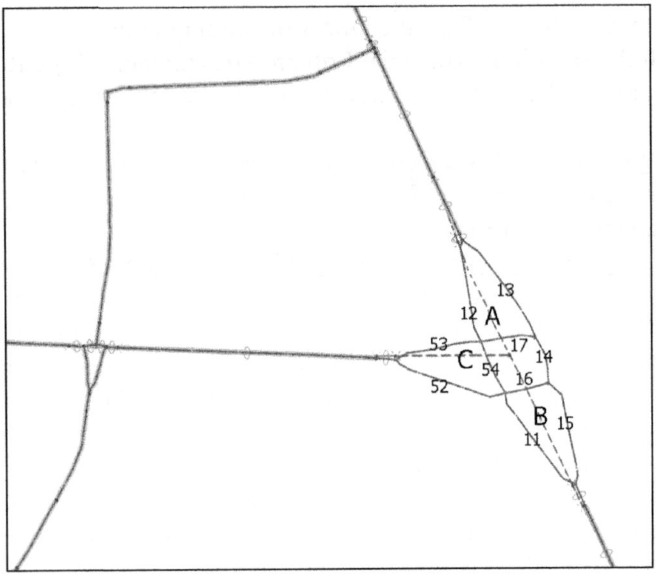

Fig. 10. A case of missing matches produced by the Road Matcher tool.

5 Conclusions and Future Work

In this paper we define a conceptual data model for representing "segmented properties" and two implementation approaches: the *structural implementation* and the *dynamic implementation*. Moreover, the procedure to transfer segmented properties from one implementation to the other one is presented. In particular, we show that the integration of different networks, when segmented properties are included, becomes more complex and that the geometric conflation does not solve completely the integration problem, but a consistent amount of work is necessary in order to integrate also the segmented properties.

Finally, we illustrate a methodology applied in a practical project carried out in Lombardy Region, where it was necessary to transfer segmented properties from an old road network onto the updated road network.

Future work can be focused on: the design of a tool that implements automatically all the methods for transferring segmented properties described in the previous section; the study of metadata that can describe the accuracy of the transferred segmented properties and the study of advanced methods for comparing segmented properties coming from different sources.

References

Blasby, D., Davis, M., Kim D., and Ramsey, P. (2003): GIS Conflation using Open Source Tools, White paper, The Jump Project.

Cobb, M.A., Chung, M.J., Foley III, H., Petry, F.E., and Shaw, K. B. (1998): A Rule-based Approach for the Conflation of Attributed Vector Data, GeoInformatica, Vol. 2, N. 1, pp. 7-35.

Duckham, M., and Worboys, M. (2007): Automated Geographical Information Fusion and Ontology Alignment. In Spatial Data on the Web: modeling and management, Springer Verlag, pp. 109-132.

Dueker, K.J., and Butler, J.A. (1998): GIS-T Enterprise Data Model with Suggested Implementation Choices, URISA Journal, Vol. 10, N. 1, pp 12-36.

Hage, C., Jensen, C.S., Pedersen, T.B., Speicys, L., and Timko, I. (2003): Integrated data management for mobile services in the real world. In Proceedings of the 29th VLDB, Endowment (Berlin, Germany, 2003). pp. 1019-1030.

Hariharan, R., Shmueli-Scheuer, M., Li, C., and Mehrotra, S. (2005): Quality-driven approximate methods for integrating GIS data. In Proceedings of the 13th ACM GIS'05 (Bremen, Germany, 2005), pp 97-104.

Lynch, M.P., and Saalfeld, A.J. (1985): Conflation: Automated Map Compilation – A Video Game Approach, In Proceedings of the 7th Auto-Carto, Falls Church, VA.

Scarponcini, P. (2002): Generalized Model for Linear Referencing in Transportation. Geoinformatica , Vol. 6, N. 1, 35-55.

Walter, V., and Fritsch, D. (1999): Matching Spatial Data Sets: a Statistical Approach, International Journal for Geographical Information Science, Vol. 13, No. 5, pp. 445-473.

Detecting Hotspots in Geographic Networks

Kevin Buchin[1], Sergio Cabello[2], Joachim Gudmundsson[3], Maarten Löffler[1], Jun Luo[4], Günther Rote[5], Rodrigo I. Silveira[1], Bettina Speckmann[6], Thomas Wolle[3]

[1] Department of Information and Computing Sciences, Utrecht University, The Netherlands, {fbuchin, loffler, rodrigo}@cs.uu.nl
[2] Department of Mathematics, Inst. for Math., Physics and Mechanics, Ljubljana, Slovenia, sergio.cabello@imfm.uni-lj.si
[3] NICTA Sydney, Australia {fjoachim.gudmundsson, thomas.wolle}@nicta.com.au
[4] Shenzhen Institute of Advanced Technology, Chinese Academy of Sciences, China, jun.luo@sub.siat.ac.cn
[5] Institut für Informatik, Freie Universität Berlin, Germany rote@inf.fu-berlin.de
[6] Department of Mathematics and Computer Science, TU Eindhoven, The Netherlands, speckman@win.tue.nl

Abstract. We study a point pattern detection problem on networks, motivated by geographical analysis tasks, such as crime hotspot detection. Given a network N (for example, a street, train, or highway network) together with a set of sites which are located on the network (for example, accident locations or crime scenes), we want to find a connected subnetwork F of N of small total length that contains many sites. That is, we are searching for a subnetwork F that spans a cluster of sites which are close with respect to the network distance.

We consider different variants of this problem where N is either a general graph or restricted to a tree, and the subnetwork F that we are looking for is either a simple path, a path with self-intersections at vertices, or a tree. Many of these variants are NP-hard, that is, polynomial-time solutions are very unlikely to exist. Hence we focus on exact algorithms for special cases and efficient algorithms for the general case under realistic input assumptions.

M. Sester et al. (eds.), *Advances in GIScience*, Lecture Notes in Geoinformation and Cartography, DOI 10.1007/978-3-642-00318-9_11,

1 Introduction

Consider the following scenario: You are given a detailed map of the road network of an area together with the exact locations of all crimes committed during the last year. Your job is to determine the area of the network with the greatest concentration of crimes. To do so, you will want to find many crimes that are somehow "close". But finding crimes whose locations are close with respect to the Euclidean distance might not give you the right answer—the crimes need to be close with respect to the road network. In other words, you need to find a comparatively "small" part of the network which contains the locations of many crimes. This is usually referred to as a crime *hotspot*.

The problem of detecting crime hotspots has received a lot of attention in recent years (see for example (Celik et al., 2007; Levine, 2005; Ratcliffe, 2004; Ratcliffe and McCullagh, 1998; Rich, 2001)). Crime hotspots are relevant to both crime prevention practitioners and police managers: They allow local authorities to understand what areas need most urgent attention, and they can be used by police agencies to plan better patrolling strategies.

Most problems of this type have been almost exclusively considered in the fields of geographic data mining (Miller and Han, 2001) and geographical analysis (Okabe et al., 2006; O'Sullivan and Unwin, 2002). Many different variants have been studied. The data set can be a point set (each point indicating the location of a crime) or a crime rate aggregated into regions such as police beats or census tracts. Even though both provide useful information, for the purpose of finding hotspots, the precise locations of the crimes are required. Existing methods also differ in the shape of the hotspot. For example, a well-known technique, the "Spatial and Temporal Analysis of Crime", outputs areas of higher crime rate as standard deviational ellipses (Illinois Criminal Justice Information Authority, 1996). However, in urban areas, most human activities, including the criminal ones, are georeferenced to the street network, and any measure of proximity should take the network connectivity and network distances into account, rather than using the Euclidean distance.

In this paper we address the problem of finding hotspots in networks from an algorithmic point of view. We model the problem as follows. The input network N is a connected graph with positive edge lengths. The connected subnetwork F of N which we are searching for is a *fragment* of N, that is, a connected subgraph of N: the edges of F are contained in edges of N (they are either edges of N or parts of edges of N). The *length* of a fragment F is the sum of its edge lengths. Together with N, we are given a set S of *sites* (locations of interest), which are located on the edges or vertices of N.

Generally, we are looking for a fragment of small length that should contain many sites (for an example see Fig. 1). These sites then form a cluster with respect to the network distance. More formally, we consider the following problem:

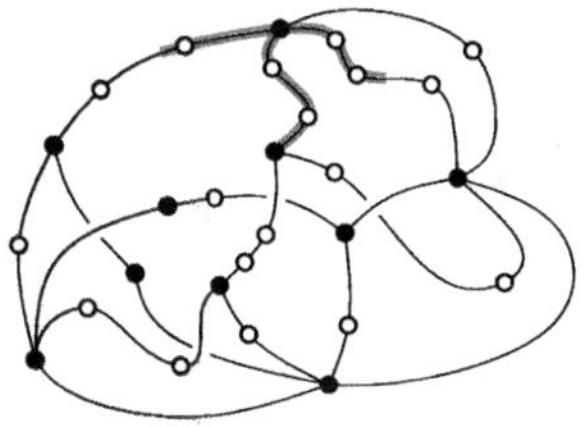

Fig. 1. A network with sites (white circles), nodes (black circles), edges and a fragment (gray).

> *Most Relevant Fragment.* Given a network N with m edges, a set S of n sites on N, and a positive real value d. Find a fragment F of N (from a particular class of graphs) of length at most d that contains the maximum number of sites.

Not surprisingly, the most general problem where N is a graph and the fragment F is a graph, a tree, or even a path, can easily be shown to be NP-complete, that is, polynomial-time solutions are very unlikely to exist. Hence we investigate under which realistic input assumptions the general problem becomes tractable. Furthermore, we present exact and efficient algorithms for special (simple) cases which we believe to be interesting also from a practical point of view, since they can form a solid foundation for effective heuristics that solve the general case.

Notation. We consider various variants where N is either a tree or a graph and F is either a simple path, a path with self-intersections at vertices (but no duplicate edges), or a tree. (Note that if F is allowed to be a general graph then the optimal solution will always be a fragment F which is a tree). We denote each variant by the pair of symbols NF, where N and F is one of four codes: **G** stands for a general graph, **T** for a tree, **Ps** for a simple path (without repeated vertices), and **Pi** for a path with possible self-intersections at vertices. For example, **TPs** denotes the instance of the problem where N is a tree and F is a simple path.

Throughout the paper we assume that the sites are given in sorted order along the edges of N, otherwise sorting the sites would force a lower bound of $\Omega(n \log n)$ for the time complexity of our algorithms.

Results. Recall that we are given a network N with m edges together with n sites on N. We are looking for a fragment of length at most d which contains the maximal number of sites. The simplest case when N is a path can trivially be solved in $O(n + m)$ time by sweeping a path of length d along N. We first discuss two more challenging variants where N is a tree. Here optimal and efficient solutions based on dynamic programming exist. In particular, in Section 2 we consider **TPs**: N is a tree and F is a simple path. In this case we can find the most relevant fragment in $O(n + m)$ time and $O(n + m)$ space. In Section 3 we discuss **TT**: both N and F are trees. Here we can find the most relevant fragment in $O(mn + n^2)$ time.

In Section 4 we study several realistic input assumptions under which efficient algorithms exist for the general problem when N is a graph. If we assume a bound on the maximum vertex degree and on the length of the smallest edge in N—both assumptions are satisfied in general street networks—problems **GP** and **GT** can be solved in polynomial time. The same holds for networks N of bounded treewidth.

Related work. Spatial analysis has been studied intensively in GIS for decades (Fotheringham and Rogerson, 1994) and it has been used in many other areas such as sociology, epidemiology, and marketing (Stillwell and Clarke, 2005). Many spatial phenomena are constrained to network spaces, especially when they involve human activities. For example, car accidents tend to happen only on roads and gas stations are also usually located along roads. There is an ample body of work concerning spatial network analysis and network restricted clustering (Aerts et al., 2006; Spooner et al., 2004; Steenberghen et al., 2004; Yamada and Thill, 2007). Like many spatial analysis methods, most spatial network analysis uses statistical methods such as the network K-function method (Spooner et al., 2004). As already mentioned, the problem of finding crime hotspots has received a lot of attention itself (Celik et al., 2007; Levine, 2005; Ratcliffe, 2004; Ratcliffe and McCullagh, 1998; Rich, 2001). A large part of the existing methods look for hotspots of a particular shape (like an ellipse). Others instead output a *crime map*, dividing the map into a grid and showing the different crime intensities at every grid cell (Ratcliffe, 2004). Although popular in practice, these methods in general do not provide guarantees on the output quality or running time.

On the more algorithmic side, the problems studied in this paper are related to the *orienteering problem* (Golden et al., 1987) (also known as *bank robber problem* (Awerbuch et al., 1998)), as well as to the k-MST and k-TSP problems. In the graph version of the orienteering problem one is given a graph with lengths on edges and rewards on nodes, and the goal is to find a path in the graph that maximizes the reward collected, subject to a hard limit on the length of the path. Many variants of the orienteering problem have been studied (Arkin et al., 1998; Awerbuch et al., 1998; Blum et al., 2007; Chekuri et al., 2008; Chen and Har-Peled., 2006). Even though most of them look for a path, versions where the subgraph sought is a cycle or tree have also received some attention (see for example (Arkin et al., 1998)).

2 TPs: Looking for a Simple Path *F* in a Tree *N*

In this section we assume that the network N is a tree T. We first show in Section 2.1 that we can in fact assume that T is a rooted tree where each internal vertex has two children. Here we also introduce the notation used in this section and state a useful lemma. In Section 2.2 we then show how to find the most relevant fragment in linear time and space.

2.1 Preliminaries

We assume for simplicity of exposition that no site lies on a vertex of T. Sites at vertices complicate the algorithm a little but are no fundamental problem. Select an arbitrary vertex of T as a root, denoted by v_{root}. We transform the input tree into a tree where each internal vertex v has precisely two children, denoted by v_ℓ, v_r (see Fig. 2): a vertex with $t \geq 3$ children can be replaced by a path of $t-1$ degree-three vertices with zero-length edges between them. Vertices with a single child can be eliminated by simply merging the two incident edges. A fragment in the original network corresponds to a fragment of the same length in the new network, and vice versa.

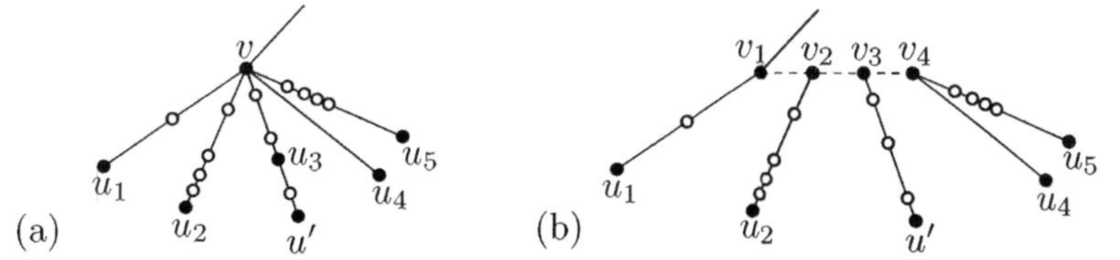

Fig. 2. Transforming the input tree (a) into a rooted tree where each internal vertex has two children (b). The dashed edges in (b) have length zero.

We preprocess T so that the distance $d_T(v, v')$ can be obtained in constant time for any query pair of vertices v, v' in T. This can be done in linear time by building a data structure for lowest common ancestor queries (Bender et al., 2005) and storing for each vertex its distance from the root.

For any pair of sites a, b in the tree T, let $\pi_T(a, b)$ denote the unique path in T that connects them. For each vertex v of T, let $T(v)$ denote the subtree of T rooted at v, and let $m(v)$ be the maximum number of sites from S contained in any path from v to a leaf of $T(v)$. For any edge vu, where v is the parent of u, let $T(vu)$ be the subtree consisting of $T(u)$ plus the edge vu, and let $m(vu)$ be the maximum number of sites from S contained in any path from v to a leaf of $T(vu)$. Let $n(F)$ denote the number of sites of S contained in a fragment F of T and let $n(e)$ denote the number of sites of S along the edge e. Note that $m(vu) = n(vu) + m(u)$. The following bounds, whose proof we omit for lack of space, will be useful to analyze our algorithms.

Lemma 1.

$$\sum_{\substack{u \in V(T) \\ u \text{ not a leaf}}} \min\{m(uu_r), m(uu_\ell)\} = O(n),$$

and

$$\sum_{\substack{u \in V(T) \\ u \text{ not a leaf}}} n(T(uu_r)) \cdot n(T(uu_\ell)) = O(n^2).$$

2.2 Finding the most relevant path

In this section we use dynamic programming to find a path in T of total length at most d that covers the maximum number of sites of S. The approach requires linear time and space.

For each interior vertex v we compute lists $P(v), P(vv_r), P(vv_\ell)$. The list $P(v)$ has $m(v)$ elements. The jth element is (a pointer to) a site $p \in S$ with the property that the path $\pi_T(v, p)$ is a path of minimum length among the paths contained in $T(v)$ that have one endpoint in v and contain j sites of S. Analogously, the list $P(vv_\ell)$ has $m(vv_\ell)$ elements, storing the minimum-length paths in $T(vv_\ell)$ that have one endpoint in v, and similarly for $P(vv_r)$.

The only way in which the length of these lists changes is by adding elements at the front. Thus, we store each list as an *extensible array*, but we store the elements in reverse order: the jth element of a list of length m is stored in array position $A[m - j]$. Standard techniques can be used to implement such arrays with constant access time and amortized constant time for extending them by one element (Cormen et al., 2001, Section 17.4). The total space is linear in the total number of added elements. The arrays are reused for different lists to achieve overall linear time and space.

We process the tree bottom-up and maintain a value $k_{\max}$ that equals the number of sites of S in the best path of length d so far. Initially $k_{\max} = 1$. When v is a leaf, we allocate an empty list $P(v)$ and set $m(v) = 0$. Consider an internal vertex v. Its two children v_r, v_ℓ have already been processed. We aim for a time bound of $O(n(vv_\ell) + n(vv_r) + \min\{m(vv_r) + m(vv_\ell)\})$ for processing v.

(i) We construct $P(vv_\ell)$ and $P(vv_r)$. $P(vv_\ell)$ is obtained by adding the ordered sequence of $n(vv_\ell)$ sites of S on the edge vv_ℓ to the beginning of the list $P(v_\ell)$. The list $P(v_\ell)$ is destroyed in this operation. We construct $P(vv_r)$ similarly, and the total amortized running time is $O(1 + n(vv_\ell) + n(vv_r))$.

(ii) We find the best path contained in $T(v)$ that intersects vv_r but not vv_ℓ. We look for a path of length $k_{\max} + 1$ by simultaneously scanning $P(vv_r)$ with a shifted copy of itself. Formally, we start with $j = 1$, and while $j \leq n(vv_r)$ and $j + k_{\max} \leq m(vv_r)$ do:
   (a) if the distance between the jth site of $P(vv_r)$ and the $(j + k_{\max})^{\text{th}}$ site of $P(vv_r)$ is at most d, then we increment $k_{\max}$ by one.
   (b) otherwise, we increment j by one.

   The same approach can be used to find the best path among those contained in $T(v)$ and intersecting vv_ℓ but not vv_r. Case (b) happens at most $n(vv_r) + n(vv_\ell)$ times, and each time that case (a) occurs, the value $k_{\max}$ is incremented by one. Therefore, this task takes $O(1 + \Delta + n(vv_r) + n(vv_\ell))$ time, where Δ is the increment in the value of $k_{\max}$.

(iii) We find the best path that intersects both vv_r and vv_ℓ as above: we simultaneously scan the lists $P(vv_\ell)$ and $P(vv_r)$, looking for a path with $k_{\max} + 1$ sites and incrementing $k_{\max}$ whenever we find such a path. Assume w.l.o.g. that $m(vv_\ell) \leq m(vv_r)$; the other case is symmetric. Start with $j = m(vv_\ell)$, and while $j \geq 1$ and $k_{\max} - j + 1 \leq m(vv_r)$ do:
   (a) if the distance between the jth element of $P(vv_\ell)$ and the $(k_{\max} - j + 1)^{\text{st}}$ element of $P(vv_r)$ is at most d, then we increment $k_{\max}$ by one.
   (b) otherwise, we decrement j by one.

   Case (b) happens at most $\min\{m(vv_r), m(vv_\ell)\}$ times, and each time that case (a) occurs, the value $k_{\max}$ is incremented by one. Therefore, this task takes $O(1 + \Delta + \min\{m(vv_r) + m(vv_\ell)\})$ time, where Δ is the increment in the value $k_{\max}$.

   The operations of steps (ii) and (iii) together have now taken care of all paths in $T(v)$ that are not contained in one of the subtrees $T(v_\ell)$ or $T(v_r)$.

(iv) Finally, we compute $P(v)$, as follows. Assume without loss of generality that $m(vv_\ell) \leq m(vv_r)$; the other case is symmetric. We will re-use the list $P(v_r)$ to represent the list $P(v)$. For each $j = 1, \ldots, m(vv_\ell)$, the jth element of $P(v)$ is simply the minimum of the jth element of $P(vv_r)$ and the jth element of $P(vv_\ell)$. The elements beyond the $m(vv_\ell)^{\text{th}}$ element are left unchanged. This pairwise comparison of the two lists takes $O(1 + \min\{m(vv_r), m(vv_\ell)\})$ time.

After processing each vertex v of T, we have computed the optimum value $k_{\max}$. Of course, the pair of sites defining the optimum path can be retrieved if we remember the relevant pair of sites each time we increment $k_{\max}$. At each vertex v we spend $O(1+\Delta(v)+n(vv_r)+n(vv_\ell)+\min\{m(vv_r)+m(vv_\ell)\})$ time, where $\Delta(v)$ is the increment that $k_{\max}$ takes when processing vertex v. The sum of $\Delta(v)$ over all vertices v is the final value of $k_{\max}$, and therefore is bounded by n. The sum of $n(vv_r) + n(vv_\ell)$ over all vertices v is n, since each site is counted once in the sum. The sum of $\min\{m(vv_r) + m(vv_\ell)\}$ over all vertices v is $O(n)$ because of Lemma 1. We summarize.

Theorem 1. *Given a tree network with m vertices, a set S of n sites along its edges, and a value d, we can find in $O(n+m)$ time and $O(n+m)$ space a path fragment that has length at most d and contains the maximum number of sites from S.*

3 TT: Both *N* and *F* are Trees

In this section we again assume that the input network is a tree T. We use the notation and transformation described in Section 2.1 and can hence assume that T is a rooted tree where each internal vertex v has precisely two children.

We use dynamic programming and process the vertices of T bottom-up. For each internal vertex v we compute a list $L(v)$, such that from $L(v)$ we can compute the optimal solution where v is the highest vertex in T. The jth entry, $L(v)[j]$, of $L(v)$ stores the length of the smallest tree fragment of $T(v)$ containing v and covering j points of S. If there is no such tree fragment of $T(v)$ then we set $L(v)[j] = \infty$. We also set $L(v)[0] = 0$ to simplify some formulas below. For each leaf v, the tree $T(v)$ contains no sites of S, and $L(v)$ will be empty. When all the leaves have been processed we continue bottom-up. Consider an interior vertex v for which the lists $L(v_r), L(v_\ell)$ of its children v_r, v_ℓ have already been computed. We compute $L(v)$ as follows:

(i) For each child u of v we build a list $L(vu)$ from $L(u)$ with the following property: The jth entry of $L(vu)$ stores the length of the smallest tree fragment of $T(vu)$ containing v and covering j sites. The list is constructed as follows. Consider the points $s_1, s_2, \ldots, s_{n(vu)}$ along the edge vu ordered from v to u. For $j = 1, \ldots, n(vu)$, add the jth entry to $L(vu)$ containing the distance between v and s_j. Then, for $j = n(vu), \ldots, n(T(vu))$ we set the jth entry of $L(vu)$ to be $|vu|+L(u)[j-n(vu)]$, where $|vu|$ denotes the length of the edge vu. The total time to compute the lists $L(vv_r), L(vv_\ell)$ is $O(n(T(v))) = O(n)$.

(ii) The lists $L(vv_r)$ and $L(vv_\ell)$ are used to construct $L(v)$, as follows. For each integer $j = 1, \ldots, n(T(v))$ we set

$$L(v)[j] = \min\{\, L(vv_r)[a] + L(vv_\ell)[b] \mid 0 \le a \le n(T(vv_r)), 0 \le b \le n(T(vv_\ell)),\ a + b = j\,\}.$$

This procedure constructs the list $L(v)$ using time

$$O\big(n(T(vv_r)) + n(T(vv_\ell)) + n(T(vv_r)) \cdot n(T(vv_\ell))\big) = O(n + n(T(vv_r)) \cdot n(T(vv_\ell))).$$

Each vertex v of T is processed once and requires $O(n + n(T(vv_r)) \cdot n(T(vv_\ell)))$ time. The sum of $O(n)$ over all vertices is $O(mn)$. The sum of $n(T(vv_r)) \cdot n(T(vv_\ell))$ over all vertices is $O(n^2)$ (see Lemma 1). Hence, we can construct the lists $L(v)$ for all vertices v of T in $O(mn + n^2)$ time.

We describe now how to find the most relevant tree fragment of length at most d in T. First, we compute the most relevant tree fragment that does not contain any vertex of T, and therefore is a path. This can be done in $O(n + m)$ time by finding optimal solutions contained in each edge of T. Next, for each vertex v, we use $L(v)$ to find the most relevant tree fragment that has v as highest vertex. Taking the best among these solution gives the optimal solution. If a tree fragment has v as highest vertex, then it is contained in $T(v_{\text{parent}}v)$, where v_{parent} denotes the parent of v. (We can handle the case $v = v_{\text{root}}$ by adding a dummy parent to v_{root}.) Let $s_1, \ldots, s_{n(v_{\text{parent}}v)}$ be the points of S on the edge vv_{parent}, ordered from v to v_{parent}. We construct a list $M(v)$, where the jth entry stores the length of the smallest tree fragment of $T(v_{\text{parent}}v)$ that has v as highest vertex and contains j points of S, using:

$$M(v)[j] = \big\{\, L(v)[a] + |vs_b| \;\big|\; 0 \le a \le n(T(v)), 0 \le b \le n(vv_{\text{parent}}),\ a + b = j\,\big\}.$$

Constructing $M(v)$ takes $O(n(T(v)) \cdot n(vv_{\text{parent}}) = O(n \cdot n(vv_{\text{parent}}))$ time for a vertex v of T, which sums up to $O(n^2)$ time over all vertices v of T. The largest number of sites contained in a tree fragment with v as highest vertex is given then by the unique index j_v satisfying $M(v)[j_v] \le d$ and $M(v)[j_v + 1] > d$.

Theorem 2. *Given a tree network with m vertices, a set S of n sites along its edges, and a value d, we can find in $O(mn + n^2)$ time using $O(n)$ space a tree fragment that has length at most d and contains the maximum number of sites from S.*

4 GP and GT: Exact Algorithms

While the general problem considered in this paper is NP-hard, in many applications we have additional information and/or restrictions on the network and the fragment, which make polynomial-time solutions possible. Here we discuss two such scenarios. In Section 4.1 we bound the maximum vertex degree of N as well as the length of the smallest edge in N and in Section 4.2 we consider networks N of bounded treewidth. In both cases we describe fixed-parameter tractable algorithms.

4.1 Limiting vertex degree and edge length

Real-world road networks are unlikely to contain high degree vertices or very short edges (with respect to the length d of the fragment). Let D be the maximum vertex degree of N, and let s be the length of the shortest edge in N. If we assume that both D and the fraction $f = d/s$ are constant, then we can solve **GP** and **GT** in time polynomial in n and m.

To solve **GP** when f and D are small, we can simply enumerate all possible paths, and then choose the best one. The optimal path consists of one partial edge of N, then a sequence of complete edges, and then another partial edge. We call the part consisting of complete edges the *skeleton* of the path. The skeleton can consist of at most f edges. The number of skeleton paths of at most f edges is $O(m \cdot D^f)$, since we can start at any vertex, and at each new vertex we have at most D choices of how to proceed. We compute all of these, and then look for the best path that has that skeleton.

To find the best path using a given skeleton, we have to append two partial edges to its endpoints that cover the largest amount of sites, while their length remains bounded by d minus the length of the skeleton. To be able to do this, we pre-compute for each edge two lists with the distance to the k-th point on the edge, as seen from one endpoint. This takes linear time in total. Then, for a given skeleton, we guess an adjacent edge to both of its endpoints, and then find the best combination of partial edges on those two edges. Note that both edges may be the same edge, in which case the two partial edges can overlap, but when this is the case we can simply take the whole edge. There are D^2 choices for the adjacent edges per skeleton. Finding the best partial edges takes time linear in the number of sites on those edges, which is in the worst case $O(D^2 n)$. Multiplying this time by the number of skeleton paths we obtain the following result.

Theorem 3. *On graphs with degree at most D and smallest edge length s,* **GP** *can be solved in $O(nm \cdot D^{d/s+2})$ time.*

We can use a similar approach for **GT**. A solution again consists of a number of complete edges of N and a number of partial edges. The complete edges are all connected and form a *skeleton tree.* We enumerate all skeleton trees of length at most d. A skeleton tree can have at most f edges. The number of skeleton trees containing a given "root" vertex can be bounded by the number of D-ary trees, which is known to be

$$\binom{Dk}{k-1}/k = \frac{\binom{Dk}{k}}{(D-1)k+1} \approx \frac{D^{Dk+1/2}}{(D-1)^{(D-1)k+3/2}}/\sqrt{k^3} \leq e^k \cdot D^k,$$

if they contain k vertices (and $k-1$ edges), cf. (Beineke and Pippert, 1971), see also (Rote, 1997) for an elementary proof. (The approximation uses Stirling's formula, and $e \approx 2.718$ is Euler's constant.) Summing this over all possible sizes $k = 1, \ldots, f+1$, we conclude that there are no more than $O(m \cdot (eD)^{f+1})$ skeleton trees.

Now, all edges that are adjacent to a given skeleton tree might be used partially in a solution. Some of these edges are connected to the skeleton by only one endpoint, and some by both endpoints. Therefore, as a preprocessing step, we compute for each edge e in N three lists. The first list stores, for each integer k, the shortest possible path, starting at the left endpoint of e, within e, that contains k sites (so just the distance to the k-th site from the left). The second list stores the same information, but starting from the right endpoint. The third list stores the shortest pair of paths, starting at the left and right endpoints of the edge, that contains k sites. We can easily compute all of these lists in quadratic time.

With this information, we can solve the problem for a given skeleton tree by considering the correct lists for all adjacent edges (depending on which endpoints are in the skeleton tree). We need to find the best combination of partial edges, which can be done in $O(mn + n^2)$ time with an algorithm very similar to that in Section 3. Since we do this for each skeleton tree, the total running time is $O((m^2n + mn^2) \cdot (eD)^{f+1})$.

Theorem 4. *On graphs with degree at most D and smallest edge length s,* **GT** *can be solved in $O((m+n)mn \cdot (eD)^{d/s+1})$ time.*

Note. The running times for a single path or tree in the above proofs are overly pessimistic, since they allow that *all* m edges and *all* n sites enter the calculation. By a more thorough examination of the edges and sites that are actually relevant for each path or tree, the running time of Theorem 3 can be improved to $O(n \cdot D^{d/s} + m \cdot D^{d/s+1})$, and the running time of Theorem 4 can be improved to $O(m(eD)^{d/s+1} + n^2 \cdot (eD)^{2d/s+2})$. Details will be given in the full paper.

4.2 Networks of bounded treewidth

A *tree decomposition* is a mapping of a graph into a tree and the *treewidth* of a graph measures the number of graph vertices mapped onto any tree node in an optimal tree decomposition. It is NP-hard to determine the treewidth of a graph, but many problems on graphs are solvable in polynomial time if the treewidth of the input graph is bounded (see e.g. (Bodlaender, 2007)). Here we sketch an algorithm for **GT** on a network N of bounded treewidth.

Formally, a *tree decomposition* of a network $N = (V, E)$ is a pair (T, X) with $T = (I, F)$ a tree, and $X = \{X_i \mid i \in I\}$ a family of subsets of V, called *bags*, one for each node of T, such that

- $\bigcup_{i \in I} X_i = V$.
- for all edges $\{v, w\} \in E$ there exists an $i \in I$ with $\{v, w\} \subseteq X_i$.
- for all $i, j, k \in I$: if j is on the path in T from i to k, then $X_i \cap X_k \subseteq X_j$.

The *width* of a tree decomposition $((I, F), \{X_i \mid i \in I\})$ is $\max_{i \in I} |X_i| - 1$. The *treewidth* $tw(N)$ of a network N is the minimum width over all tree decompositions of N. A tree decomposition (T, X) is *nice*, if T is rooted and binary, and the nodes are of four types:

- *Leaf nodes* i are leaves of T and have $|X_i| = 1$.
- *Introduce nodes* i have one child j with $X_i = X_j \cup \{v\}$ for some vertex $v \in V$.
- *Forget nodes* i have one child j with $X_i = X_j \setminus \{v\}$ for some vertex $v \in V$.
- *Join nodes* i have two children j_1, j_2 with $X_i = X_{j_1} = X_{j_2}$

Using nice tree decompositions often makes it easier to develop and describe algorithms for graphs of bounded tree-width. Any tree decomposition can be converted into a nice tree decomposition of the same width in linear time (Kloks, 1993).

We construct a network N' from N by adding the sites of S as vertices to N'. N' has $|V| + n$ vertices and $n + m$ edges. We refer to a vertex of N' that originated from N or S as *network vertex* or *site vertex*, respectively.

The general approach is as follows. We assume that we are given a nice tree decomposition (T, X) of N' of width $tw(N)$. With each bag i of T, we associate a table containing certain information. These tables represent partial solutions for the subnetwork N_i' of N' that corresponding to the subtree of T rooted at i. More specifically, in the tables we keep track of subforests in N_i', of their lengths and of the number of site vertices they contain. Such a subforest might have vertices in common with X_i. These vertices are represented by an *interface*, which is a set of disjoint subsets of X_i. An interface of a forest tells us which vertices of X_i are involved in the forest, and it also tells us which vertices belong to the same tree of that forest.

Our algorithm employs dynamic programming on (T, X). We start at the leaves, and for an internal node i of T, we compute the table of i using the tables of the children of i. For that, we combine the information of compatible interfaces from the children of i. The resulting running time is exponential in the treewidth, but polynomial in the size of the input.

Theorem 5. *Given a graph network with m edges whose treewidth is bounded by some constant, a set S of n sites along its edges, and a value d, we can find in $O((m+n)\,n^2)$ time a tree fragment that has length at most d and contains the maximum number of points from S.*

5 Conclusions and Open Problems

We studied a network analysis problem motivated by crime hotspot detection. Our approach focused on finding cases for which polynomial-time solutions are possible. Specifically, we showed that if the network N is a tree, efficient algorithms exist to solve the problem: we gave a linear-time algorithm for the **TPs** variant and a simpler $O(mn+n^2)$-time algorithm for **TT**. Furthermore, we gave exact polynomial-time algorithms for the realistic cases in which the maximum degree of the vertices and the minimum edge length in N are bounded, and for networks N of bounded treewidth. Although our algorithms are efficient from a theoretical point of view, the practical suitability of them could only be determined in experiments.

Various extensions of this work are possible. First of all, can we give an effective heuristic for the general problem based on our exact and efficient solutions to special cases? For example, we could consider to test various spanning trees of an input network and overlay the solutions to arrive at a global solution. Judging the quality of such an approach requires an extensive experimental evaluation. Second: how about a setting where there are two types of sites, for example cars and accidents? Then we would be interested in a short fragment where the ratio between cars and accidents is small—a question which is related to so-called ratio-clustering.

Acknowledgments

The authors thank Matya Katz for helpful discussions. This research was initiated during the GADGET Workshop on Geometric Algorithms and Spatial Data Mining, funded by the Netherlands Organisation for Scientific Research (NWO) under BRICKS/FOCUS grant number 642.065.503.

NICTA is funded by the Australian Government as represented by the Department of Broadband, Communications and the Digital Economy and the Australian Research Council through the ICT Centre of Excellence program.

K. Buchin and J. Luo were supported by Netherlands Organisation for Scientific Research (NWO) under BRICKS/FOCUS grant number 642.065.503. M. Löffler and R. I. Silveira were supported by the Netherlands Organisation for Scientific Research (NWO) under the project GOGO. B. Speckmann was supported by the Netherlands Organisation for Scientific Research (NWO) under project no. 639.022.707.

References

Aerts, K., Lathuy, C., Steenberghen, T., and Thomas, I. (2006). Spatial clustering of traffic accidents using distances along the network. In *Proc. 19th Workshop Intern. Cooperation on Theories and Concepts in Traffic Safety*.

Arkin, E. M., Mitchell, J. S. B., and Narasimhan, G. (1998). Resource-constrained geometric network optimization. In *Proc. 14th Symp. Computational Geometry*, pages 307–316.

Awerbuch, B., Azar, Y., Blum, A., and Vempala, S. (1998). New approximation guarantees for minimum-weight k-trees and prize-collecting salesmen. *SIAM Journal on Computing*, 28(1):254–262.

Beineke, L. W. and Pippert, R. R. (1971). The number of labeled dissections of a k-ball. *Math. Ann.*, 191:87–98.

Bender, M. A., Farach-Colton, M., Pemmasani, G., Skiena, S. S., and Sumazin, P. (2005). Lowest common ancestors in trees and directed acyclic graphs. *Journal of Algorithms*, 57(2):75–94.

Blum, A., Chawla, S., Karger, D. R., Lane, T., Meyerson, A., and Minkoff, M. (2007). Approximation algorithms for orienteering and discounted-reward TSP. *SIAM Journal on Computing*, 37(2):653–670.

Bodlaender, H. (2007). Treewidth: Structure and algorithms. In *Proc. 14th Colloquium on Structural Information and Communication Complexity*, number 4474 in LNCS, pages 11–25.

Celik, M., Shekhar, S., George, B., Rogers, J. P., and Shine, J. A. (2007). Discovering and quantifying mean streets: A summary of results. Technical Report 07-025, University of Minnesota - Comp. Science and Engineering.

Chekuri, C., Korula, N., and Pál, M. (2008). Improved algorithms for Orienteering and related problems. In *Proc. 19th ACM-SIAM Symp. Discrete Algorithms*, pages 661–670.

Chen, K. and Har-Peled., S. (2006). The orienteering problem in the plane revisited. In *Proc. 22nd Symp. Computational Geometry*, pages 247–254.

Cormen, T. H., Leiserson, C. E., Rivest, R. L., and Stein, C. (2001). *Introduction to Algorithms*. MIT Press and McGraw-Hill, 2nd edition.

Fotheringham, S. and Rogerson, P. (1994). *Spatial Analysis and GIS*. Taylor and Francis, London.

Golden, B., Levy, L., and Vohra, R. (1987). The orienteering problem. *Naval Research Logistics*, 34:307–318.

Illinois Criminal Justice Information Authority (1996). STAC user manual.

Kloks, T. (1993). *Treewidth*. PhD thesis, Utrecht University.

Levine, N. (2005). Crime mapping and the Crimestat program. *Geographical Analysis*, 38:41–56.

Miller, H. and Han, J., editors (2001). *Geographic Data Mining and Knowledge Discovery*. CRC Press.

Okabe, A., Okunuki, K., and Shiode, S. (2006). Sanet: A toolbox for spatial analysis on a network. *Geographical Analysis*, 38(1):57–66.

O'Sullivan, D. and Unwin, D. (2002). *Geographic Information Analysis*. Wiley.

Ratcliffe, J. H. (2004). The hotspot matrix: A framework for the spatio-temporal targeting of crime reduction. *Police Practice and Research*, 5:05–23.

Ratcliffe, J. H. and McCullagh, M. J. (1998). Aoristic crime analysis. *International Journal of Geographical Information Science*, 12:751–764.

Rich, T. (2001). Crime mapping and analysis by community organizations in hartford, connecticut. *National Institute of Justice: Research in Brief*, pages 1–11.

Rote, G. (1997). Binary trees having a given number of nodes with 0, 1, and 2 children. *Séminaire Lotharingien de Combinatoire*, B38b:6 pages.

Spooner, P. G., Lunt, I. D., Okabe, A., and Shiode, S. (2004). Spatial analysis of roadside Acacia populations on a road network using the network k-function. *Landscape Ecology*, 19:491–499.

Steenberghen, T., Dufays, T., Thomas, I., and Flahaut, B. (2004). Intra-urban location and clustering of road accidents using GIS: a Belgian example. *International Journal of Geographical Information Science*, 18:169–181.

Stillwell, J. and Clarke, G. (2005). *Applied GIS and Spatial Analysis*. Wiley.

Yamada, I. and Thill, J. (2007). Local indicators of network-constrained clusters in spatial point patterns. *Geographical Analysis*, 39:268–292.

Estimating Spatio-Temporal Distribution of Railroad Users and Its Application to Disaster Prevention Planning

Toshihiro Osaragi

Department of Mechanical and Environmental Informatics,
Graduate School of Information Science and Engineering,
Tokyo Institute of Technology
2-12-1-W8-10 O-okayama, Meguro-ku, Tokyo 152-8552, Japan

Abstract. To develop measures for minimizing human damage from a devastating earthquake, it is important to understand the characteristics of the population and its spatio-temporal distribution in an urban area. In the present paper, a model is constructed that simulates the route selection behavior and the transfer choices of railroad users using a geographic information system. The spatio-temporal distribution of users is estimated by applying the model to the Tokyo metropolitan area, using data collected in a person-trip survey. Some numerical examples using the proposed model are shown for detailed disaster prevention planning. In particular, the number and the spatio-temporal distribution of people with difficulty returning home are discussed.

Keywords: railroad user, spatio-temporal distribution, disaster prevention planning, person trip survey, difficulty returning home

1 Introduction

In recent years, interest in disaster prevention planning for a devastating earthquake directly below Tokyo has grown, and numerous investigations and studies have been completed that estimate fatalities from such an event. Most previous estimates of human casualties have been estimated based on static population distributions, such as the daytime population distribution or nighttime population distribution obtained from the national census or other sources. However, the actual population distribution varies

M. Sester et al. (eds.), *Advances in GIScience*, Lecture Notes in Geoinformation and Cartography,
DOI 10.1007/978-3-642-00318-9_12,

hourly, and the degree and spatial distribution of human damage are closely linked to the time when the disaster occurs. In particular, the temporal variation in the number of railroad users is extremely large in Tokyo metropolitan area, and cannot be ignored. Consequently, the purpose of the current research is to build a model for estimating the spatio-temporal distribution of railroad users using data extracted from the *Tokyo Metropolitan Area Person Trip Survey* of 1998 (hereafter referred to as "*PT data*"), and thereby develop a detailed understanding of the temporal and spatial variations in the number of railroad users.

The proposed model is noted for its capability to estimate the information relating to specific attributes and destination data for railroad users, as well as the capacity to estimate the spatio-temporal movements of individual types of railroad users. For example, the time, location, and elementary school children traveling a long distance to school by train can be identified. The model can track elderly residents that leave their residence to shop at department stores a distance from their homes. These age groups would require special care and attention in the event of a large earthquake. To establish sophisticated disaster prevention planning for a specific location, the individual attributes of the population must be considered.

The advanced model developed in the current study examines the spatio-temporal distribution of people that would encounter difficulty in returning home after a disaster. In addition to the number of individuals with difficulty returning home, the people that remain within the city are also considered in the model. To demonstrate the ability of the constructed model to estimate detailed attributes and anticipated information, the current study includes individuals with a high possibility of remaining in the city in need of support, and discusses the spatio-temporal distribution of these residents in case of disaster.

2 Methods of Estimating The Spatio-Temporal Distribution of Railroad Users

2.1 Previous Research Relating to the Distribution of Railroad Users

As part of transportation planning, civil engineers have previously conducted research about railroad users (Morichi et al. 2001, Iwakura et al. 2000, Harada et al. 2002). These studies are focused on alleviating rush hour congestion in urban transport regions, and provide the elemental research that supports policy options for leveling out the volume of railroad users at peak times in high traffic time zones (Iwakura et al. 2000). The analyses include the effects of a flex-time system, peak-pricing, and charges per used time zone. However, the users' detailed attributes have not been previously evaluated because the research has focused on the number of railroad users during morning rush hour. Furthermore, although methods of forecasting spatio-temporal user demand have been developed, the forecasting of time-specific user demand has not been sufficiently addressed (Harada et al. 2002).

In contrast, Toriumi et al. (2008) examine impacts on railroad users in the event of an earthquake with an epicenter directly below the Tokyo metropolitan area, using the model developed by Taguchi (2005). Although Toriumi et al. (2008) examine each railroad's temporal movements, their methods differ from the method employed in the current study. That is, their examination is limited to railroad users with commuting passes, using the *Metropolitan Transport Census* (2000), and detailed user attributes such as gender, age and occupation and purpose of movement cannot be obtained through their method.

2.2 Features of Proposed Model

A large number of route choice models have been developed over the last few years for route choice in public transit systems and path choice within transit stations. Most of these approaches are based on discrete choice models (Bovy and Stern 1990, Chorus et al. 2007). In the present paper, a model was constructed to estimate the dynamically varying spatio-temporal distribution of railroad users based on their detailed attributes provided in the PT data — i.e. to answer the questions "what types of people (attributes such as age and sex) use the railroad, at what time, in which sectors (position coordinates), and for what reasons (purpose)?" More specifically, position and time information for the departure/arrival railroad stations are extracted from PT data. Railroad in the present paper includes not only urban heavy railroad but also other kind of public transportation such as subway/metro/underground and light rail transit (LRT). Based on that information, time specific position information is estimated for railroad users by recreating spatial movements on railroad lines using Geographic Information System (GIS) network analyses. Figure 1 provides the details of estimating the spatio-temporal distribution of railroad users using the present model.

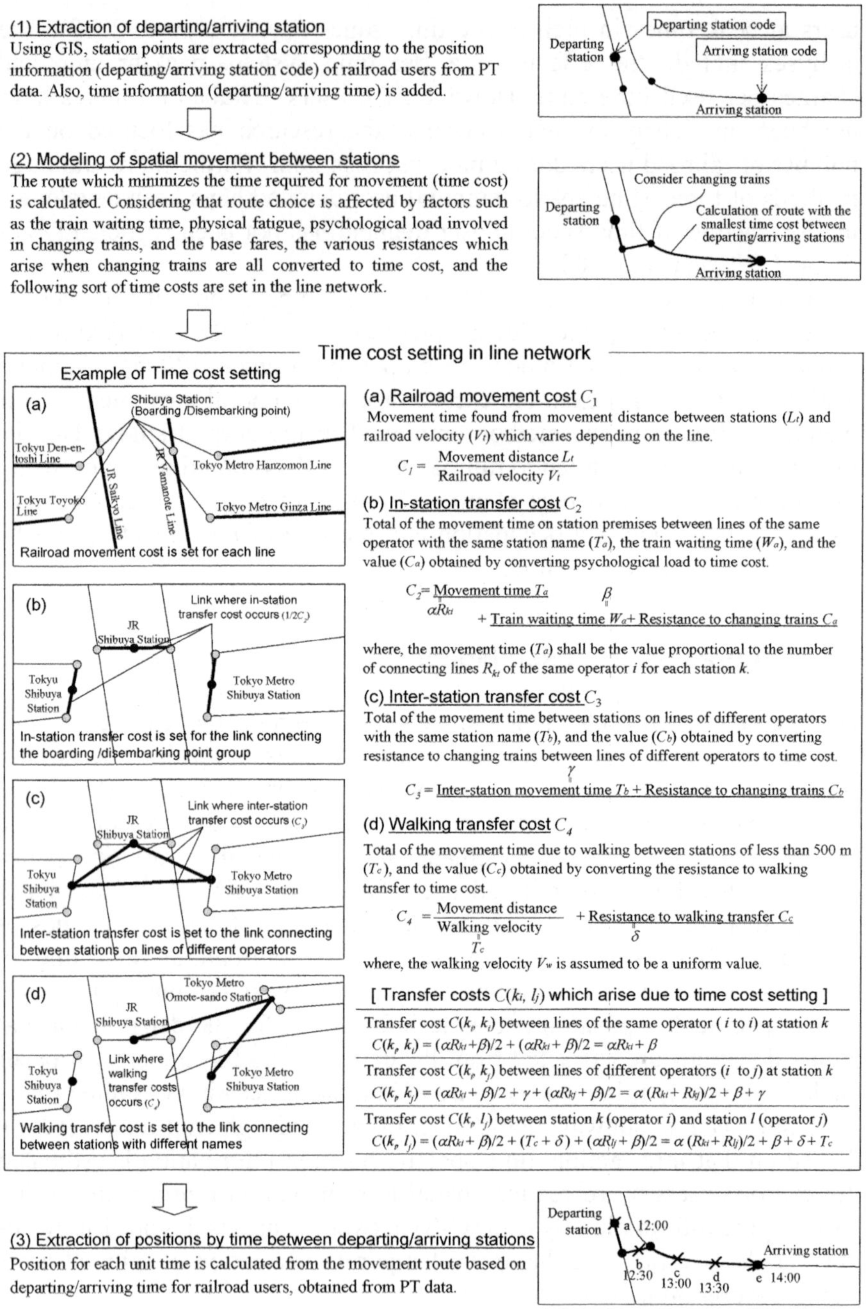

Fig. 1. Model for estimating spatio-temporal distribution of railroad users

Table 1. Outline of person trip survey (PT data)

Regions subject to survey:	Tokyo, Kanagawa, Saitama, Chiba and Southern Ibaraki Prefectures
Survey time and date:	One weekday, excluding Monday/Friday in October–December of 1998
Object of survey:	Persons living within the region, extracted/selected from persons at least 5 years old
Sampling:	Random sampling based on census data (1,235,883 persons from 32,896,705 persons)
Valid data:	883,044 samples (mean weighting coefficient is around 37.3 (= 32,896,705/ 883,044))
Content of data:	Personal attributes, departure/arrival time and location, purpose of trip, means of trip, etc.
Purpose of trip:	One of the 18 purposes for each trip
Means of trip:	One of the 15 means for each unlinked-trip, one or more means for each trip

A part of the information provided in the PT data is listed in Table 1. Using PT data, the railroad user position and time information for the departure/arrival stations can be determined. However, the accuracy of time information is not so high and time necessary for transit is unknown. Therefore, the route and transfer stations taken by the user between the departure and arrival stations cannot be directly determined. Thus, the model selects the route that "minimizes the cost of movement (*time-cost*)" as the spatial movement of railroad users. Route selection by railroad users may be affected by factors such as waiting time before train arrival, physical fatigue, the psychological burden of changing trains, and the base fare payments that arise from transferring among lines under different management (operators). Although changing trains is generally accomplished at one station or between stations with the same name, there are users that change trains by walking to a station located within close proximity of another (walking transfer). The various resistances that arise from needing to transfer between trains are converted to time-cost values and incorporated into the model. The following discussion outlines the specific method for describing time-cost.

For movement along a single train line, movement time is estimated based on the distance between stations and the railroad velocity, and considered to be the "*railroad movement cost.*" However, train velocity will vary among train lines. Thus, a model describing the railroad velocity is developed using the average distance between stations for each line, as shown in Figure 2.

To accommodate transferring between lines, the movement time on the station premises, the waiting time for train arrival, and user resistance to changing trains (i.e., the psychological burden of changing trains) are converted to time-cost values, and the total values are taken to be the "*in-station transfer cost.*" However, movement time at a station depends upon the size of the station, which is indicated by the number of connected lines. The in-station transfer cost is set for links connecting the representative boarding/disembarking points for lines under the same operator with the same station name (corresponding to different platforms). The representative points are assigned as the center of gravity for a group of boarding/disembarking points, which correspond to the station representative points for each operator.

Furthermore, the movement time between stations of the same name with different operators, and resistance to changing trains between operators (base fare resistance) are converted to time-cost values, which are then totaled as the "*inter-station transfer cost.*" The inter-station transfer cost is set for links that connect between representative station points for lines with the same station names under different operators.

In addition, walking time between stations with different names and resistance to walking transfers (resistance that results from the necessity of walking a long distance) are also converted to time-cost values, and those values are totaled to represent the "*walking transfer cost.*" A walking transfer cost value is set for links connecting representative station points where the distance between stations with different names is 500 meters or less.

The waiting time cost might vary for passengers, depending on their departure times and their egress time from transit lines. Also the walking-transfer resistance might be higher for elderly people. Although all the parameters used in the model might vary between different trip characteristics and age-groups, they are assumed to be constant for the model's simplicity.

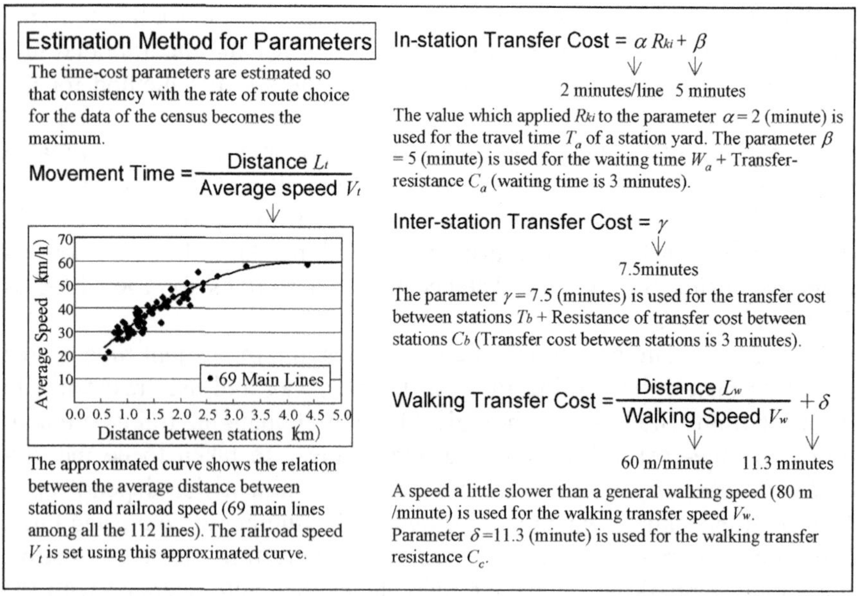

Fig. 2. Time-cost parameter for railroad networks

2.3 Estimation of Model Parameters

The Metropolitan Transport Census data provides information concerning the use of transfer stations. The information is derived from commuters' monthly-passes, by which they can travel on the specific lines and transfer on the specific stations. Based on these data, transit choices are identified and are constructed as measured values (referred to as the "*real value*") using the following method. After applying the method, all transfer routes between Station *A* and Station *B* are extracted, and represented as R_i ($i = 1, 2, \ldots, n$). Furthermore, the probability of choosing any one of these routes is represented as C_i ($i = 1, 2, \ldots, n$) ($\sum_i C_i = 1.0$). In sum, for multiple routes available as user choices for movement between Station *A* and Station *B*, and for which the collective time-costs are approximately equivalent, the ratios of choosing these routes are similar or equivalent. In contrast, when a particular route has a minimum collective time-cost, the ratio of choosing the particular route is close to 1.0.

The route estimation model is calibrated using the following method. Railroad user movement routes are estimated based on the above model, and the model parameter is estimated using the steepest descent method to enhance coincidence with the movement route choice ratio. The data for movement among stations on the same line are not included in the calibration of the model, because they do not include the information about transfer stations and do not contribute to the calibration.

The estimated values of the time-cost parameters after model calibration are shown in Figure 2. Given that resistance to changing trains at one station is low, whereas resistance to changing trains between different stations and walking transfers is high, users are tolerant of transfers among trains under the same operator, and tend to select routes that do not require additional initial fares or walking transfers.

The estimated model ran only once to simulate the spatial distribution of railroad users on a typical day. Its spatial distribution for a given time can be dynamically presented by using the results of simulation.

2.4 Validation of the Estimated Model

The estimates generated by the calibrated model show that more than eighty-two percent (82.9%) of transfer movement routes were correctly extracted from the Metropolitan Transport Census data; when the data were weighted and converted to reflect user numbers, ninety-four percent (94.0%) of the routes were correctly estimated. To confirm compatibility with other statistical data, the following validations were completed.

2.4.1 Total Users per Day at Main Stations

Movement routes are calculated using data for all railroad users (number of trips: 497,835). Based on this, a comparison of the estimated value of the number of users per day for stations with multiple connected lines (155 stations) and stations with one connected line (918 stations) with the statistical values reported in the *Urban Transport Yearbook* (Institution for Transport Policy Studies 2003). The results are shown in Figure 3. There is a strong correspondence between the estimates derived from the model and the statistical values, indicating that the model can reproduce the actual spatial movement with comparatively good precision.

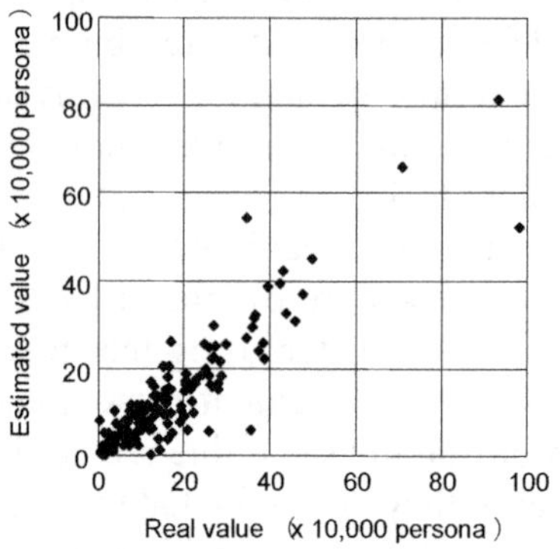

Fig. 3. Fitting of the estimated model

2.4.2 Elapsed Time Required for Trips

Validation of elapsed time for trips was conducted by comparing the time estimated by the model and based on the PT data. The average margin of error for the actual time minus estimated time is – 3.6 minutes, and the standard deviation is 11.3 minutes. Although standard deviation is somewhat large, the actual time and the estimated time are compatible and reflect accurate results.

3 Spatio-Temporal Distribution of Railroad Users

3.1 General Railroad User Characteristics

The number of persons moving at 8:00 is 10 times higher than the number at 12:00 and 2 times higher than the number at 18:00 (Figure 4). In the event of a disaster, users moving on train lines may begin evacuation later than persons present at the station premises. The danger for users will increase when the system is moving passengers at high-density and high-speed. Disaster countermeasures must consider the passenger volume and the concentration points on train lines at approximately 8:00 and after 18:00.

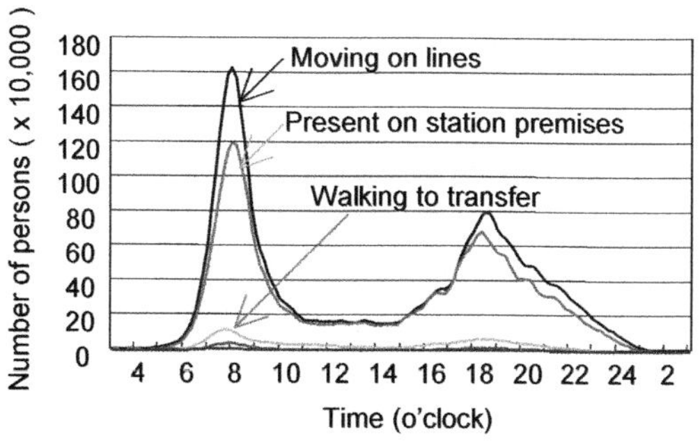

Fig. 4. Railroad users by usage phase

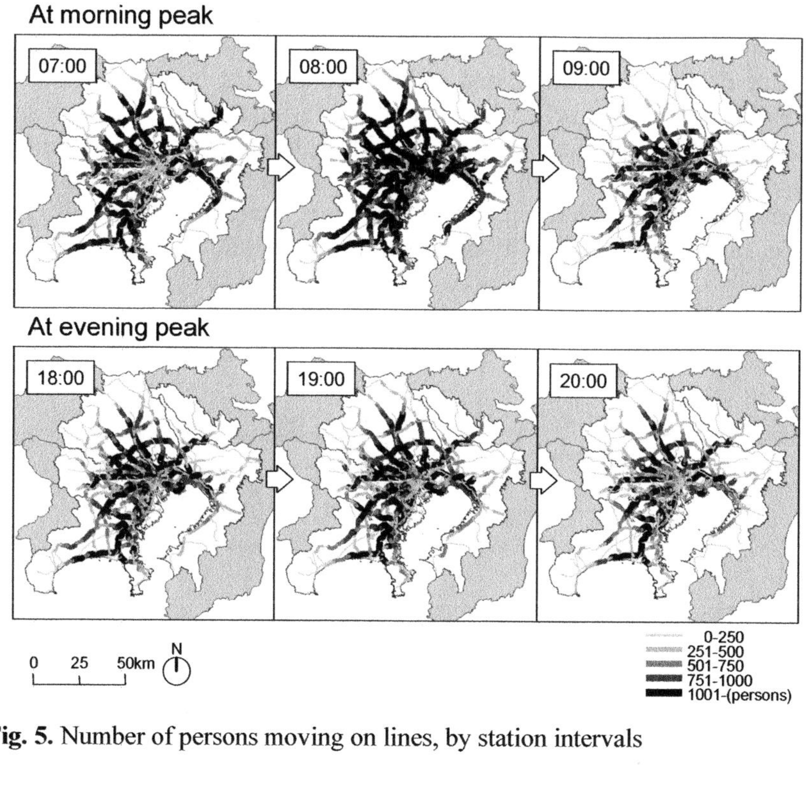

Fig. 5. Number of persons moving on lines, by station intervals

The spatial distribution of users by station interval (Figure 5) indicates that the distribution moves from the suburbs toward the city center from 7:00 to 9:00, and reverses from 18:00 to 20:00. If an earthquake were to occur during these peak time periods, the damage would be more extensive. Measures for reducing railroad usage during peak hours (including recommendations such as staggered commuting times) would not only improve comfort, but would be an important aspect of disaster prevention.

3.2 Personal Attributes of Railroad Users

The spatio-temporal distribution of railroad users is evaluated based on personal attributes. The percentage of male users is overwhelmingly high in the early morning and late night, and the percentage of male and female users is the same during the daytime (Figure 6, left panel). In addition, housewives, househusbands, and the unemployed show an increased use during the daytime because of excursions such as shopping. A nighttime increase can also be seen for pupils, children, and kindergartners, who account for about half of the total (Figure 6, right panel). Although only a small number of users are in the age 5–15 group, there is a temporary peak in usage during the morning and evening commuting hours. For users 70 years old or older, the percentage is highest during the daytime because of the timing of their outings (Figure 7). Although the number of individuals using the train lines is small during the day and in the nighttime hours, the percentage of young children and elderly users increases during these time periods, and their vulnerability during a disaster should be considered.

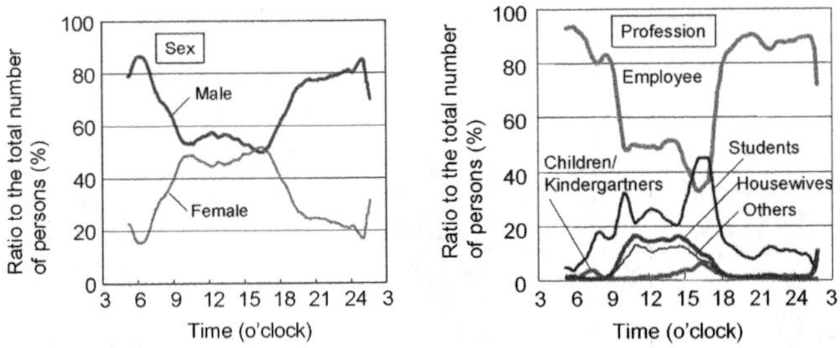

Fig. 6. Ratio of male/female and professions of railroad users

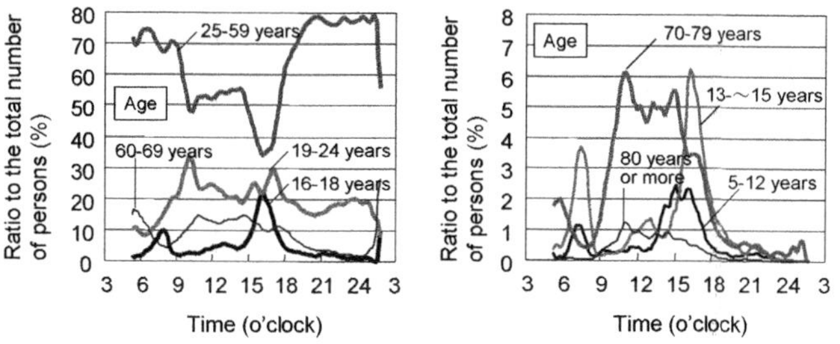

Fig. 7. Ratio of railroad user ages

3.3 Purpose of Railroad User Movements

The spatio-temporal distribution of railroad users is separated by user travel purpose (Figure 8). During the morning and evening hours, the percentage of commuters is high. However, during the daytime, the percentage of shopping is higher, and this purpose accounts for half of the total use at 15:00. Unlike persons commuting to work or school, there is a high probability that persons with purposes such as shopping do not following a determined route, and the event of a daytime earthquake, a large percentage of people may be affected by the earthquake in unfamiliar areas.

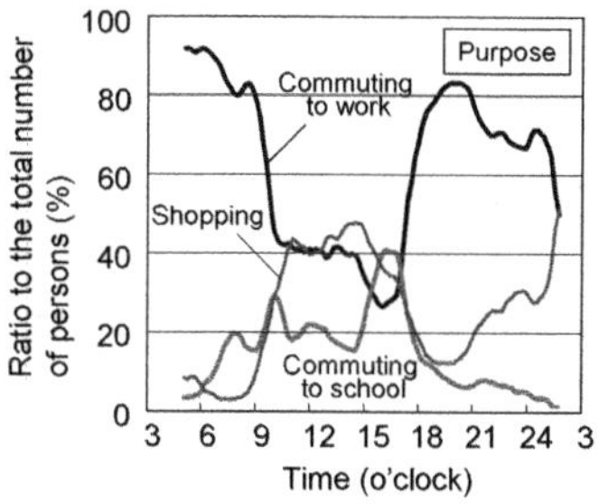

Fig. 8: Ratio of railroad user purposes

The spatial distribution shown in Figure 9 reveals a close resemblance between the morning and evening hours based on the spatial distribution of users commuting to or from work or school. In contrast, the distribution is different in the morning and evening for persons with purposes such as shopping. Furthermore, users commuting to school are distributed over a wide geographic range in both the morning and evening time periods, indicating the wide distribution of high schools, universities, and other schools across the Tokyo metropolitan area. Persons that commute to and from work also travel over a broad geographic region using railroads.

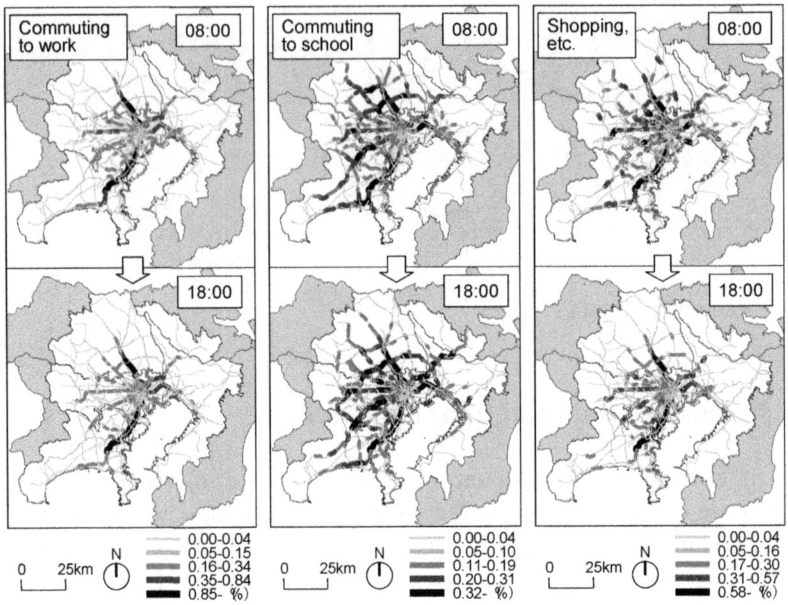

Fig. 9. Spatio-temporal distribution based on railroad user purposes: percentage of persons with each purpose as a fraction of total persons with a purpose on railroad lines

4 Spatio-Temporal Distribution of Persons with Difficulty Returning Home

4.1 Method for Estimating the Spatio-Temporal Distribution of Persons with Difficulty Returning Home

Previous research on the difficulty of residents returning home in the event of a disaster evaluated the city's measures for damage prevention, and considered the history and experiences of residents during previous catastrophic disasters (Nakabayashi 1985, 1995). The Tokyo Metropolis published a report regarding damage forecasts for earthquakes occurring directly below the city (Tokyo Metropolis, 1997). In the Tokyo metropolitan area, a large number of people are traveling for variety of purposes within a one and half hour's distance from home using rapid transit railways. In the event of a major earthquake, public transportation including buses is expected to be paralyzed and unavailable for transport, leaving an extremely large number of people in the city and with difficulty returning home. In the event of a major earthquake, automobiles including buses must stop on roadsides for preventing collateral disasters. Many roads are expected to be damaged and blocked by collapsed buildings and traffic signals will be turned off. The government assumes therefore that there is no alternative way, excepting walking, to get home. This is the same in areas where a dense network of bus route is available.

The spatio-temporal distribution of persons with difficulties returning home was estimated by anticipating which users remain without transportation home in the event of an earthquake based on the data contained in Tokyo Metropolis (1997), as shown in Figure 10. More specifically, the ability or difficulty in returning home was determined in accordance with the distance required to return home, which was set as the distance to the center coordinates of his/her residential zone (a spatial unit of PT data) based on the time-specific position coordinates of the railroad users calculated in the previous chapter. The time-specific position coordinates for low-use time periods are represented as the center coordinates of the residential zone. In actuality, the distance that a person can traverse by walking home differs depending on the time of earthquake occurrence and the degree of damage. However, at present an individual's evaluation of whether the return home is feasible cannot be anticipated. Therefore, a simple historical method is used. In addition, the figures in the current study focus only on railroad users, and address markedly fewer individuals than studies focusing on all persons who are outside their homes during a disaster. However, many of the individuals outside their homes that are traveling a farther distance are railroad users, and this estimate is important to include in disaster planning.

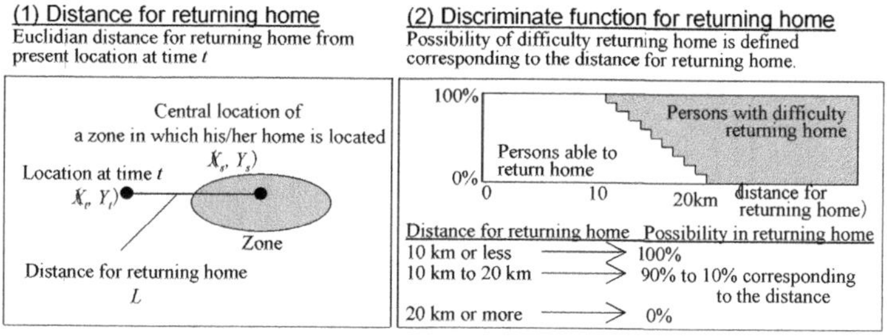

Fig. 10. Definition of persons with difficulties returning home

4.2 Persons with Difficulty Returning Home

The number of persons with difficulty returning home abruptly increases between 7:00 and 9:00, and gradually decreases from 18:00 to 24:00 (Figure 11, left panel), with a maximum (approximately 4.2 million users) reached at approximately 14:00. Riders at the peak account for approximately 45% of all railroad users who would have difficulty returning home in the event that an earthquake interrupts transportation. Figure 11 (right panel) shows that many people are in the process of traveling to a destination in the morning and evening. Particularly the evening, accommodations would need to be provided for travelers experiencing difficulty returning home. The spatial distribution (Figures 12 and 13) confirms that users that would experience difficulty returning home are distributed along railroad lines in the morning, and spread out from the city center to the west side during the day. At night, many remain in the city center.

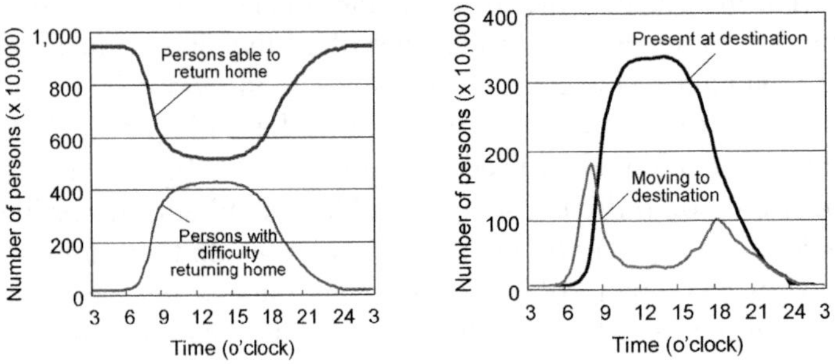

Fig. 11. Persons able to return home and persons with difficulty returning home, by movement segment

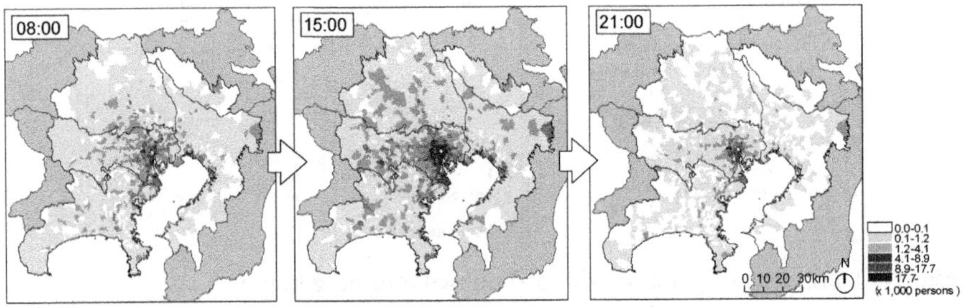

Fig. 12: Spatio-temporal distribution of persons with difficulty returning to homes in each sub-zone, by present location

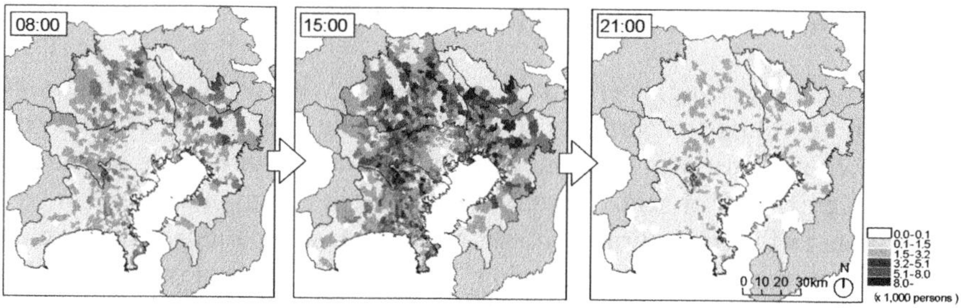

Fig. 13. Spatio-temporal distribution of persons with difficulty returning to homes in each sub-zone, by location of residence

4.3 Spatial Distribution by Personal Attributes

The percentage of persons with difficulty returning home in the age group 5–18, which includes pupils, children, and kindergartners, temporarily increases at approximately 8:00. During the day, the percentage of persons over age 80, housewives and househusbands, and the unemployed increases (Figures 14 and 15). In the event of an earthquake during the time period of commuting to school in the morning, emergency response to students, children, kindergartners, and other underage persons in the age group 5–18 will be crucial.

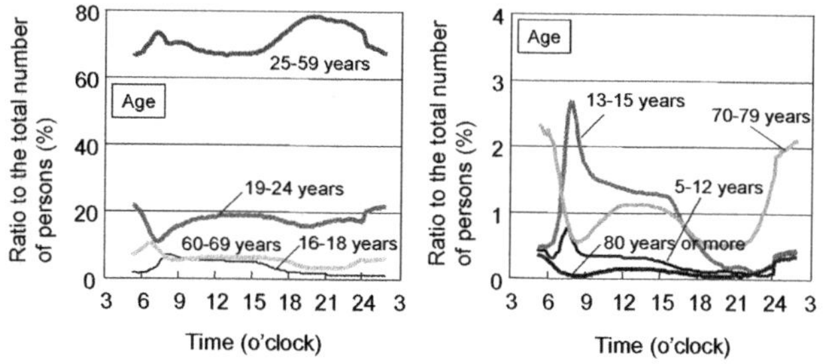

Fig. 14. Ratio of persons with difficulty returning home, by age

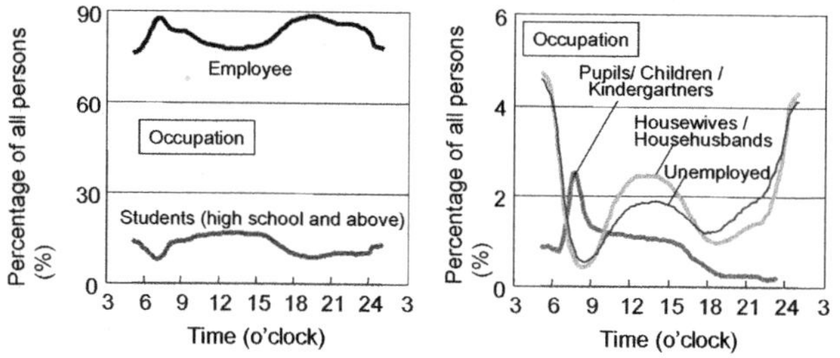

Fig. 15. Number of persons with difficulty returning home, by occupation

4.4 Spatial Distribution by Purpose for Travel

During the day, a high percentage of persons with difficulties returning home are traveling for purposes such as shopping (Figure 16). A high percentage of persons with difficulties returning home during the day are at their place of work or school. In contrast, at night a high percentage of persons with difficulty returning home are found at locations outside their home for purposes such as shopping. The spatial distribution (Figure 17) shows that persons with difficulty returning home and the purpose of commuting to work tend to concentrate in the city center, while persons with the purpose of commuting to school are distributed over a broad geographic range throughout the day. Persons who travel with a purpose, such as shopping, tend have a wide geographic distribution in the morning, and be concentrated in the city center in the evening. Although individuals that experiences difficulty in returning home are concentrated in the city center (Figure 12), when evaluated by purpose, distinguishing characteristics are evident.

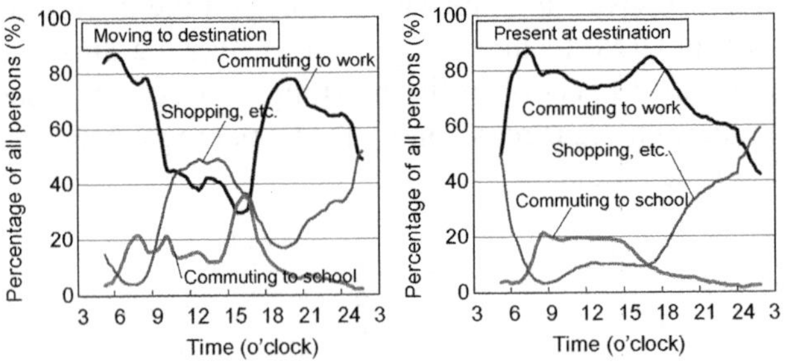

Fig. 16. Ratio of persons with difficulty returning home, by purpose

These distinguishing characteristics, such as the distribution over a wide geographic area vary with time. For example, persons with the purpose of commuting to kindergarten or school involve additional users because the young must be met by their parents or guardians, who may not be at the particular location at other times of the day. Similarly, persons with purposes such as shopping who do not belong to any organization may be traveling to different locations in particular time periods. Consequently, the need for emergency measures to assist individuals with difficulties returning home is not limited to the city core, but is spread throughout the metropolitan area depending on the time of day.

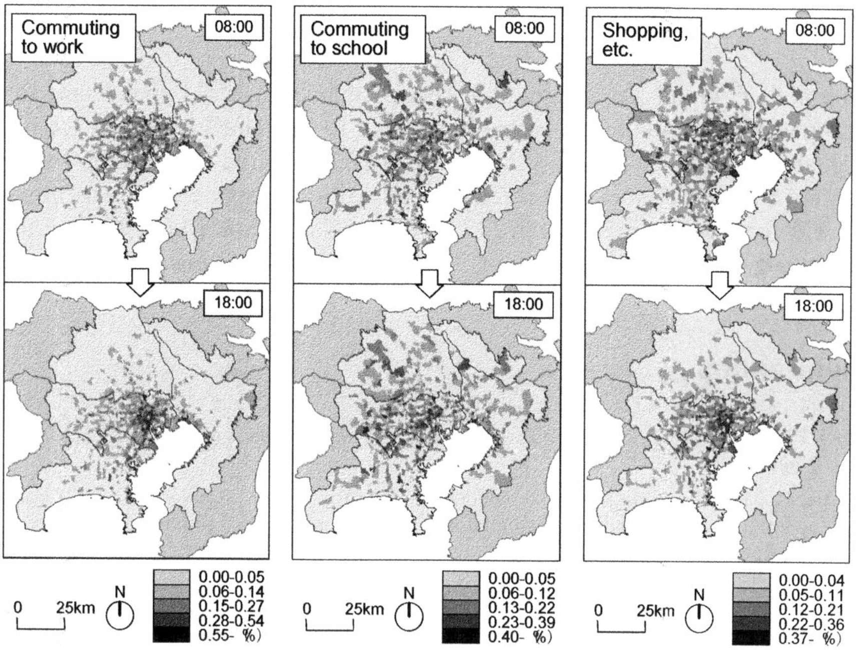

Fig. 17. Spatio-temporal distribution of persons with difficulty returning home, by zone and by purpose

5 Summary and Conclusions

We constructed a model for estimating the spatio-temporal distribution of railroad users using PT data, and developed an understanding of the time variation and the variation in location of railroad users by personal attributes and movement purposes. Using this model, it became possible to understand factors that previously could not be understood such as the potential for human damage, and the specific profile of persons affected by an earthquake, which vary greatly with time and location. By combining the profiles of affected persons and their spatio-temporal distribution, it is possible to plan for disaster assistance and to allow for more detailed disaster prevention planning suited to particular time periods and locations. As an application of the spatio-temporal distribution of railroad users estimated by the proposed model, the spatio-temporal distribution of persons with difficulty returning home was examined.

Acknowledgements

The author would like to give his special thanks to Ms. Ikuko Otani for computer-based numerical calculations, and to Mr. Hideki Oshima for his valuable comments at the 14th annual conference of the Geographic Information Systems Association in 2005. This work was sponsored by a Grant-in-Aid for Scientific Research (17310093). The author would like to acknowledge the valuable comments from anonymous referees.

References

Bovy P, Stern E (1990) Route choice: Wayfinding in transport networks. Kluwer Academic Publishers

Caspar C, Molin E, Arentze T, Hoogendoorn S, Timmermans H, Van Wee B (2007) Validation of a multimodal travel simulator with travel information provision, Transportation Research Part C 15: 191-207

Harada C, Iwakura S, Morichi S (2002) Analysis and modeling of commuters' departure time in urban railway network, The Infrastructure Planning Committee, JSCE, 26, CD-ROM

Institution for Transport Policy Studies (2003) Urban transport Yearbook, Institution for Transport Policy Studies

Iwakura S, Watanabe S, Doi A (2000) Development of peak-hour train speed model for dynamic commuter travel forecasting, The Infrastructure Planning Committee, JSCE, 17: 709-714

Morrichi S, Iwakura S, Morishige T, Itoh M, Hayasaki S (2001) Tokyo metropolitan rail network long-range plan for the 21st century, Transportation Research Board 80th Annual Meeting, Washington, D.C.

Toriumi S, Kawaguchi Y, Taguchi A (2008) Estimation of the damage of railway commuter traffic caused by forthcoming earthquake in Tokyo metropolitan area, Operations research as a management science research, 53-2: 111-118

Nakabayashi I (1985) Study for figuration of damage in a metropolis during an earthquake disaster (4) estimation of difficulty to arrive home from each office, Summaries of Technical Papers of Annual Meeting, Architectural Institute of Japan F: pp 361-362

Nakabayashi I (1992) Development of estimation method on obstructed homeward commuters after earthquake disaster, Comprehensive Urban Studies: pp 35-75

Taguchi A (2005) Time dependent traffic assignment model for commuter traffic in Tokyo metropolitan railway network, Transactions of the Operations Research Society of Japan, 48: 85-108

Tokyo Metropolis (1997) Report on the damage assessment of earthquake directly under Tokyo, Tokyo Metropolis

Rasterizing Census Geography: Definition and Optimization of a Regular Grid

Himanshu Mathur, Stefania Bertazzon

Department of Geography, University of Calgary,
2500 University Dr. NW, Calgary, AB, T2N 1N4, Canada
mathur.himanshu@gmail.com; bertazzs@ucalgary.ca

Abstract. The results of first and second order analyses tend to be affected by the characteristics of the spatial units at which data are sampled. This paper discusses the definition of a regular grid superimposed on a set of given irregular census units, and the subsequent redistribution of the census variables to the newly defined grid cells. A statistical criterion guides the definition and optimization of the grid: through an objective function, the method aims at preserving the global spatial autocorrelation measured for a salient variable on the original census units. Several aspects of the grid positioning and population redistribution are critically discussed. Ultimately, the proposed method constitutes a valuable alternative to the spatial heterogeneity that affects many empirical spatial data.

1 Introduction

Data sampled at different spatial units (e.g., postal code level or census tracts) are a common occurrence in the social sciences. The results of first and second order analyses tend to be affected by the characteristics of the spatial units at which data are sampled (spatial heterogeneity, or the varying shape and size of spatial units). The related aggregation/disaggregation problems have been widely researched under the heading of MAUP (Modifiable Areal Unit Problem); scale, the other critical factor in MAUP is one of the growing research areas in geography. Technical advances in geomatics have dramatically increased the availability of data on a raster support. Such technological progress is slowly being matched by conceptual and computational advances, which are allowing for increasingly sophisti-

M. Sester et al. (eds.), *Advances in GIScience*, Lecture Notes in Geoinformation and Cartography,
DOI 10.1007/978-3-642-00318-9_13, © Springer-Verlag Berlin Heidelberg 2009

cated spatial analyses in a raster data model. While, traditionally, only the physical and natural sciences have adhered to the raster paradigm, arguably, the whole range of geographical analysis applications is slowly migrating to a raster data model. The consequences of such a shift are likely to be significant for the social sciences and human geography research, which have traditionally depended on a vector model and developed a host of concepts and techniques within that model. The availability of satellite data for use in the social sciences, the ever present MAUP, and the computational advances in the analytical capabilities within the raster model are strong incentives to consider ways of "rasterizing" vector data, and specifically census data.

This study is part of a larger analysis of health and socio-economic variables in the city of Calgary (Gavrilova et al., 2005; Bertazzon and Olson, 2008). Health and medical geography analyses induce an additional problem, in that, due to confidentiality issues, medical data are typically released at the postal code level: the analyst is thus forced to "merge" census data and postal code data, making arbitrary choices which may bias their results. To maintain the generality of our conclusions, the present analysis is implemented on the "total population" variable, as recorded by Census Canada. The goal of the present work is to experiment with a regular grid, superimposed on the city, and to determine an "optimal" grid for first and second order analysis. The main contributions of this work are the definition of criteria to optimize the grid, the discussion of an array of experiments with variable grid size and shape, and the definition of guidelines to repeat the process for other cities and case studies.

Geographers have a long tradition of studying data for areal units. The best example would be studying spatial objects such as zones, regions or countries. As Chapman (1977) put it, "Geography has consistently and dismally failed to tackle its entitation problems, and in that more than anything else lies the root of so many of its problems" (page 7). The effect of the selection of areal units on analysis is termed the modifiable areal unit problem (MAUP). If relations between variables change with the selection of different areal units, the reliability of results is in question. The use of small areal units tends to provide unreliable rates because the population used to calculate the rate is small; conversely, larger areal units provide more stable rates but may mask meaningful geographic variation (Nakaya, 2000).

The MAUP had been most prominent in the analysis of socio-economic and epidemiological data (Openshaw and Alvandies, 1999; Wong et al., 1999; Nakaya, 2000). The effects of the MAUP can be divided into two components: the scale effect and the zonation effect (Amrhein 1995). The scale effect is the variation in numerical results that occurs due to the number of zones used in an analysis. For example, the difference in numerical results between mortality rates by municipality and health area in British Columbia is a scale effect. The zonation effect is the variation in numerical results arising from the grouping of small areas into larger units. For example, using Canadian census data, numerical differences in employment rates between census tract data and its enumeration area would be a zonation effect.

There exist a potentially infinite number of different options for aggregating the data. Openshaw (1984) calculated that if one was to attempt to aggregate 1,000 objects into 20 groups, you would be faced with 101,260 different solution combinations. Numerous administrative boundaries exist, such as enumeration districts (or dissemination areas), wards, counties, census tracts, etc. Regular, often square, grids are common, though polygons have been used in other studies of crime distribution (Hirschfield et al., 1997). Although there are a large number of different spatial objects and ways in which a large geographical area can be sub-divided, the choices of areal units tend to be dominated by what is available rather than what is best.

As a result, research into the MAUP has been primarily empirical, focusing on the effects of aggregation on various statistics computed from a specific dataset. For example, Openshaw and Taylor (1979) examine correlation coefficients using an Iowa electoral dataset; Fotheringham and Wong (1991) study multiple regression parameters using Buffalo census data; Amrhein and Reynolds (1996)—one of the papers in the special issue of *Geographical Systems* that focuses on the MAUP—and Amrhein and Reynolds (1997) study the effects of aggregation on univariate statistics and make a tentative link between a spatial statistic and the relative change in variance. Recognition of spatial patterns is a fundamental requirement for landscape ecology, and various spatial autocorrelation statistics, such as the Moran Coefficient, are often employed as a tool for this task (Jelinski and Wu, 1996; Qi and Wu, 1996); hence it is important to know how spatial statistics are affected by aggregation as well. Reynolds and Amrhein (1998) show that the variables with same Moran's Coefficient can have very different spatial structures, although the possibilities decrease as the Moran's Coefficient approaches the maximum allowed by the spatial structure.

Spatial dependence exists whenever a variable has a regular distribution over space and its value at a given location depends on values of the same variable at other locations. Autocorrelation statistics are basic descriptive statistics for any data that are ordered in a sequence (Griffith, 1984); they summarize information about how the values of one variable are arranged in space (Odland, 1988). Our focus will be on the neighbourhood view of spatial data, in which data are observed for a given discrete set of fixed locations. In the traditional approach to spatial autocorrelation, the overall pattern of dependence in the data is summarized into a single indicator, such as the Moran's I or Geary's c (Cliff and Ord, 1981; Haining, 1990; Getis, 2008).

Gehlke and Biehl (1934) observed that the size of the correlation coefficient increases with aggregation. They grouped 252 census tracts in Cleveland into larger units of approximately the same area with contiguity restrictions (tracts should be adjacent). They calculated the correlation between two sets of variables, male juvenile delinquency and median monthly income. Then they compared these results with random groupings of data without contiguity restrictions. Gehlke and Biehl (1934) found that random data aggregations have no systematic effect on correlations. They also demonstrated that the variation in the size of the correlation coefficient was related to the size of the units involved: the larger the size of the units, the greater the correlation coefficient. These findings raised more questions than they answered. They concluded that "a relatively high correlation might occur at census tracts level (which has larger sets of units), when the traits so studied were completely dissociated in the individuals or families of those traits". Their findings also raised the question: Is geographical area an entity possessing traits or merely one characteristic of a trait itself?

Yule and Kendall (1950) noted that any value of the correlation coefficient from 0 to 1 between wheat and potato yields can be produced merely by choosing an appropriate size of the unit of area for which the yields are measured. Any "real" correlation between the two variables under consideration (wheat and potato yields) is illusory.

Openshaw (1977) re-examined the effect of aggregation in the late 1970s. He is one of the first researchers to re-emphasize the importance of the aggregation problem and relative insignificance of the scale problem. He compared scale and aggregation effects on the correlation between the number of early- and mid-Victorian houses in South Shields. The conclusion that he drew is that Yule and Kendall (1950) were correct, but that they had clearly underestimated the severity of the problem. Different variables can be affected by aggregation in different ways. It is not that all the variables show an increase in magnitude of correlation coefficient as scale changes; some may remain constant while some decrease.

A recent interesting implementation is the SPARTACUS (System for Planning and Research in Towns and Cities for Urban Sustainability) project, initiated in 1996 by a consortium of regional planning agencies and consultants in Europe, to analyze the implications of urban land use and transportation policies. The disaggregation process begins with zone-level household and employment data for the base year and for the forecast year for all policy scenarios tested. If no information on the distribution of population and employment within the zone is available, data can be allocated equally to each grid cell within the zone (represented in the GIS as a polygon). If some information on the distribution within the zone is available, the disaggregation can be performed with greater accuracy. This is done by assigning a weight to each grid cell, which is used to allocate total population or employment proportionally. If sufficient data are available, population can be given different weights by socioeconomic group. Disaggregation of data in the SPARTACUS was performed using microsimulation techniques. Each single activity (such as place of residence or work) was assigned a raster cell by applying Monte Carlo Simulation.

2 Background and Case Study

The case study for this implementation is the city of Calgary, which is largely characterized by a regular street network (Manhattan grid); however, the natural features and topography (rivers, hillsides), the presence of "anomalies" (reservoirs, parks, commercial areas), and the urban development of the past years (cul-de-sacs, crescents) represent important departures from the grid structure. These features, coupled with the land use and age of urban development in the city, also result in a variable unit size, even where regular structures are apparent. There are 1321 dissemination areas and 181 census tracts (Fig. 1) within Calgary for which the population is published.

The settlement pattern in the city can be thought of as shaped by a series of overlaying structures, some being irregular grids, some others irregular grids deriving from a regular grid with some departures: topography and hydrology (including parks and reservoirs, Indian reserves, airport, etc.) present an irregular pattern; townships and ranges present the most regular pattern; land use parcels present substantial departures from a regular pattern; postal codes (both at the 3-digit scale and at the 6-digit scale) present an irregular pattern departing from a regular design; likewise is the geometry of the census units at their various sales (subdivisions, tracts, enumeration areas).

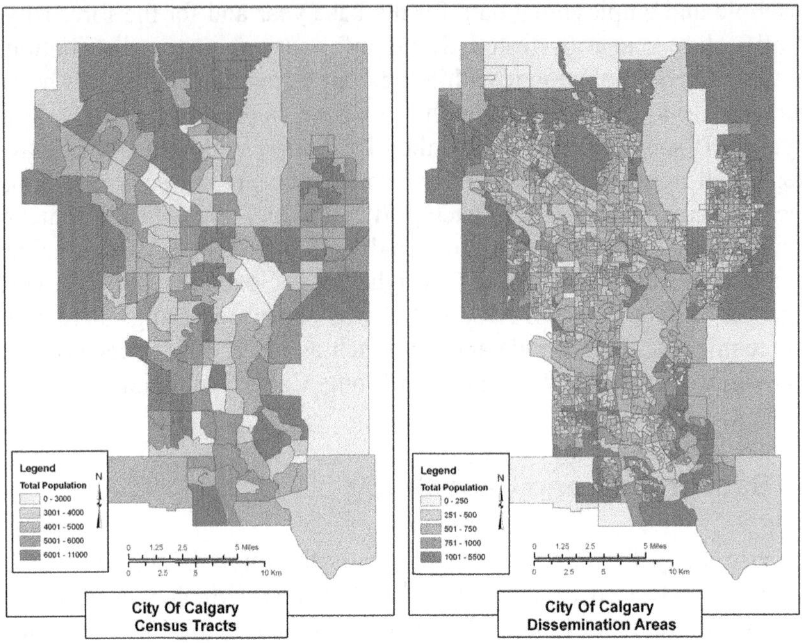

Fig. 1. Calgary population, 2001: dissemination areas and census tracts

3 Methodology

Aggregated demographic datasets are associated with analytical and cartographic problems due to the arbitrary nature of areal unit partitioning. This paper describes a methodology for generating a grid-based representation of population that is aimed at mitigating these problems. This grid-based technique is termed as rasterizing the census geography. This technique preserves the global spatial autocorrelation of the dataset, which means that ideally the association of nearest neighbours across each grid cells does not change globally; realistically however, its changes are minimized. It will never be possible to fully reconstruct the detail of the spatial structure from the aggregated census data, but some spatial disaggregation should be possible either by the use of data collected for grid cells or by interpolating from the population-weighted centroid locations available for census dissemination areas. The raster (grid cell) representation produced in this way is implicitly a density surface of the variable concerned.

The rasterization process developed in this project was carried out in ArcView 3.2. Two scripts written in Avenue (for ArcView 3.2) were used to carry out the rasterization process: Grid Maker, an avenue script used in ArcView 3.2 to generate grid(s) of varying sizes (Lascov, 2005); and Data Partitioner, used to calculate the population for a designed polygon in a boundary theme (Dou, 2002).

In this project, a method to cope with the problem is proposed, which is based on manipulating the newly created spatial units in order to preserve as much as possible the spatial dependence observed in the original spatial units. This manipulation takes the form of determining the sample grid size of the population distribution using permutations of the data that preserve the global spatial autocorrelation in the data. This approach adjusts the sample grid size (N x M) of the census population data so that the adjusted value (N x M) measures the "equivalent" amount of independent information in the sample. The proposed approach aims at mending/reducing the MAUP by using a non-arbitrary grid and implementing a "guided" selection of the square unit area or grid size. It also tries to explore a possible solution to this aggregation/disaggregation problem by the approach of conserving the global spatial autocorrelation of the variable under consideration (total population).

3.1 Grid Definition

The geometric shape of the city of Calgary (Canada) can be approximated by a rectangle, with edges roughly orthogonal to the cardinal directions, and a length of about 30 km in the N–S direction and about 25 km in the E–W direction (Fig. 1). The grid definition therefore begins by drawing one such rectangle over the city. The rectangle is subsequently subdivided into experimental arrays of grid cells varying in number (size) and dimension. All the arrays are constituted of rectangular cells. Whereas different shapes could theoretically have been tested[1], the patterns of the underlying Calgary geometries—whenever some regularities can be detected—are fundamentally rectangular (Section 2).

[1] Obvious candidates would be hexagonal or triangular shapes, owing to their ability for completely filling the space. Other possible choices might be circular or elliptic units.

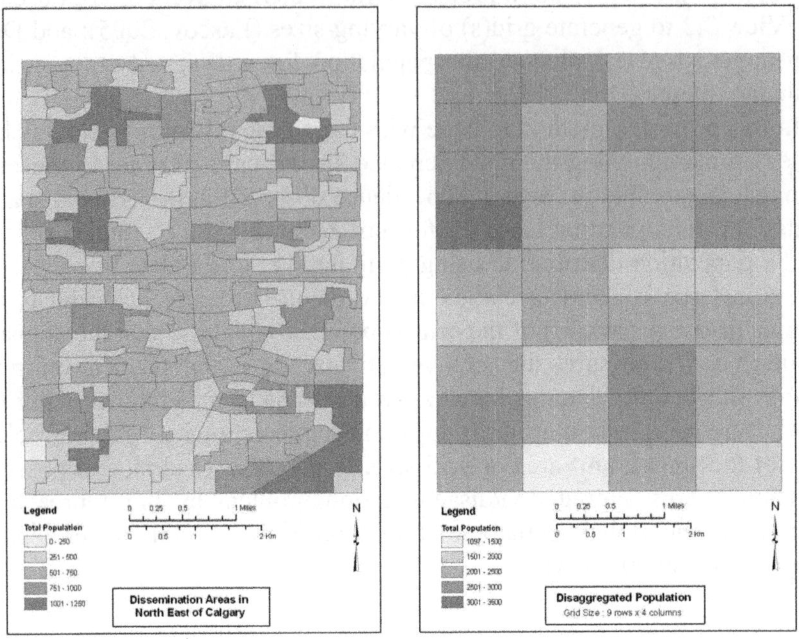

Fig. 2. Example of rasterization of a northeast area

Rasterizing the population can be intuitively understood by comparing the two images in Fig. 2, which presents the census units in one of the more regular parts of the city, in an area in the northeast of Calgary.

Before proceeding to the application of the statistical criterion, a few geometric refinements shall be discussed: the main steps are: land use filter; cell size experiments; population redistribution; new database creation; robustness to positional shift.

Land-Use Filter

The census data is overlaid on the land-use map, and only areas designated as residential land-use are selected. Land cover information, derived from the land use map of Calgary (Fig. 3) was used to filter out non-populated zones from the set of census dissemination areas.

The land-use categories were broadly grouped into "residential" and "non-residential". The group of parcel defined as residential included the classes: Residential – High Density, Residential – Medium Density, Residential – Low Density. The grid definition and population redistribution phases are only performed on this set of parcels, and no distinction is made among the various density classifications.

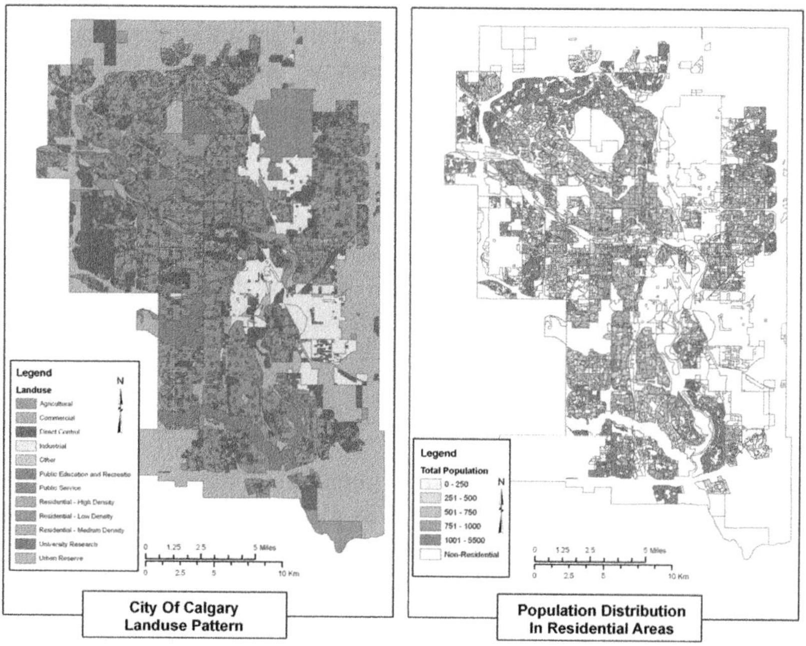

Fig. 3. Land use map and filtered population distribution

Grid Cell Overlay

Grids containing cells of various numbers and dimensions were experimentally overlain on the study area. Even though the grid must, to an extent, be arbitrary, the following criteria were used to guide the experiments:

1. The number of cells in a grid should be approximately the same as the number of the given dissemination areas. This criterion produces several rectangular grids of various dimensions:

$$Nr * Nc = Total\ No.\ of\ cells \tag{1}$$

Where:

Nr = Number of rows,
Nc = Number of columns

A close examination of the grids thus obtained reveals that, following this criterion, the spatial autocorrelation for the regular grid tends to be higher than in the given spatial units. This is due to the fact that the redistribution of population within the grid-cells smoothes the overall distribution of population over the grid (Openshaw, 1988).

2. The grid size can also be determined by visually analyzing the study area and matching the boundaries of the original dissemination area with the newly formed grid-like structure. By following this method, grids of different cell sizes should be overlain on different parts of the city. The provision of a flexible grid that dynamically mimics the different parts of the city is an ambitious goal, beyond the scope of this work, and will be discussed as a future line of research.

Population Redistribution

After overlaying the grid on the study area, the population of each cell of the newly formed grid is calculated. The algorithm used for redistributing population among newly formed grid cells uses the formula:

$$Pop_n = Area_n * (Pop_i / Area_i) \tag{2}$$

Where:

Pop_n	=	Population of the newly formed grid cells
Pop_i	=	Population of the original dissemination areas
$Area_n$	=	Area of the newly formed grid cells
$Area_i$	=	original dissemination areas

In this experiment the *data partitioner* tool is used to divide the population among newly formed spatial units, according to Eq. 2. An alternative method was considered, that would take into account the population density of the whole study area. According to this method, the population density is multiplied by the area of each grid cell to obtain the new population distribution for each cell. The previous method is preferred over this one, because the latter distributes the whole population homogeneously over the study area which might give inconsistent results.

Database Creation

A new database is created after the population is divided among the grid cells. This database consists of the following fields:

1. Area ID: Unique area code for each of the newly formed dissemination areas.
2. Area: Total area of the grid cell (depends upon the grid size).
3. Easting: Denotes easting of the centroid of the dissemination area.
4. Northing: Denotes northing of the centroid of the dissemination area.
5. Population: The new population for each dissemination area.

Grid Refinement: Removal of Irrelevant Cells

As exemplified in Fig. 3, defining a regular grid over the city results implies—regardless of the grid shape and size—the definition of a series of empty cells, i.e., cells whose entire area is not classified as residential. These cells are simply removed by the grid, and excluded from the statistics calculation. A more complex situation occurs in all cases when a cell is only partially classified as residential. For these problematic cells, two different methods are applied and integrated. The first method is called Density of Population (DoP), the second one Area inside Polygon (AiP).
The most straightforward method is the AiP. A threshold portion of the cell is defined; a threshold of 25% was chosen. Consequently, cells that are classified as residential for less than 25% of their area are removed, and their population is distributed evenly among the neighbouring cells. However, this method causes cells containing a small but densely populated area to be removed, but this would imply the loss of a large portion of population; conversely, large residential areas within a cell do not necessarily imply high population. In order to correct for the deficiencies of the AiP method, the DoP method is defined as follows. A test grid is overlain on the city; using the data partitioner tool according to the formula in Eq. 2, the population is redistributed from census units to grid cells. Cells with 0 population are removed. A population threshold is set depending on the grid size; for example, for a 20 x 20 grid, a hypothetical threshold of 2% of the total population can be considered. Consequently, all cells with population <2% are marked for possible deletion. At this stage, the two methods (AiP and DoP) are integrated, and only cells whose population and residential area are below their respective thresholds are removed.

Given a cell c_{ij},

$$\text{if} \quad RA(c_{ij}) < 25\% \quad \text{and} \quad TP(c_{ij}) < 2\% \quad \Rightarrow \quad c_{ij} \text{ is removed.}$$

Where:

$RA(c_{ij})$ is the residential area of cell c_{ij},
$TP(c_{ij})$ is the total population of cell c_{ij}.

In light of the arbitrariness of these thresholds, both thresholds were set to 0, so that no population is lost, and only cells with 0 population are removed.

3.2 Calculation of Moran's I

Since the goal of this work is the provision of a flexible tool for socio-economic analysis, and the focus is on the spatial relationships of socio-economic variables, the main criterion is the convergence of analytical results obtained at the dissemination area level, and analyses conducted at the "new grid" level. The reference variable for the implementation of this criterion is the *total population* (total number of residents, for each spatial unit) and the relationship of interest is spatial dependence.
The software used to calculate Moran's I is GEODA. In this, the inputs are a shape file and a weight matrix. The weights are calculated by the queen contiguity method, which computes nearest neighbour in such a way that all cells with a common edge or corner are defined as neighbours. Moran's I for the original dissemination area is calculated as I_0. For the population distribution at the dissemination area level, the value of the Moran's coefficient is 0.3725, while at the census tract its value is 0.1108, using the parameters just defined. The values of Moran's I for different grids are calculated as $I_{G(m \times n)}$., where m and n represent the numbers of rows and columns of each grid G. A residual is calculated as in Eq. 3.

$$\varepsilon = (I_0 - I_{G(m \times n)}) \tag{3}$$

A residual plot represented al for all computed values of $G(m \times n)$ represents the gradual change in the residual for varying numbers of rows and columns, after each iteration of the rasterization and population redistribution processes. The grid size that produces the lowest residual value is selected as the best grid.

Robustness to Positional Shift

Once a set of best grid candidates has been identified according to the objective function, a final test is conducted on the set of best candidates, to assess their robustness to positional movement. Each grid is tested by moving it along the horizontal and vertical axes in the four cardinal directions (east, west, north, south) and re-computing their absolute residuals. By analyzing the response of absolute residuals to the movement of the grid, a measure of robustness for each grid is formulated.

4 Results and Discussion

Grid sizes ranging from 4 rows by 4 columns to 60 rows by 60 columns were tested in the objective function (Eq. 3), for the city of Calgary, at the dissemination area level and at the census tract level, respectively. These grids are denominated "gross grids" as they contain all the cells in the bounding rectangle, including those with 0 population. Subsequently, each of these grids is refined according to the procedures described in 0, in order to remove cells with no residential areas and cells with 0 population: this operation results in a set of grids, denominated "net grids". For each m x n pair, the number of cells in the net grid (valid cells) is always lower than the number of cells in the gross grid.

At the DA level (1372 original units), grids ranging from 14 x 10 to 66 x 50 were tested, resulting in a set of grids ranging from 140 to 3,300 cells. Moran's I was calculated only for the valid cells of each net grid and for each grid the absolute residuals were computed. Fig. 4 summarizes the results for a sample of cell sizes and dimensions.

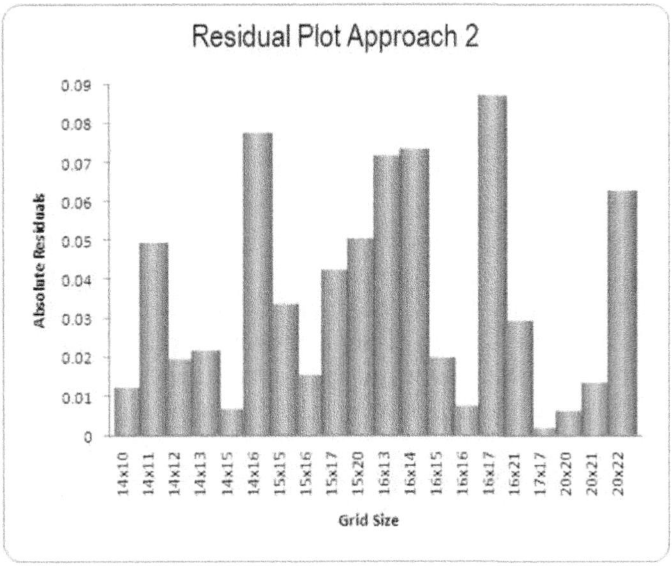

Fig. 4. Absolute residuals as a function of cell size and dimension, DA level

As shown in Fig. 4, the lowest residuals are displayed by grids of size 14 x 15, 16 x 16, 17 x 17, 20 x 20 and 27 x 20. Even though these are empirical results, a 2-dimensional representation of the residuals confirms that a depression, which may be interpreted as a "local minimum" is present in the area around these values. It may be worth noting that most of the best candidates are "square grids", i.e., grids where the number of rows equals the number of columns: grids of size 14 x 15[2], 16 x 16, 17 x 17 and 20 x 20. The single observed anomaly to this pattern is the 20 x 27 grid, which, in addition to departing from the square dimensions, contains a number of cells considerably higher than the grids in the square set. It may also be observed that the finer grids produce the highest residuals, due to an oversmoothing effect that increases the spatial dependence in the regular distribution, thus departing from its observed values and increasing the absolute residuals. The best candidates at this scale produce regular patterns ranging between 210 and 400 cells[3].

At the census tract level (182 original units), grids ranging from 5 x 6 to 50 x 50 cells were tested, resulting in regular sets ranging from 30 to 2500 cells. The best candidates at this scale are grids of dimensions 6 x 12, 6 x 14, 7 x 6, 8 x 6, 9 x 7 and 14 x 6. The regular sets produced by the best candidates range between 72 and 84. This set of candidates appears to be less regularly distributed than the set for the dissemination area level, and to produce a less clear depression, failing to indicate a distinct "local minimum".

It may be inferred that at both scales, the best candidates produce a range of much fewer spatial units than the original patterns: a range of 210–400 units compared to the original 1372 at the dissemination area level, and a range of 72–84 units compared to the original 182 at the census tract level. One important, consequent observation is the best range at the dissemination area level is very close, in number of units, to the number of census tract units. This result shall be explored further, but it appears that the methodology applied at the dissemination area level produces not only a distinct local minimum, within a range of consistently square grids, but also a range of number of units whose lower limit coincides almost exactly with the number of units at the census tract level (210 vs. 181). This result suggests that the grid derived at the dissemination area level may produce results that are meaningful also at the census tract level, or at least comparable with analyses at that scale. For all these reasons, the following discussion will be limited to the grids derived at the dissemination area level. Table 1 summarizes the characteristics of the best candidate grids at this scale.

---

[2] This is very close to a square grid.
[3] Not considering the 20 x 27 candidate.

Table 1. Best candidate grids, DA level

Grid size	Gross grid cells	Net grid cells	Computed Moran's I	Residuals I(DA)=0.3725
14x10	140	89	0.3601	0.0124
14x15	210	125	0.3793	-0.0068
15x16	240	137	0.3880	-0.0155
16x16	256	151	0.3801	-0.0076
17x17	289	169	0.3705	0.0020
20x20	400	213	0.3789	-0.0064

As shown in Table 1, all these grids present very low absolute residual values; as such all of them constitute good candidates, and none appears clearly superior to the other ones. The final test of robustness to positional shift was conducted on this set of candidates. The set of grids in Table 1 was therefore tested by moving each of them along the horizontal and vertical axes by 1 km in each of the four cardinal directions and re-computing their absolute residuals, as shown in Table 2.

Table 2. Positional robustness of the grids

Grid size	Horizontal Shift		Vertical Shift		Average Moran's I	Average Residual
	West direction	East direction	North direction	South direction		
14x10	0.3268	0.0918	0.3036	0.1910	0.2283	0.1442
14x15	0.3815	0.3054	0.2628	0.2418	0.2979	0.0746
15x16	0.3149	0.2237	0.3134	0.3914	0.3109	0.0617
16x16	0.3741	0.3208	0.3732	0.4352	0.3758	0.0033
17x17	0.3820	0.4366	0.3588	0.3921	0.3924	0.0199
20x20	0.4601	0.4278	0.4322	0.4985	0.4547	0.0822
20x21	0.4591	0.4091	0.4536	0.4437	0.4414	0.0689

Even though the 17 x 17 grid presents the lowest absolute residual (Table 1), the 16 x 16 grid appears the most robust to positional movement. While the residuals in Table 1 are all very close in magnitude, the robustness test shows a significant superiority (by approximately 10 orders of magnitude) of the 16 x 16 grid over the other candidates. Additional robustness tests were conducted for the entire set of calculated grids, and for smaller movements, as shown in Fig. 5.

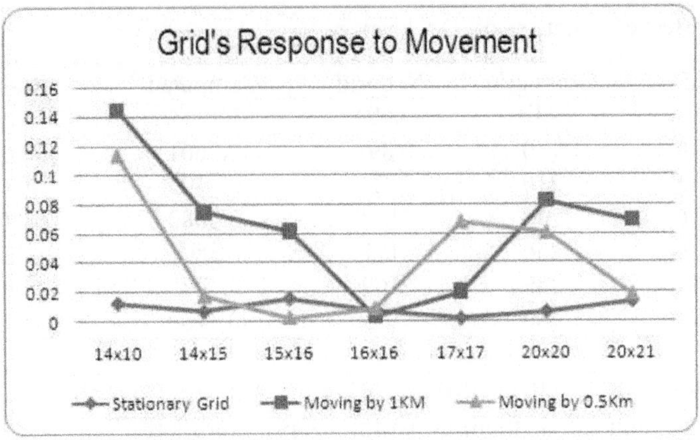

Fig. 5. Robustness test for several grids and movement magnitudes

The summary shown plot in Fig. 5 corroborates the conclusion that even if the 17 x 17 grid may have the lowest absolute residual, it is the 16 x 16 that is more robust to positional shifting. Given the unavoidable arbitrariness of the origin of the overlaid grid, it can be concluded that the 16 x 16 partition, formed by 151 cells, is the most reliable best-fitting approximation of the population distribution. The grid, and the consequent population redistribution, is shown in Fig. 6.

The regular pattern represented in Fig. 6 provides a new database that can be used for statistical analysis on the census variables.

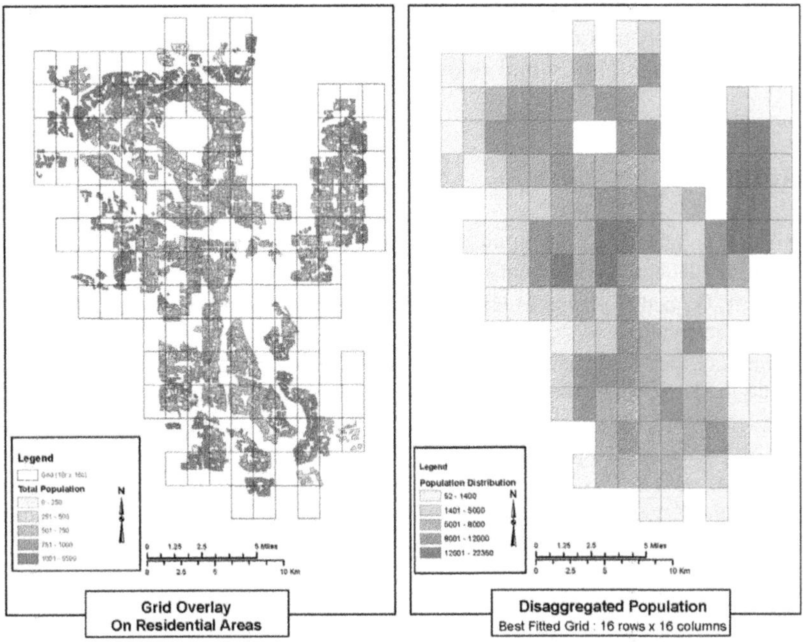

Fig. 6. Best-fitting grid

5 Conclusion

This work offers an alternative to the spatial heterogeneity that often affects spatial data. The contribution of this work consists of the development of a methodology for the rasterization of census units, which can be optimized according to some set criteria. The construction of an optimal grid may offer a practical answer to many analytical and planning problems. Additionally, in the process of constructing the grid, the population distribution is split into pixels, which provides a homogeneous raster to merge and analyze census data and remotely sensed data. In this perspective, a complementary approach to this project may be the use of remotely sensed imagery and its integration with census data. In this case, the pixel size is given and the target grid known; it is the population redistribution that would follow the main lines of the work discussed here.

An important line of future enquiry that will complement this work requires a broad range of analysis on the constructed grid. Specifically: first and second order analysis shall be conducted, all else being equal, on the original census data and on the constructed grid. A careful comparison of the statistical properties and the conceptual meaning of both analyses will aid assessing the validity and relevance of the proposed approach.

References

Amrhein, C.G. (1995). Searching for the elusive aggregation effect: Evidence from statistical simulations. Environment & Planning A, 27(1):105.

Amrhein, C.G., and H. Reynolds (1996). Using spatial statistics to assess aggregation effects. Geographical Systems, 3:143–158.

Amrhein, C.G., and H. Reynolds (1997). Using the Getis statistic to explore aggregation effects in Metropolitan Toronto census data. The Canadian Geographer, 41(2):137–149.

Anselin, L. (1994). Exploratory spatial data analysis and Geographic Information Systems. In M. Painho, ed., New Tools for Spatial Analysis, pp. 45–54. Luxembourg: EuroStat.

Bertazzon, S., and S. Olson (2008) Alternative Distance Metrics for Enhanced Reliability of Spatial Regression Analysis of Health Data. Gervasi et al. (Eds.): Lecture Notes in Computer Science 5072, Proceedings of the International Conference on Computational Science and its Applications, Part I, pp. 361–374.

Chapman, G.P. (1977). Human and Environmental Systems: A Geographer's Appraisal. New York: Academic Press.

Cliff, A.D., and J.K. Ord (1981). Spatial Processes: Models and Applications. London: Pion Ltd.

Dou, J. (2002). ArcScript "Data Partitioner". ESRI Script Library.

Fotheringham, A.S., and D.W.S. Wong (1991). The Modifiable Areal Unit Problem in multivariate statistical analysis. Environment and Planning A, 23:1025–1045.

Gavrilova, M., S. Bertazzon, A. Hoang, and H. Mathur (2005). Dynamic grid-based approach for analysis of multi-dimensional data. In Proceedings of the 4th ISPRS Workshop on Dynamic and Multidimensional GIS (DMGIS'05). GMVAG 2005, London, UK, ISPRS Press, Vol. XXXVI, Part 2, W29, pp. 28–36.

Gehlke, C., and K. Biehl (1934). Certain effects of grouping upon the size of the correlation coefficient in census tract material. Journal of American Statistical Association, 29:169–170.

Getis, A. (2008). A history of the concept of spatial autocorrelation: A geographer's perspective. Geographical Analysis, 40(3):297–309.

Griffith, D.A. (1984). Reexamining the question "Are locations unique?" Progress in Human Geography, 8:82–95.

Haining, R. (1990). Spatial Data Analysis in the Social and Environmental Sciences.

Hirschfield, A. (1997). Crime pattern analysis, spatial targeting and GIS: Evaluating the Safer Merseyside Partnership Initiative. International Seminar on the Mapping and Analysis of Geo-referenced Crime Data.

Jelinski, D.E., and J. Wu (1996). The modifiable area unit problem and implications for landscape ecology. Landscape Ecology, 11(3):129–140.

Lascov, O. (2005). ArcScript "Grid Maker". ESRI Scripts Library.

Nakaya, T. (2000). An information statistical approach to the modifiable areal unit problem in incidence rate maps. Environment & Planning A, 32(1).

Odland, J. (1988). Spatial Autocorrelation. Newbury Park: Sage Publications.

Openshaw, S., and P.J. Taylor P.J. (1979). A million or so correlation coefficients: Three experiments on the modifiable areal unit problem. In N. Wrigley, ed., Statistical Methods in Spatial Sciences, pp. 127– 144. London: Pion.

Openshaw, A. (1984). The Modifiable Areal Unit Problem, CATMOG 38. GeoAbstracts, Norwich.

Openshaw, S., and S. Alvanides (1999). Zone design for planning and policy analysis. In J. Stillwell, S. Geertman, and S. Openshaw, eds., Geographical Information and Planning, pp. 299–315. Berlin: Springer.

Openshaw, S. (1977). A geographical solution to scale and aggregation problems in region building, partitioning, and spatial modeling. Transactions of the Institute of British Geographers, 2 (new series):459–472.

Qi, Y., and J. Wu (1996). Effects of changing resolution on the results of landscape pattern analysis using spatial autocorrelation indices. Landscape Ecology, 11(1):39–49.

Reynolds, H., and C.G. Amrhein (1998). Some effects of spatial aggregation on multivariate regression parameters. In D. Griffith, C. Amrhein, and J.-M. Huriot, eds., Econometric Advances in Spatial Modelling and Methodology: Essays in Honour of Jean Paelinck. Dordrecht: Kluwer.

Wong, D.W.S., H.A. Lasus, and R.F. Falk (1999). Exploring the variability of segregation index D with scale and zonal systems: An analysis of thirty U.S. cities. Environment and Planning A, 31:507–522.

Yule, G., and M. Kendall (1950). An Introduction to the Theory of Statistics. New York: Charles Griffin and Co., Ltd.

Towards Semantic Interpretation of Movement Behavior

Miriam Baglioni[1], José Antônio Fernandes de Macêdo[2,3], Chiara Renso[4], Roberto Trasarti[4], Monica Wachowicz[5]

[1] Department of Computer Science, University of Pisa, Italy
baglioni@di.unipi.it
[2] Database Laboratory, École Polytechnique Fédéral de Lausanne, Switzerland
jose.macedo@epfl.ch
[3] Department of Computer Science, Federal University of Ceará, Brazil
jose.macedo@lia.ufc.br
[4] KDDLAB, ISTI-CNR, Pisa, Italy
{chiara.renso, roberto.trasarti}@isti.cnr.it
[5] TechnicalUniversity of Madrid, Spain
m.wachowicz@topografia.upm.es

Abstract. In this paper we aim at providing a model for the conceptual representation and deductive reasoning of trajectory patterns obtained from mining raw trajectories. This has been achieved by means of a semantic enrichment process, where raw trajectories are enhanced with semantic information and integrated with geographical knowledge encoded in an ontology. The reasoning mechanisms provided by the chosen ontology formalism are exploited to accomplish a further semantic enrichment step that gives a possible interpretation of discovered patterns in terms of movement behaviour. A sketch of the realised system, called Athena, is given, along with some examples to demonstrate the feasibility of the approach.

1 Introduction

A flood of data related to moving objects is available today, and it will expand in the near future, particularly due to the automated collection of positioning data from mobile phones and other location-aware devices. This

M. Sester et al. (eds.), *Advances in GIScience*, Lecture Notes in Geoinformation and Cartography,
DOI 10.1007/978-3-642-00318-9_14,

flood of data enables a novel class of applications, where the discovery of consumable, concise, and applicable knowledge is a key step. The presence of a large number of location-aware wirelessly connected mobile devices presents a growing possibility to devise location based services exploring spatio-temporal data representing the footprints of moving objects, which we call raw trajectories. They represent the spatial track of a moving object in a fixed time interval.

There are currently interesting practical opportunities for exploring various kinds of spatio-temporal relationships, including trajectory patterns. From the point of view of the application, trajectory patterns can be considered as the spatio-temporal evidence of movement behaviour. For example, in a traffic management application, a trajectory pattern could represent a congestion or could disclose a frequent origin-destination path of working activities. Clearly, the discovery, representation and analysis of trajectory patterns challenge the research community with respect to methods for aggregating, generalising and explaining those patterns. In this scenario, data mining techniques play a fundamental role due to the fact that data mining is the application of specific algorithms for extracting patterns from data. However, data mining algorithms recently developed for mining raw trajectories have mainly produced trajectory patterns which are difficult to be interpreted in an application domain. The main problem has been the difficulties to correlate those patterns with movement behaviour in order to improve our knowledge, for example how to avoid congestion or making things easy for pedestrians.

Movement behaviour is a particularly complex process, since several factors can affect the movement itself, including the nature of the moving object, its motivation and the geographic environment where the object moves through. Our main research premise is that a standard pattern extraction algorithm, despite being highly specialised, cannot cope with such a complexity in terms of trajectory patterns interpretation, understanding and evaluation. This paper proposes the development of ontologies to associate semantics to raw trajectory data, providing a unique definition to various kinds of movement behaviour. Indeed, in this context, we propose an ontology-based approach for the semantic interpretation of trajectory patterns as movement behaviour. Indeed, as defined by Dodge et al. in (Dodge et al., 2008), a movement behaviour depends on the context since each movement takes a particular meaning when happens in a geographical environment. For example, a specific trajectory pattern showing a group of trajectories moving slowly in a city may be identified as a lane forming due to a traffic light or a traffic jam, depending on the contextual geographical knowledge (for example if the moving objects are moving in dense populated neighbourhood, or on a motorway).

The basic idea is to define an approach to incrementally enriching raw trajectory data with context geoinformation. The first step is the definition of semantic trajectory as sequence of stops (movement suspension) and moves (the actual movement). A further enrichment step exploits the knowledge capabilities provided by the ontology to integrate context geoinformation with the semantic trajectory patterns. Eventually, a reasoning step allows to infer new knowledge (e.g. the inferred movement behaviour) that can be added back to the ontology. For example, semantic trajectories patterns showing that typically stops are happening in tourist places can be associated with the movement of tourists. In contrast, semantic trajectory patterns stopping at offices in the morning and residential areas in the evening can be identified as a home-work routing behaviour.

The remainder of this paper is organised as follows: Section 2 describes our main contribution by proposing a semantic enrichment process. Section 3 recalls a concise definition of ontology. In Section 4 semantic trajectories are introduced and their conceptual representation. Section 5 discusses the representation of some mining patterns over semantic trajectories. Section 6 introduces the semantic trajectory patterns integrated ontology that gives the basic reasoning structure for the extracted models. Section 7 sketches some details of the experimental system we set up to test the feasibility of the introduced approach. Section 8 presents some related work. Finally, Section 9 draws some conclusions and open issues.

2 Semantic Enrichment Process

The analysis of raw trajectories strongly relies on the contextual geographical knowledge. For example, the analysis of urban movement can result in meaningless results if each raw trajectory is not associated to the moving object (e.g. car, dog, person, etc), features in the geographical space (e.g. building, street segments, hospital, etc), landmarks (e.g. Eiffel Tower, Everest, etc) and events (e.g. football matches, accidents, etc). The conceptual representation of a trajectory via its association to several semantic concepts is essential to provide the means for disclosing valuable information about movement behaviour.

From a user perspective, representing trajectories using geographical knowledge increases their understanding, but also helps users to query about movement behaviour using their own terminology. For example, it is more natural to a urban planner write the query: "*Give me all tourist activities.*", than formulate some complex query that expresses the meaning of tourist activities. Let's assume, for example, that tourist activities are identified by trajectories having tourist behaviour as people stopping in tourist places (such as Museums) and stopping also in accommodation places. Based on this definition, the following query should be issued: "*Give me all trajectories that contains stops located at a tourist place and also contains stops in accommodation places* .". In the former query the concept *tourist activities* is a main behaviour concept while in the latter query this concept is implicit in the query. Furthermore, the query itself can refer to concepts (tourist place, accommodation place) that are the "semantic generalisation" of "atomic" concepts stored in a knowledge base (museums, hotels, ...). The corresponding query could be rephrased as *"Give me all trajectories which stop their movement inside a museum or near a monument (o any other geographical object defined as tourist place) and stop their movement inside a Hotel or a B&B (o any other geographical place defined as accommodation place)"*. It is clear that going from such kind of query on raw trajectory data to the former desired query "give me all tourist activities", an incremental process should have built where semantics and reasoning play a central role.

The research challenge relies on developing an approach that employs formal ontologies to enable an enrichment process that augments the semantics of both raw trajectory data and mined trajectory patterns.

Ontologies have certainly become a research topic in several disciplines, ranging from philosophy, geography, geomatics up to machine learning and artificial intelligence. The definition given by (Gruber, 2008) is used to define ontology as "a technical term denoting an artifact that is *designed* for a purpose, which is to enable the modeling of knowledge about *some* domain, real or imagined". Such ontologies determine what can be represented and what can be inferred about a given domain, using a specific formalism of concepts. The main objective of our current research is to exploit a process that can give the user a semantic interpretation of trajectory patterns in terms of movement behavior. The process we propose is based on the abstraction of a semantic model of trajectories, the application of data mining algorithms on them for generating trajectory patterns, and a final further enrichment step that consists of exploiting an ontology reasoning engine for inferring movement behavior. Figure 1 illustrates the flow of such an enrichment process.

An important added value of relying on an ontology formalism is that it comes with an embedded reasoning engine that gives an increased expressive power for movement behavior interpretation, putting together the different levels of knowledge of an application domain. This is further discussed in the next section.

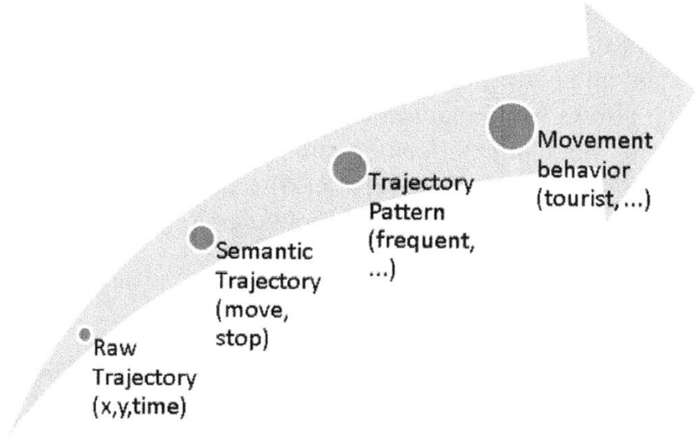

Fig. 1. The trajectory semantic enrichment process

3 Ontology Languages

Ontology languages are formal languages used to construct ontologies. They allow the encoding of knowledge about specific domains and often include reasoning facilities that support the processing of that knowledge. Among all the ontology languages, we considered Web Ontology Language (OWL), that is a well known standard arisen from the Semantic Web and it is now a W3C recommendation (OWL, 2004). An interesting feature of OWL is that it relies upon a family of languages known as Description Logics (DL) that provide a deductive inference system based on a formal well founded semantics (Baader et al., 2003). The basic components of DL are *concepts (classes)* and *roles (properties)*, termed as TBox, and *individuals (instances)*, termed as ABox. Concepts describe the common properties of a collection of individuals and roles are binary relations between concepts. Furthermore, a number of language constructs, such as intersection, union and role quantification, can be used to define new concepts and roles, by means of *axioms*. The main reasoning tasks are *classification* and *satisfiability*, *subsumption* and *instance checking*. Classification is the computation of a concept hierarchy based on subsumption, whereas instance checking verifies that an individual is an instance of a concept. This latter task is mainly used to check to which class each trajectory pattern is instance of.

Currently, OWL has three standard sublanguages of increasingly expressive power that are: (a) *OWL Lite*, which is the syntactically easiest version and can define hierarchies and simple constraints; (b) *OWL DL*, which allows the maximum expressiveness while retaining computational completeness and corresponds to Description Logics, and finally, (c) *OWL Full*, which allows for maximum expressiveness and the syntactic freedom of RDF with no computational guarantees. A further sub-language of OWL, called OWLPRIME (OWLPRIME, 2008), is implemented in the Oracle 11g Semantic Technologies, the technology used to implement the Athena system, described later in this paper.

4 The Semantic Representation of Trajectories

The trajectories captured by positioning systems (e.g. GPS and GSM equipments) are usually represented as a sequence of <*sample point, time*> pairs, called *raw* trajectory. The main characteristic of data acquired by those mechanisms is that they rigorously reveal the geometric facet of a trajectory, but suffers from a lack of semantics representing the movement behavior of the moving object.

A trajectory, as defined in (Spaccapietra et al., 2008), is the *user defined record of the evolution of the position of an object that is moving in space during a given time interval in order to achieve a given goal.* Moving objects do not necessarily move continuously during a trajectory. Consequently, trajectories may themselves be semantically segmented by defining a temporal sequence of time sub-intervals where alternatively the object position changes and stays fixed. We call the former the *moves* and the latter the *stops*. We can then see a trajectory as a sequence of moves going from one stop to the next one (or as a sequence of stops separating the moves). Yet, identifying stops (and moves) within a trajectory depends on the application requirements. Formal definitions of stops and moves are given in (Spaccapietra et al., 2008), whereas the semantic conceptualisation of a trajectory has been introduced in (Alvares et al., 2007), and in (Wachowicz et al., 2008). In Figure 2 we show our ontology-based representation of stops and moves, in terms of concepts and their relationships.

More precisely, boxes represent the main concepts whereas arrows represent relationships between two concepts. Every Trajectory is composed of stops (*trajCompOfStop*) and almost every stop is connected to other two stops by two moves. Therefore, we have represented the concepts of *Stop* and *Move* and four explicit relations connecting them, namely, *fromStop*, *toStop*, *inMove* and *outMove*. Only the first and the last stop are connected with a single move. Furthermore, every stop is connected to an interval (*stop_Has_Time*) that represents the time of the stop.

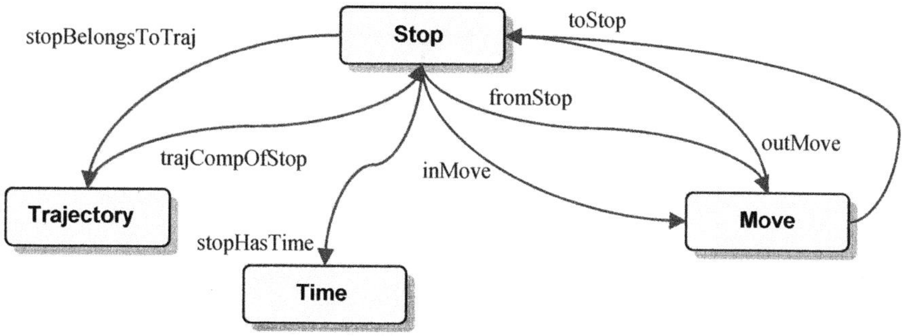

Fig. 2. The semantic trajectory representation

5 Mining Semantic Trajectories

In this section, our objective is to demonstrate how semantic trajectories can be associated with trajectory patterns and later with movement behavior. The examples given below might look straightforward, but it is important to point out they are used to illustrate the feasibility of our approach.

The problem of finding frequent patterns can be stated as: given a data set D and a user specified minimum support (*minsupp*), the task of frequent pattern discovery is to find all sets of items with support at least *minsupp*.

Frequent patterns (Agrawal et al., 2003) can be applied to both stops and moves independently. Frequent Stops (Moves) means finding sets of stops (or moves) that are most frequent. More details of patterns on semantic trajectories can be found in (Alvares et al., 2007), and in (Bogorny et al., 2008).

The following example represents frequent stop patterns:

```
Hotel[weekday],Museum[weekday]   (s=0.21)
```

This pattern is a mining expression and states that people that stay in a hotel during the week also go to (in any order) a Museum with a support of 21%.The conceptual representation is depicted in Figure 3. A *FrequentStopPattern* is defined as a set of *Stop*, by means of the relation *fpContainsStop*.

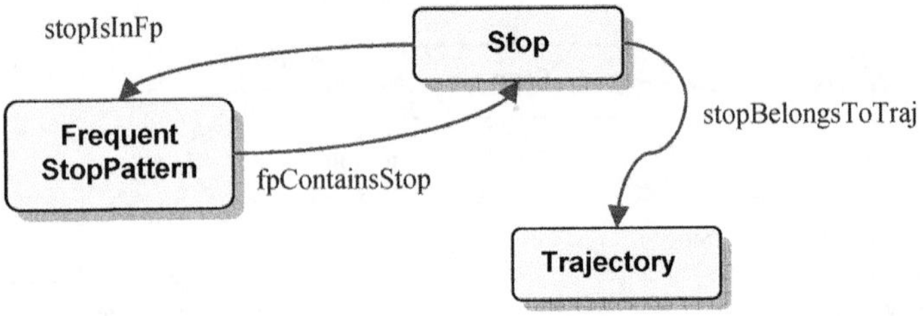

Fig. 3. Trajectory patterns representation

Similarly, in (Bogorny et al., 2008) authors have defined frequent moves and sequential stops and moves. Besides, it is important to point out that several other mining algorithms have been defined in the literature to extract patterns from both raw and semantic trajectories (Nanni et al., 2008). However, we focus here only on frequent patterns since the proposed approach could be applied to other patterns as well.

6 Reasoning on Trajectory Patterns

Once semantic trajectories and mining models have been defined, the next step is to express them using an ontology formalism, to combine them with domain knowledge, in order to define complex movement behavior.

Indeed, domain knowledge is composed of geographical knowledge about where the trajectory movement takes place using the application domain of reference, such as tourist or traffic management applications. Here, we focus on an urban human movement. The geographical knowledge is based on an urban ontology that describes the places where people move through (e.g. museums, hotels, universities, theatres, etc). Figure 4 shows an example of the ontology obtained by combining the three aspects described so far. Indeed, the core part of the ontology is the semantic trajectory, appearing on the upper side of the picture. The central part of the ontology figure shows the geographical knowledge, linked to the Stop concept by means of the *is_at* property, stating that a stop happens at a specific city place. Notice that the taxonomy defining interesting city places classifies two different types: tourist and accommodations places. Each of these semantic concepts represents a number of geographical features belonging to the application domain, such as monuments, hotels, museums and so on.

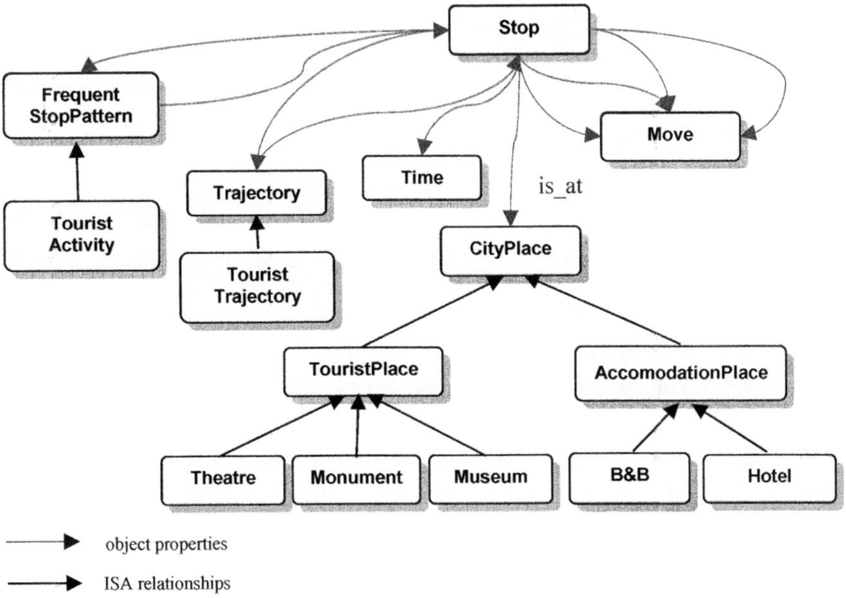

Fig. 4. The integrated ontology

New application concepts are described by means of *axioms*. Each axiom is a combination of logical operators that implicitly describes a class of objects. In the following example, we are interested in characterising trajectory patterns according to a movement behaviour by giving a possible interpretation of mined patterns (in the ontology are represented as *instances*) respect to that knowledge. In the following, we show two examples of axioms, expressed in OWL syntax, to specify a typical tourist behavior.

```
TouristActivity ≡
fpContainsStop some (Stop and (is_at some Tou-
ristPlace)) and fpContainsStop some (Stop and (is_at
some AccomodationPlace))
```

The above axiom defines a tourist activity as a frequent pattern that contains stops that are located at a *tourist place* (e.g. Museum, Monument …) and contains stops located at *accommodation places* (e.g. Hotel, B&B,…). Similarly, we can exploit this definition for characterizing individual trajectories, thus defining the concept of *tourist trajectory.*

```
TouristTrajectory ≡
trajCompOfStop some (Stop and (is_at some Tou-
ristPlace)) and trajCompOfStop some (Stop and (is_at
some AccomodationPlace))
```

The added value of having such an ontology-based approach, allows us to define axioms in terms of high-level semantic concepts, abstracting away from the geometric coordinates of the geographical features. Indeed, in this approach, each stop is treated as a semantic concept (e.g. Museum, Monument) instead of using the spatial coordinates. Moreover, the domain ontology gives a further abstraction level on top of semantic patterns, i.e. the *AccomodationPlace* class that subsumes hotels and B&B. Therefore, we can reason in terms of *AccomdationPlace* instead of each specific subclass (*Hotel*) or specific instance (*CentrumHotel*). As a further analysis step, the availability of a reasoning engine allows to build an automatic system for the characterisation of trajectory patterns in terms of people behaviour.

7 System Implementation

In order to better understand the feasibility of this approach, we have implemented a system where trajectory patterns are first extracted by means of a frequent pattern algorithm, then they are classified in the appropriate behavior classes, according to a semantic trajectory ontology and related axioms.

As a first step, we built a prototypical ontology using Protégé (Protégé, 2008) editor with Pellet reasoner (Pellet, 2008). Here, stop and moves and patterns data were imported from a database directly into Protégé. The obvious drawback of this approach is that importing very large raw trajectory data sets is not efficient, since both Protégé and Pellet are not scalable for large datasets (i.e. large collection of individuals). Indeed, we have to point out that the use of a formal ontology as a solution for enriching trajectory mining patterns is not a "silver bullet". In fact, the knowledge representation by means of a formal ontology brings a fundamental advantage of having ready-to-use inference system (i.e. reasoner) providing useful reasoning services for granted. However, reasoning on individuals (i.e. ABox reasoning) are not scalable in any of the Description Logics language flavours and both trajectory and data mining data pose a big challenge in this respect. Actually, reasoning scalability is an important open issue investigated in the Description Logic research domain.

For this reason, we investigated the use of Oracle11g Semantic Technologies (Oracle, 2008) as an ontology storage system and associated reasoner. The advantage of using Oracle technologies is that it is optimized for handling large datasets of individuals. Oracle semantic technologies store ontologies, represented as RDF triples, in relational tables. Both defined and inferred knowledge is stored, which results in faster processing of queries as all inferences are readily available. Oracle reasoning services are executed on demand and include traditional inference classification, satisfiability, subsumption, and instance checking. The drawback is that the Description Logic language that Oracle implements, called OWLPRIME, is a subset of OWL DL and has some limitations in expressiveness. To overcome these OWLPRIME limitations, Oracle provides *rules* that allow the designer to complement the basic OWLPRIME reasoning with more sophisticated and application dependent inference mechanisms. The rules can be added to a rule base that can be used conjointly with the ontology during semantic query execution.

7.1 Architecture

The Athena architecture overview is illustrated in Figure 5. Here, the user directly poses a query using the ontology concepts where trajectories/patterns are classified by the reasoner. The ontology is populated by instances coming from relational tables storing semantic trajectories, patterns and geographical features.

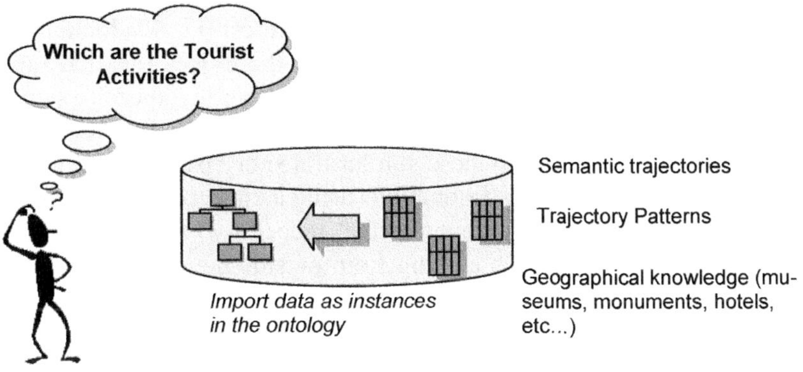

Fig. 5. The Athena Approach

More in detail, we used a dataset of trajectories coming from GPS installed on cars moving in the Milan area[1]. This dataset has been stored as raw trajectories in the moving object database Hermes (Pelekis and Theodoridis, 2006), which is based on Oracle. From them, we extracted a collection of stops computed over a simplified geographical domain containing Museums, Theatres, Hotels, Bed and Breakfast, Monuments. From these database tables, we developed a Java procedure described as follows:

- starts from a dataset of raw trajectories and geographical places and computes the semantic trajectories stored as table of stops;
- runs the frequent pattern algorithm on stops table (here we used PatternList (Bonchi et al., 2006) and stores back the results in the frequent patterns tables;
- imports the ontology from Protégé to Oracle 11g;
- reads the database tables and translates them into RDF triples;
- insert the triples in the ontology as instances;
- runs the reasoner to infer new triples.

7.2 Real Case Scenario

The objective of the analysis is to understand tourist movements in Milan. A first simple analysis is to visualise all tourist trajectories, i.e. trajectories that contain stops at tourist places and accommodation places (see step 3 and 4 below). A further objective is to understand the movement behavior among tourists. Starting from the semantic trajectories computed by system, expressed as sequences of places, we exploit a data mining algorithm to discover the trajectory frequent stop patterns. Among these trajectory patterns, we can choose a specific pattern of interest (i.e. MadonninaB&B, Duomo and Castello at step 5 below) to find the trajectories that support this patterns. This operation can be done by joining together the semantic trajectory and raw trajectory data using the IDs (see step 6 below). A further analysis step is to group these similar trajectory patterns in subgroups depending on their common destination (using a clustering algorithm (Rinzivillo et al., 2008)), thus discovering that indeed two types of movement behavior are present: tourists coming from outside the city and moving to the centre, and tourists moving from the city centre to outside (see step 7 below).

This analysis process is illustrated in details below:

---

[1] This dataset has been collected by *Octotelematics* and has been given for internal use inside the consortium of the European Project GeoPKDD (http://www.geopkdd.eu). It counts 17000 trajectories of one week.

1 – The objective of the analysis is to discover tourist common behaviour in the city. First of all we select all the geographical places that are interesting for the analysis (corresponding to the ontology classes) such as Hotel, Bed and Breakfast, Museum, etc. and the original raw trajectories in Milan. The following picture shows the Milan trajectories and some colored areas, corresponding to geographical objects of interest for the analysis, encoded as classes in the integrated ontology shown in Fig. 4

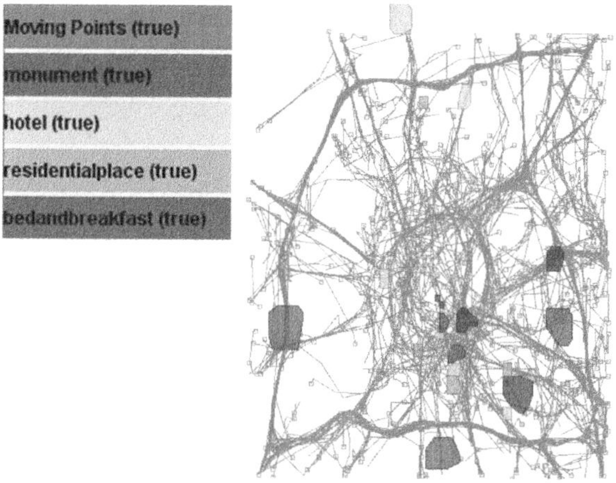

2 – The execution of Athena computes semantic trajectories, data mining patterns, and imports them, along with geographical objects, in Oracle11g Semantics Technologies. A map file is provided to define the correspondence between ontology concepts and database tables. When everything is loaded, the system calls the reasoner over the ontology and the imported instances in order to infer new facts (e.g. tourist activities).

```
Execute Athena(Map_file).
```

3 – The inferred facts are stored in the ontology, therefore we can query[2] the system to select the trajectories that are kind of *tourist* according to the definition in the ontology. Here we discovered that trajectories with ID *Trajectory192*, *Trajectory345* have been classified by the ontology as *tourist*.

```
SELECT m, r FROM table
```

[2] Parts of queries are omitted due to query readability. The rest is the specification of the name of the semantic model created, the type of rules used and the aliases: *SEM_Models('MODELGEOPKDD') ,SEM_rulebases('owlprime') ,SEM_ALIASES(SEM_ALIAS('','http://www.owl-ontologies.com/GeoPKDDOnto.owl#')),null))*.

```
(SEM_MATCH('(?m ?r :tourist),[…]'))
```

Result:

Trajectory192

Trajectoy345

…

4 –Tourist trajectories can be joined with the raw trajectories exploiting the ID. Indeed, the geometric coordinates allows us to draw them on a map. Below in yellow the subset of the raw geometric trajectories showing a tourist behavior, based on the definition of the described urban ontology.

5 – To discover common behavior, we exploit the Frequent Stop Patterns data mining algorithm, and we select one of the resulting computed patterns (id=47). This pattern is characterised by people stopping at a specific B&B, namely MadonninaB&B, and two different monuments, Duomo and Castello.

```
SELECT distinct(s.p) from
TABLE
(SEM_MATCH('(?m?r:FrequentStopPattern)',[…]'))
t,
TABLE(SEM_MATCH('(?m:fpContainsStop?p),[…]')) s
WHERE t.m = s.m and GET_ID (t.m)=47
```

Result:

MadonninaB&B

Duomo

Castello

6 – In order to understand which are the raw trajectories that support this specific tourist frequent pattern, we can, through the use of the Stop ID, Trajectory ID, and the *StopBelongsToTraj* ontology property, join the pattern with raw trajectories, shown in the picture below. This is a powerful capability of the Athena system since it allows us to combine, in the same query, semantic information with raw geometric data.

7 – A further data mining analysis step can be run using the found tourist trajectories, to have a better understanding of common tourist behavior. Here we used a density-based clustering algorithms on trajectories. The result shows two very well defined groups of trajectories: from outside the city to the center (A) and vice versa (B). This can suggest to the user that the first group describes tourists in their first day of holidays and the second group describes tourists in their last day of holidays.

(A) (B)

This example, despite its simplicity, shows the feasibility of the proposed ontology-based semantic enrichment process over raw trajectory data, exploiting the synergy between geographical knowledge, automated reasoning, data mining to perform very complex analysis of movement behaviour processes. It is interesting to notice that a different kind of analysis is possible with this methodology. Indeed, finding patterns that have not been classified under any ontology class defining a specific behaviour, will result in an "unknown" behaviour. This in turn may call for a further in-depth analyses based on the iterating steps of the knowledge discovery process, possibly arriving to a better understating of new unexpected movement behaviour.

8 Related Work

This work is a continuation of a previous preliminary work on characterizing semantic trajectories by means of ontologies (Baglioni et al., 2008). Current work extends previous one by exploiting mining algorithms in defining trajectory patters. Furthermore, we developed the Athena system to concretize the approach.

Several approaches have been proposed for the development of formal ontologies. They may differ from each other in terms of methods for defining concepts, for identifying relations between these concepts and for building reasoning tasks (Teller et al., 2007). In particular, (Caglioni and Rabino, 2007) suggests that semantic models are unique sources for deriving ontologies because they already include appropriate definitions of concepts. However, most of the semantic data models developed so far have considered moving objects with an absolute representation of space, and have proposed query languages and data mining functions that support their manipulation in space and time (Guting and Shneider, 2005). Within these semantic models, a moving object is usually represented using its location as a function of time. A different formal model has been proposed by (Noyon et al., 2007) where the basic primitives are the relative positions and relative velocities of the neighboring objects and regions, as perceived by an observer acting in the environment.

9 Conclusions and Future Work

In this paper we presented an approach to provide the interpretation of movement behavior. This approach exploits a formal ontology as a representation and reasoning mechanism that enables semantic interpretation of raw trajectories and their mining patterns. The obtained ontology is immersed in a domain ontology representing geographical knowledge. This allows getting the maximum semantics in term of encoded application geographical knowledge. Indeed, this ontological approach allows us to reason in terms of semantic concepts. This added value, when applied to mining patterns, produces an automated interpretation of them within an application domain, which we have called as semantic enrichment process.

Ongoing research work is towards the development of a more robust and complete Athena system. First, we are investigating the expressiveness of OWLPRIME combined with Oracle user-defined rules in order to get maximum expressivity while retaining a good computational efficiency. Second, we aim at improving our system by integrating more data mining algorithms. Indeed, the proposed approach can be applied also to raw trajectories and associated mined patterns. Obviously, in this case the ontology misses the semantic trajectory part and no relations can be expressed between trajectory and pattern, thus offering less expressiveness.

Another challenging issue that could be investigated is how to exploit the semantic ontology in the steps of a knowledge discovery process, especially in the pre-processing step. The aim is to derive a suitable raw trajectory data in order to avoid extracting uninteresting patterns. A further challenge will be the combination of ontology-based semantic enrichment process with visual analytics techniques. Within this context, we are planning to improve the integration of Oracle 11g with visualization techniques to provide users with a more friendly and effective interaction. This could offer better support for user evaluation of inferred movement behavior.

Acknowledgments

This work has been partially supported by GeoPKDD EU Project IST-6FP-014915, http://www.geopkdd.eu/.

References

Agrawal, R., Imielinski, T. and Swami, A. (2003) Mining association rules between sets of items in large databases, Proc. ACM SIGMOD Int. Conf. on Management of Data, pp. 207-216

Alvares, L. O., Bogorny, V., Kuijpers, B., Macedo, J., Moelans, B. and Vaisman, A. (2007) A Model for Enriching Trajectories with Semantic Geographical Information, Proc. of the ACM 15th International Symposium on Advances in Geographic Information Systems (ACM-GIS'07).

Baader, F., Calvanese, D., McGuinness, D. L., Nardi, D. and Patel-Schneider, P. F. (2003) The Description Logic Handbook: Theory, Implementation, Applications. Cambridge University Press, Cambridge, UK.

Baglioni, M., Macedo, J., Renso, C. and Wachowicz, M. (2008) An ontology-based approach for the semantic modelling and reasoning on trajectories, SeCoGIS 2008, Barcelona, Spain, October 20-23, LNCS Vol. 5232.

Bogorny, V., Kuijper B. and Alvaresz, L. O. (2008) ST-DMQL: a Semantic Trajectory Data Mining Query Language, IJGIS to appear.

Bonchi, F., Giannotti, F., Orlando S., Lucchese, C., Perego, R. and Trasarti, R. (2006) ConQueSt: a Constraint-based Querying System for Exploratory Pattern Discovery. Proceedings of the 22nd IEEE International Conference on Data Engineering (ICDE'06).

Caglioni, M. and Rabino, G. (2007) Ontologies of urban systems. In Joint Congress of the European Regional Science Association and ASRDLF.

Dodge, S., Weibel, R. and Lautenschultz, A-K (2008) Towards a Taxonomy of Movement Patterns, Information and Visalization, 7, 240-252

Gruber, T.R. (2008) Ontology. Entry in the *Encyclopedia of Database Systems*, Ling Liu and M. Tamer Özsu (Eds.), Springer-Verlag.

Güting, R. and Schneider, M. (2005) Moving objects databases. Morgan Kaufmann.

Nanni, M., Kuijpers, B., Korner, C., May, M. and Pedreschi, D. (2008) Spatiotemporal data mining. In F. Giannotti and D. Pedreschi, editors, Mobility, Data Mining, and Privacy: Geographic Knoweledge Discovery. Springer-Verlag.

Noyon, V., Claramunt, C. and Devogele, T. (2007) A Relative Representation of Trajectories in Geogaphical Spaces, GeoInformatica, Vol. 11, N. 4, December.

Oracle (2008) Oracle Semantic Technologies http://www.oracle.com/technology/tech/semantic_technologies/index.html

OWL (2004) W3C Consortium. The web ontology language. http://www.w3.org/TR/owlfeature

OWLPRIME (2008) Owl Prime http://www.w3.org/2007/OWL/wiki/OracleOwlPrime

Pelekis, N. and Theodoridis, Y. (2006) Boosting Location-based Services with a Moving Object Database Engine. In Proc. of Int. ACM SIGMOD/PODS Workshop on Data Engineering for Wireless and Mobile Access, pp. 3-10.

Pellet (2008) Pellet Reasoner http://pellet.owldl.com/

Protege (2008) Protégé-OWL editor http://protege.stanford.edu /overview/ protege-owl.html

Rinzivillo, S., Pedreschi, D., Nanni, M., Giannotti, F., Andrienko, N. and Andrienko, G. (2008) Visually–driven analysis of movement data by progressive clustering Information Visualization, v.7 (3/4), pp. 225-239

Spaccapietra, S., Parent, C., Damiani, M. L., de Macedo, J. A., Porto, F. and Vangenot, C. (2008) A conceptual view on trajectories. Data & Knowledge Engineering, vol. 65, number 1, pp. 126-146.

Teller, J., Lee, J. and Roussey, C. (2007) Ontologies for Urban Development. Studies in Computational Intelligence, v. 61, Springer.

Wachowicz, M., Macedo, J., Renso, C. and Ligtenberg, A. (2008) The role of a multi-tier ontological framework in reasoning to discover meaningful patterns of sustainable mobility. In: Geographic data mining and knowledge discovery, H. J. Miller and J Han (eds.), 2nd edition, Taylor and Francis (in press).

Three-Valued 9-Intersection for Deriving Possible Topological Relations from Incomplete Observations

Yohei Kurata

SFB/TR 8 Spatial Cognition, Universität Bremen
Postfach 330 440, 28334 Bremen, Germany
ykurata@informatik.uni-bremen.de

Abstract. Topological relations, which concern how two objects intersect, are one of the most fundamental and well-studied spatial relations. Typically, topological relations are distinguished by the presence or absence of pairwise intersections between several parts of two objects. However, when an observation is not complete, it is often impossible to determine the presence or absence of some intersections. In order to support such uncertain situations, this paper introduces *three-valued 9-intersection matrix* (*3-9i matrix*), which records the presence, absence, or indeterminate state of the nine types of intersections between the objects. We can use the 3-9i matrix to describe the spatial arrangement of objects in various scenarios of incomplete observations. In addition, from the pattern of the 3-9i matrix we can computationally determine a set of topological relations that may hold between the objects. We assess the degree of topological information derived from the 3-9i matrix for 14 sets of topological relations between simple lines, regions, and bodies embedded in $\mathbb{R}^1$, $\mathbb{R}^2$, $\mathbb{R}^3$, $\mathbb{S}^1$, and $\mathbb{S}^2$.

1 Introduction

Topological relations are one class of spatial relations that concern how two objects intersect with each other in a space. Topological relations have been studied extensively, as they have been considered fundamental for people's conceptualization of spatial configurations. In many studies, topological relations were distinguished based on the presence or absence of intersections between several parts of two objects (e.g., Egenhofer and Fran-

M. Sester et al. (eds.), *Advances in GIScience*, Lecture Notes in Geoinformation and Cartography,
DOI 10.1007/978-3-642-00318-9_15, © Springer-Verlag Berlin Heidelberg 2009

zosa 1991; Egenhofer and Herring 1991; Clementini and Felice 1998; Kurata 2008). However, when an observation is not complete, it is often impossible to determine the presence or absence of some of the intersections. For instance, look at Fig. 1a, which shows an old Japanese painting of a battle scene. In the center of this painting, we can see two troop units facing with each other, but a trailing cloud hides their front lines (Fig. 1b). So, we are tempted to imagine whether the battle already starts or not—or, geometrically speaking, the areas of the two units are disjoint or meeting. We sometimes have such incomplete spatial data, such as torn historical maps, aerial images with clouds or shadows, and partly-masked maps due to political reasons. Interestingly, even if the data is not complete, it is often possible to determine a small set of relations (Fig. 2a) or even a unique relation (Fig. 2b) that may hold between two objects. So, our interest is how we can systematically derive such possible relations from incomplete observations and how crisp they are.

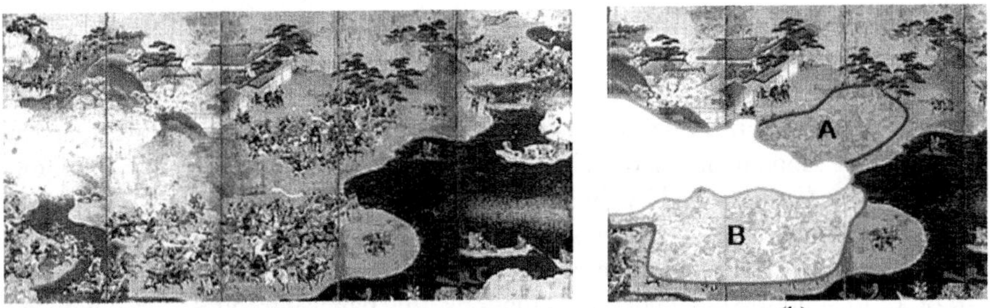

(a) (b)

Fig. 1. (a) An old Japanese painting of a battle scene, in which the spatial arrangement of two troop units is partly hidden by a cloud, as highlighted in (b)

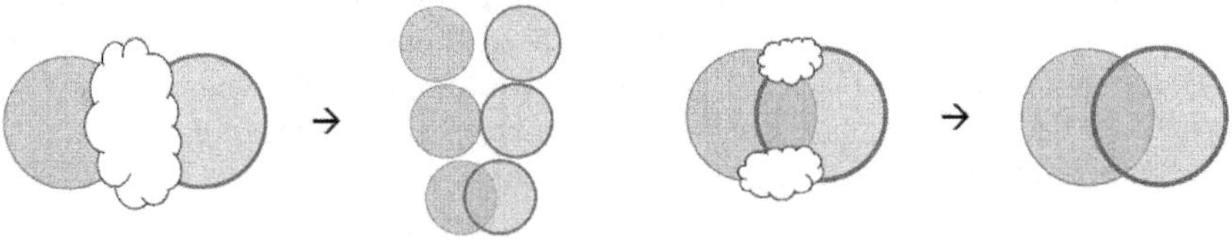

(a) (b)

Fig. 2. Mappings from partly-masked configurations of two regions to the possible topological relations between these regions

In order to characterize incompletely-observed arrangements of spatial objects from a topological viewpoint, we introduce a three-valued approach to the *9-intersection* (Egenhofer and Herring 1991). The 9-intersection is a widely-used and well-studied model of topological relations. The *three-valued 9-intersection matrix* (*3-9i matrix*) records the presence, absence, or indeterminate state of 3×3 types of intersections between two objects. We can use the 3-9i matrix to characterize uncertain spatial arrangements under various scenarios of incomplete observations. In addition, given a 3-9i matrix, we can computationally determine the set of topological relations that may hold between two objects. This technique can be used not only for topological region-region relations, but also for a variety of topological relations distinguished by the 9-intersection.

As a simpler version of the 3-9i matrix, the *three-valued 4-intersection matrix* (3-4i matrix) was already proposed by Winter (1999). His focus was to deduce quickly the topological relation between two regions, captures by such a hierarchical partition as a quadtree, by calculating the conjunction of the 3-4i matrices that represent the partial topological relations between two regions within spatial partitions. Later this approach was extended to vague regions (Winter and Bittner 1999). While their work uses the 3-4i matrix for representing a partial knowledge of a spatial arrangement, our work uses the 3-9i matrix for representing the total (but incomplete) knowledge of the spatial arrangement. Naturally, we are interested in the assessment of topological information that the 3-9i matrix can capture in various scenarios of incomplete observations, which are not yet analyzed before. Also, our target is not only region-region relations in a 2D space, but 14 sets of relations between simple lines, regions, and bodies in five types of spaces.

This research serves as a complement of *qualitative spatial calculi* (Ligozat and Renz 2004) on topological relations. Qualitative spatial calculi are a computational framework of spatial reasoning, by which we can disambiguate the spatial arrangement of multiple objects with respect to a certain relation set. To conduct such reasoning, we need preliminary rough knowledge about possible relations between pairs of the objects. Such preliminary knowledge of topological relations can be derived from incomplete observations by the technique discussed in this paper.

The remainder of this paper is structured as follows: Section 2 reviews the 9-intersection by which we distinguish topological relations. Section 3 reviews previous studies on topological relations under uncertainty. Section 4 defines the 3-9i matrix and relevant concepts. Section 5 applies the 3-9i matrix to characterize uncertain spatial arrangements under various scenarios of incomplete observations. Section 6 assesses the degree of topological information derived from the 3-9i matrices for various sets of topological relations. Finally, Section 7 concludes the discussion.

2 The 9-Intersection

The 9-intersection (Egenhofer and Herring 1991) is a model of topological relations in which the relations are characterized by certain topologically-invariant properties of intersections between the *topological parts* (*interior*, *boundary*, and *exterior*) of two objects. Under point-set topology (Alexandroff 1961), the interior of a spatial object X, denoted X°, is defined as the union of all open sets contained in X, X's boundary ∂X is defined as the difference between X's *closure* (i.e., the intersection of all closed point sets that contain X) and X°, and X's exterior X^- is defined as the complement of X's closure. The *9-intersection matrix* in Eq. 1 concisely represents the 3×3 types of intersections between two spatial objects A and B. For simplification, each element in the 9-intersection matrix is represented also by $m_{ij}(A,B)$ where $i,j \in \{1,2,3\}$. For instance, $m_{21}(A,B) = \partial A \cap B^\circ$

$$\mathrm{M}(A,B) = \begin{pmatrix} A^\circ \cap B^\circ & A^\circ \cap \partial B & A^\circ \cap B^- \\ \partial A \cap B^\circ & \partial A \cap \partial B & \partial A \cap B^- \\ A^- \cap B^\circ & A^- \cap \partial B & A^- \cap B^- \end{pmatrix} \tag{1}$$

In the most basic approach, topological relations are distinguished simply by the presence and absence of the 3×3 types of intersections. Thus, we consider a 3×3 matrix with two symbols $\emptyset$ and $\neg\emptyset$, which shows the emptiness and non-emptiness of the 3×3 elements in the 9-intersection matrix. This matrix is called the *two-valued 9-intersection matrix* (in short, the *2-9i matrix*) in this paper and denoted $\mathrm{M}^{\mathrm{II}}(A,B)$. By the patterns of the 2-9i matrix, we can distinguish, for instance, eight topological relations between two simple regions in a 2D Euclidian space $\mathbb{R}^2$ (Fig. 3) (Egenhofer and Herring 1991). Table 1 shows the numbers of topological relations between simple lines, simple regions, and simple bodies in $\mathbb{R}^1$, $\mathbb{R}^2$, $\mathbb{R}^3$, $\mathbb{S}^1$ (1-sphere), and $\mathbb{S}^2$ (2-sphere), distinguished by the patterns of the 2-9i matrix (Kurata 2008). Each topological relation set is denoted $\mathcal{TR}_{\mathbf{D}_A\mathbf{D}_B\text{-}\mathcal{S}}$, where $\mathbf{D}_A$ and $\mathbf{D}_B$ are the domains of the objects A and B (**L**:simple lines, **R**:simple regions, and **B**:simple bodies) and $\mathcal{S}$ is the class of the space that embeds A and B. For instance, $\mathcal{TR}_{\mathbf{RR}-\mathbb{R}^2}$ is the set of topological relations between two simple regions in $\mathbb{R}^2$.

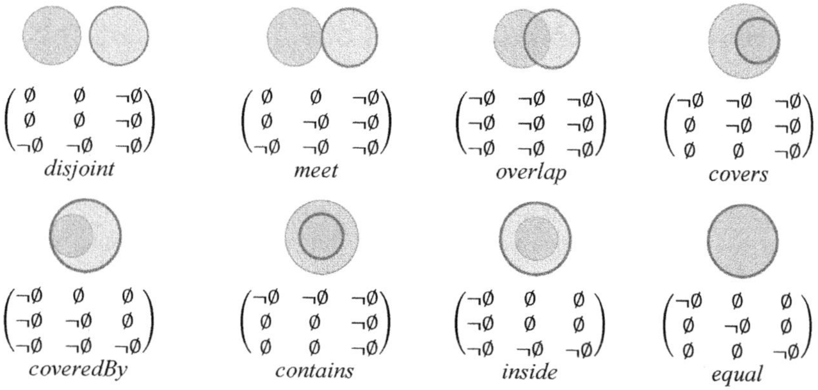

Fig. 3. Eight topological relations between two simple regions in $\mathbb{R}^2$, distinguished by the patterns of the 2-9i matrix (Egenhofer and Herring 1991)

Table 1. Numbers of topological relations between simple lines, simple regions, and simple bodies in $\mathbb{R}^1$, $\mathbb{R}^2$, $\mathbb{R}^3$, $\mathbb{S}^1$, and $\mathbb{S}^2$, distinguished by the patterns of the 2-9i matrix (Kurata 2008)

	$\mathbb{R}^1$	$\mathbb{R}^2$	$\mathbb{R}^3$	$\mathbb{S}^1$	$\mathbb{S}^2$
Line-Line	8	33	33	11	33
Line-Region	–	19	31	–	19
Line-Body	–	–	19	–	–
Region-Region	–	8	43	–	11
Region-Body	–	–	19	–	–
Body-Body	–	–	8	–	–

The 9-intersection has been studied extensively in GIS communities, because the same framework can be used for modeling a variety of topological relations. However, the 9-intersection cannot capture some topological or metrical characteristics of spatial arrangements, such as the number of intersections, and the dimension/degree of intersections, the distinction of the line's boundary subparts (i.e., start point and end point), and the distinction of the inner and outer boundary/exterior of holed regions, which may be critical for some applications. Thus, several extensions or alternative approaches of the 9-intersection have been proposed (e.g., Egenhofer and Franzosa 1995; Kurata and Egenhofer 2007; Nedas, et al. 2007).

3 Handling Topological Relations under Uncertainty

Uncertainty, which comprises such sub-concepts as *inaccuracy* (lack of correlation with reality) and *imprecision* (lack of specificity), has been an important topic in GIS studies, since geographic information is closely linked with observations of the world (Worboys 2004). Inaccuracy is triggered by measurement errors or low resolution in observations, while the imprecision results from incomplete observations or *vagueness*. For modeling vague objects whose spatial extent is not clearly definable (e.g., mountains), three-valued logic and Fuzzy set theory (Zadeh 1965) have been often applied (e.g., Yee 1987; Clementini and Felice 1996; Cohn and Gotts 1996; Erwig and Schneider 1997; Schneider 1999). On the other hand, for handling inaccuracy, probabilistic and statistical approaches have been taken (e.g., Burrough 1996; Winter 2000; Tøssebro and Nygård 2002).

Qualitative spatial relations under uncertainty are one of the key topics in the studies of uncertainty in GIS. The early studies on this topic favored three-zone approaches, which consider the area that definitely belongs to a spatial object, the area that may belong to the object, and the area that definitely does not belong to the object, respectively. Cohn and Gotts (1996) distinguished 46 relations between two regions with such indeterminate areas, namely *egg-yolks*, and they identified the mapping from egg-yolk relations to five topological relations under RCC-5 (Randell, et al. 1992). On the other hand, based on the 9-intersection (Egenhofer and Herring 1991), Clementini and Felice (1996) distinguished 44 topological relations between two regions with *broad boundaries* in which the true boundaries physically or conceptually take place. This *band*-based approach was adopted for topological line-line relations (Clementini 2005; Reis, et al. 2006), point-point relations (Lee and Flewelling 2004), and line-region relations (Tang, et al. 2006).

Recently many studies applied Fuzzy logic (Zadeh 1965) for handling spatial relations between uncertain objects. Molenaar (Zadeh 1965) discussed the topological relations between *Fuzzy regions*. Each Fuzzy region was represented by a set of points and their membership to an entity. Alternatively, Du et al. (2005) and Liu and Shi (2009) define *Fuzzy interior*, *boundary*, and *exterior* of spatial objects and described their topological relations within the framework of the 9-intersection. Similarly, Tang *et al.* (2006) adopted the framework of the 9-intersection based on the formal distinction of the interior, boundary, and exterior of *Fuzzy n-cells*.

Winter (2000) analyzed the topological relations under inaccuracy, instead of vagueness. Given an observation of two regions and its resolution, his method enables us to calculate the probability that each topological relation holds between the regions.

In this way, previous studies featured the treatment of vagueness and inaccuracy in topological relations. On the other hand, it seems that topological relations under incomplete observations have not been attracted much attention.

4 Three-Valued 9-Intersection Matrix

The two-valued 9-intersection records the presence or absence of the 3×3 types of intersections. In order to support the lack of knowledge about the presence or absence of some of the intersections, we consider a 3×3 matrix with three symbols $\{\emptyset, \neg\emptyset, ?\}$, which shows the emptiness, non-emptiness, or indeterminate state of the corresponding elements in the 9-intersection matrix (Eq. 1). This matrix is called the *three-valued 9-intersection matrix* (*3-9i matrix*). The 3-9i matrix that corresponds to the 9-intersection matrix $\mathrm{M}(A,B)$ is denoted $\mathrm{M}^{\mathrm{III}}(A,B)$. For instance, the configuration in Fig. 4a, where the presence or absence of interior-interior, interior-boundary, boundary-interior, and boundary-boundary intersections cannot be determined, is characterized by the 3-9i matrix $\begin{pmatrix} ? & ? & \neg\emptyset \\ ? & ? & \neg\emptyset \\ \neg\emptyset & \neg\emptyset & \neg\emptyset \end{pmatrix}$. Figs. 4b-c show two more examples of uncertain spatial configurations characterized by 3-9i matrices.

$\begin{pmatrix} ? & ? & \neg\emptyset \\ ? & ? & \neg\emptyset \\ \neg\emptyset & \neg\emptyset & \neg\emptyset \end{pmatrix}$ (a) $\begin{pmatrix} \neg\emptyset & \neg\emptyset & \neg\emptyset \\ \neg\emptyset & ? & \neg\emptyset \\ \neg\emptyset & \neg\emptyset & \neg\emptyset \end{pmatrix}$ (b) $\begin{pmatrix} \neg\emptyset & \neg\emptyset & \neg\emptyset \\ ? & ? & \neg\emptyset \\ ? & ? & \neg\emptyset \end{pmatrix}$ (c)

Fig. 4. Three examples of partly-masked configurations of two regions, each characterized by a 3-9i matrix

Among the eight 2-9i matrices that represent the relations in $\mathcal{TR}_{\mathrm{RR}-\mathbb{R}^2}$ (Fig. 3), $\begin{pmatrix} \emptyset & \emptyset & \neg\emptyset \\ \emptyset & \emptyset & \neg\emptyset \\ \neg\emptyset & \neg\emptyset & \neg\emptyset \end{pmatrix}$, $\begin{pmatrix} \emptyset & \emptyset & \neg\emptyset \\ \emptyset & \neg\emptyset & \neg\emptyset \\ \neg\emptyset & \neg\emptyset & \neg\emptyset \end{pmatrix}$, and $\begin{pmatrix} \neg\emptyset & \neg\emptyset & \neg\emptyset \\ \neg\emptyset & \neg\emptyset & \neg\emptyset \\ \neg\emptyset & \neg\emptyset & \neg\emptyset \end{pmatrix}$, which represent *disjoint*, *meet*, or *overlap* relations, are *consistent* with the 3-9i matrix $\begin{pmatrix} ? & ? & \neg\emptyset \\ ? & ? & \neg\emptyset \\ \neg\emptyset & \neg\emptyset & \neg\emptyset \end{pmatrix}$ in Fig. 4a; that is, $\begin{pmatrix} \emptyset & \emptyset & \neg\emptyset \\ \emptyset & \emptyset & \neg\emptyset \\ \neg\emptyset & \neg\emptyset & \neg\emptyset \end{pmatrix}$, $\begin{pmatrix} \emptyset & \emptyset & \neg\emptyset \\ \emptyset & \neg\emptyset & \neg\emptyset \\ \neg\emptyset & \neg\emptyset & \neg\emptyset \end{pmatrix}$, and

$\begin{pmatrix} \neg\emptyset & \neg\emptyset & \neg\emptyset \\ \neg\emptyset & \neg\emptyset & \neg\emptyset \\ \neg\emptyset & \neg\emptyset & \neg\emptyset \end{pmatrix}$ correspond to $\begin{pmatrix} ? & ? & \neg\emptyset \\ ? & ? & \neg\emptyset \\ \neg\emptyset & \neg\emptyset & \neg\emptyset \end{pmatrix}$ if we regard each '?' as a *wild-card* symbol (Winter 1999). This means the actual relation between the two regions in Fig. 4a is either *disjoint*, *meet*, or *overlap* (Fig. 5). Similarly, the actual relation between the two regions in Fig. 4b is uniquely determined as *overlap*, because $\begin{pmatrix} \neg\emptyset & \neg\emptyset & \neg\emptyset \\ \neg\emptyset & \neg\emptyset & \neg\emptyset \\ \neg\emptyset & \neg\emptyset & \neg\emptyset \end{pmatrix}$ is the only 2-9i matrix that is consistent with $\begin{pmatrix} \neg\emptyset & \neg\emptyset & \neg\emptyset \\ \neg\emptyset & ? & \neg\emptyset \\ \neg\emptyset & \neg\emptyset & \neg\emptyset \end{pmatrix}$. Like these examples, given a 3-9i matrix $\mathrm{M}^{\mathrm{III}}(A, B)$ that characterizes the spatial configuration of two objects A and B in a space $\mathcal{S}$, we can derive a set of relations in $\mathcal{TR}_{\mathrm{D}_A\mathrm{D}_B\text{-}\mathcal{S}}$ that may hold between A and B, called the *candidate relation set*, simply by checking the consistency between $\mathrm{M}^{\mathrm{III}}(A, B)$ and every 2-9i matrix that represents a relation in $\mathcal{TR}_{\mathrm{D}_A\mathrm{D}_B\text{-}\mathcal{S}}$. The operation to derive the candidate relation set is called *candidate-listing* and denoted $\mathrm{CL}_{\mathcal{TR}_{\mathrm{D}_A\mathrm{D}_B\text{-}\mathcal{S}}}$. In the previous example,

$$\mathrm{CL}_{\mathcal{TR}_{\mathrm{RR}-\mathbb{R}^2}}\left(\begin{pmatrix} ? & ? & \neg\emptyset \\ ? & ? & \neg\emptyset \\ \neg\emptyset & \neg\emptyset & \neg\emptyset \end{pmatrix}\right) = \left\{\begin{pmatrix} \emptyset & \emptyset & \neg\emptyset \\ \emptyset & \emptyset & \neg\emptyset \\ \neg\emptyset & \neg\emptyset & \neg\emptyset \end{pmatrix}, \begin{pmatrix} \emptyset & \emptyset & \neg\emptyset \\ \emptyset & \neg\emptyset & \neg\emptyset \\ \neg\emptyset & \neg\emptyset & \neg\emptyset \end{pmatrix}, \begin{pmatrix} \neg\emptyset & \neg\emptyset & \neg\emptyset \\ \neg\emptyset & \neg\emptyset & \neg\emptyset \\ \neg\emptyset & \neg\emptyset & \neg\emptyset \end{pmatrix}\right\}.$$

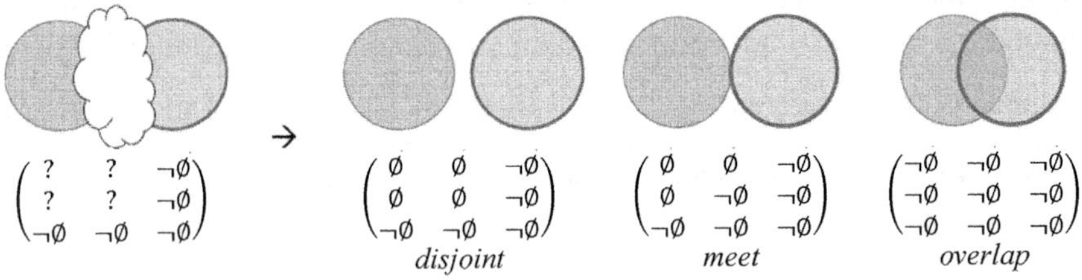

Fig. 5. A 3-9i matrix, which are mapped to three relations in $\mathcal{TR}_{\mathrm{RR}-\mathbb{R}^2}$ by the *candidate-listing* operation

4.1 Interpretable 3-9i Matrices

The 3-9i matrix may take $3^9 = 19683$ patterns. However, not all patterns have a candidate relation set. If a 3-9i matrix has a candidate relation set, the matrix is called *interpretable*.

Let us focus on $\begin{pmatrix} \emptyset & \emptyset & \neg\emptyset \\ \emptyset & \emptyset & \neg\emptyset \\ \neg\emptyset & \neg\emptyset & \neg\emptyset \end{pmatrix}$ which represents the *disjoint* relation in $\mathcal{TR}_{\mathrm{RR}-\mathbb{R}^2}$ (Fig. 3). By replacing arbitrary elements in this 2-9i matrix by '?', we can obtain $2^9 = 1024$ interpretable 3-9i matrices. Similarly, for each of the eight 2-9i matrices in Fig. 3, we can obtain 1024 interpretable 3-9i matrices. By integrating these eight sets of 1024 interpretable 3-9i matrices, we can obtain 2946 interpretable 3-9i matrices for $\mathcal{TR}_{\mathrm{RR}-\mathbb{R}^2}$.

In general, given a set of n topological relations represented by the 2-9i matrix $\mathrm{M}_k^{\mathrm{II}} = \left(m_{k\,ij}^{\mathrm{II}}\right)_{i,j\in\{1,2,3\}}$ where $k \in \{1,\cdots,n\}$, the 3-9i matrix $\mathrm{M}^{\mathrm{III}} = \left({m^{\mathrm{III}}}_{ij}\right)_{i,j\in\{1,2,3\}}$ is interpretable if and only if $\exists k \in \{1,\cdots,n\}\, \forall i,j \in \{1,2,3\}\, {m^{\mathrm{III}}}_{ij} \in \left\{m_{k\,ij}^{\mathrm{II}}, ?\right\}$.

4.2 Valid 3-9i Matrices

For some sets of topological relations, the 3-9i matrix may have the elements whose value is *fixed*. For instance, for $\mathcal{TR}_{\mathbf{RR}-\mathbb{R}^2}$, ${m^{\mathrm{III}}}_{33}$ is always $\neg\emptyset$, because the exteriors of two regions always intersect with each other in $\mathbb{R}^2$. Such *fixed elements* in the 3-9i matrix for a topological relation set $\mathcal{TR}_X$ are found simply by comparing the 2-9i matrices that represents the relations in $\mathcal{TR}_X$. Table 2 shows the position and value of the fixed elements in the 3-9i matrix for 14 sets of topological relations.

Obviously, any 3-9i matrix that has '?' at the position of a fixed element is nonsense, as its value is predetermined as $\emptyset$ or $\neg\emptyset$. Therefore, in the remainder of this paper we consider only *valid 3-9i matrices* that are interpretable and do not have '?' at the position of any fixed element. For instance, there are 1473 valid 3-9i matrices for $\mathcal{TR}_{\mathbf{RR}-\mathbb{R}^2}$. Table 2 also shows the number of valid 3-9i matrices for 14 sets of topological relations. We can see that a relation set $\mathcal{TR}_X$ has a larger number of valid 3-9i matrices as (i) the 3-9i matrices have fewer fixed elements and (ii) $\mathcal{TR}_X$ has a larger number of relations.

Table 2. The number of relations, the position and value of fixed matrix elements, and the number of valid 3-9i matrices for 14 sets of topological relations.

Relation set	$\mathcal{TR}_{\mathrm{LL}-\mathbb{R}^1}$ $\mathcal{TR}_{\mathrm{RR}-\mathbb{R}^2}$ $\mathcal{TR}_{\mathrm{BB}-\mathbb{R}^3}$	$\mathcal{TR}_{\mathrm{LL}-\mathbb{S}^1}$ $\mathcal{TR}_{\mathrm{RR}-\mathbb{S}^2}$	$\mathcal{TR}_{\mathrm{LR}-\mathbb{R}^2}$ $\mathcal{TR}_{\mathrm{LR}-\mathbb{S}^2}$ $\mathcal{TR}_{\mathrm{LB}-\mathbb{R}^3}$ $\mathcal{TR}_{\mathrm{RB}-\mathbb{R}^3}$	$\mathcal{TR}_{\mathrm{LR}-\mathbb{R}^3}$	$\mathcal{TR}_{\mathrm{LL}-\mathbb{R}^2}$ $\mathcal{TR}_{\mathrm{LL}-\mathbb{R}^3}$ $\mathcal{TR}_{\mathrm{LL}-\mathbb{S}^2}$	$\mathcal{TR}_{\mathrm{RR}-\mathbb{R}^3}$
Number of relations	8	11	19	31	33	43
Fixed elements in 3-9i matrices	$\begin{pmatrix} & & \\ & & \\ & & \neg\emptyset \end{pmatrix}$	–	$\begin{pmatrix} & & \\ & & \\ \neg\emptyset & \neg\emptyset & \neg\emptyset \end{pmatrix}$	$\begin{pmatrix} & & \\ & & \\ \neg\emptyset & \neg\emptyset & \neg\emptyset \end{pmatrix}$	$\begin{pmatrix} & & \\ & & \\ & & \neg\emptyset \end{pmatrix}$	$\begin{pmatrix} & & \\ & & \\ & & \neg\emptyset \end{pmatrix}$
Number of valid 3-9i matrices	1473	3953	463	562	2836	3016

4.3 Hierarchy of 3-9i Matrix Classes

For each topological relation set, the valid 3-9i matrices are categorized into equivalent classes with respect to the *candidate-listing* operation. For instance, under $\mathcal{TR}_{\mathrm{RR}-\mathbb{R}^2}$, $\begin{pmatrix} ? & ? & ? \\ ? & \neg\emptyset & ? \\ \neg\emptyset & \neg\emptyset & \neg\emptyset \end{pmatrix}$, $\begin{pmatrix} ? & ? & ? \\ ? & \neg\emptyset & ? \\ \neg\emptyset & ? & \neg\emptyset \end{pmatrix}$, and $\begin{pmatrix} ? & ? & ? \\ ? & \neg\emptyset & ? \\ ? & \neg\emptyset & \neg\emptyset \end{pmatrix}$ are categorized into the same class, because $\mathrm{CL}_{\mathcal{TR}_{\mathrm{RR}-\mathbb{R}^2}}\left(\begin{pmatrix} ? & ? & ? \\ ? & \neg\emptyset & ? \\ \neg\emptyset & \neg\emptyset & \neg\emptyset \end{pmatrix}\right) =$ $\mathrm{CL}_{\mathcal{TR}_{\mathrm{RR}-\mathbb{R}^2}}\left(\begin{pmatrix} ? & ? & ? \\ ? & \neg\emptyset & ? \\ \neg\emptyset & ? & \neg\emptyset \end{pmatrix}\right) = \mathrm{CL}_{\mathcal{TR}_{\mathrm{RR}-\mathbb{R}^2}}\left(\begin{pmatrix} ? & ? & ? \\ ? & \neg\emptyset & ? \\ ? & \neg\emptyset & \neg\emptyset \end{pmatrix}\right) = \{meet, overlap, coveredBy\}$. These equivalence classes of 3-9i matrices are called *topologically-equivalent 3-9i matrix classes*. For instance, there are 43 topologically-equivalent 3-9i matrix classes for $\mathcal{TR}_{\mathrm{RR}-\mathbb{R}^2}$. These classes are jointly-exhaustive and pairwise disjoint. Each class consists of 1 to 177 matrices.

Each topologically-equivalent 3-9i matrix class can be represented by the corresponding candidate relation set or, alternatively, a *representative* 3-9i matrix in the class with the fewest '?' elements. For instance, the previous class that consists of $\begin{pmatrix} ? & ? & ? \\ ? & \neg\emptyset & ? \\ \neg\emptyset & \neg\emptyset & \neg\emptyset \end{pmatrix}$, $\begin{pmatrix} ? & ? & ? \\ ? & \neg\emptyset & ? \\ \neg\emptyset & ? & \neg\emptyset \end{pmatrix}$, and $\begin{pmatrix} ? & ? & ? \\ ? & \neg\emptyset & ? \\ ? & \neg\emptyset & \neg\emptyset \end{pmatrix}$ is represented by $\{meet, overlap, coveredBy\}$ or $\begin{pmatrix} ? & ? & ? \\ ? & \neg\emptyset & ? \\ \neg\emptyset & \neg\emptyset & \neg\emptyset \end{pmatrix}$.

For each relation set, we can determine partial orders among topologically-equivalent 3-9i matrix classes based on their *crispness*. For instance, the class that consists of $\begin{pmatrix} ? & \emptyset & ? \\ ? & \neg\emptyset & ? \\ \neg\emptyset & \neg\emptyset & \neg\emptyset \end{pmatrix}$, $\begin{pmatrix} ? & \emptyset & ? \\ ? & \neg\emptyset & ? \\ \neg\emptyset & ? & \neg\emptyset \end{pmatrix}$, and $\begin{pmatrix} ? & \emptyset & ? \\ ? & \neg\emptyset & ? \\ ? & \neg\emptyset & \neg\emptyset \end{pmatrix}$, which is represented by $\{meet, coveredBy\}$, is crisper than the previous $\{meet, overlap, coveredBy\}$ class, because the possibility of *overlap* relation is excluded. Based on such crispness-based partial orders, we can determine a hierarchy of the topologically-equivalent 3-9i matrix classes. For instance, Fig. 6 shows the hierarchy of the topologically-equivalent 3-9i matrix classes for $\mathcal{TR}_{\mathrm{RR}-\mathbb{R}^2}$. Each class is represented by both its corresponding candidate relation set (icon) and its representative 3-9i matrix. We arranged these classes in eight levels, depending on the number of '?' elements in their representative 3-9i matrix. This diagram shows:

- there are 'converse' pairs of topologically-equivalent 3-9i classes, which are derived from each other by transposing their representative 3-9i matrices (or mirroring their icons);
- the converse pairs are located symmetrically across the vertical center;
- the converse pairs consist of the same number of 3-9i matrices; and

- as the topologically-equivalent 3-9i matrix class becomes crisper, the number of '?' elements in the representative 3-9i matrix decreases monotonically.

We can make similar hierarchical graphs for other sets of topological relations, even though the graphs become more complicated. A similar graph is seen in Cohn and Gotts (1996), which shows the hierarchy of 13 *clusters* of 46 egg-yolk relations. The egg-yolk relations in each cluster correspond to the same subset of five topological relations under RCC-5 (Randell, et al. 1992). These clusters, therefore, correspond to our topologically-equivalent 3-9i matrix classes. Schilder (1997) also shows a similar hierarchical graph for 82 convex relations on interval relations.

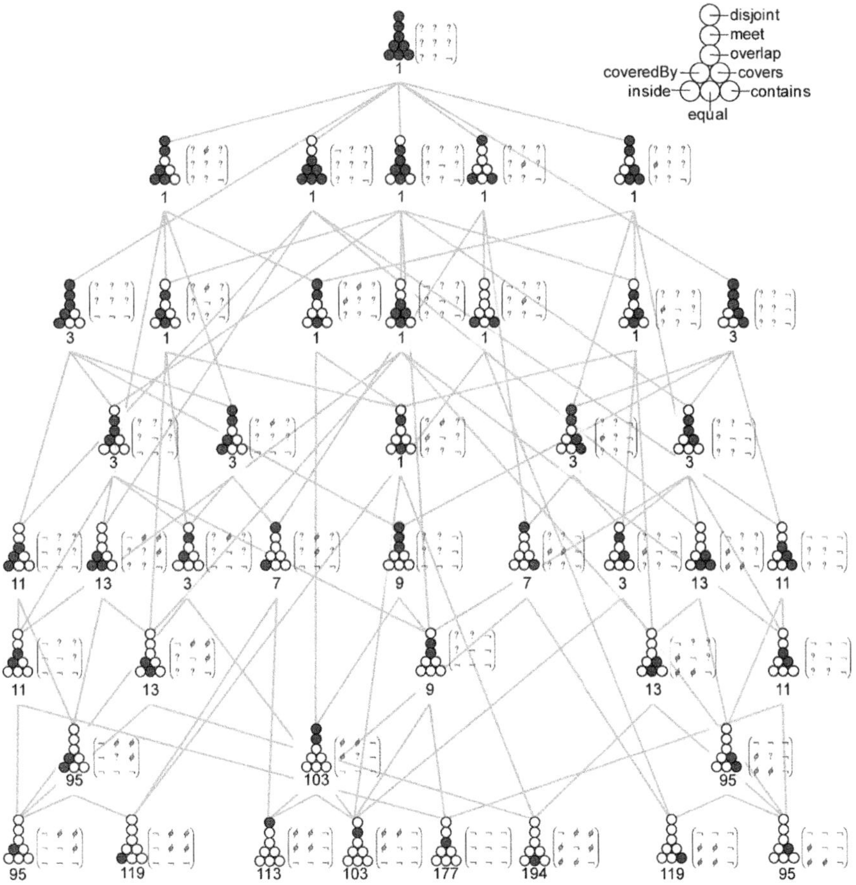

Fig. 6. The hierarchy of 43 topologically-equivalent 3-9i matrix classes for $\mathcal{TR}_{\mathbf{RR}-\mathbb{R}^2}$, together with the numbers of 3-9i matrices that form these classes

5 Expressing Uncertain Spatial Arrangements with Three-Valued 9-Intersection Matrix

This section demonstrates how the 3-9i matrix is applied to the characterization of spatial arrangements of two objects under several scenarios of incomplete observations. We then assess the degree of topological information derived from the 3-9i matrix in each scenario. Let A and B refer to the two spatial objects whose topological relation is of interest.

Scenario 1: Neutral Incomplete Observation

In this scenario, any of the 3×3 type of intersections between A and B has the risk that its presence or absence cannot be determined. This case may happen when obstacles (e.g., clouds) hide a part or whole of the interior, boundary, and exterior of each object. In this case, the 3-9i matrix may take the pattern in Eq. 2. Naturally, all valid 3-9i matrices are the target of this scenario.

$$\mathrm{M}^{\mathrm{III}} = \begin{pmatrix} \{\emptyset, \neg\emptyset, ?\} & \{\emptyset, \neg\emptyset, ?\} & \{\emptyset, \neg\emptyset, ?\} \\ \{\emptyset, \neg\emptyset, ?\} & \{\emptyset, \neg\emptyset, ?\} & \{\emptyset, \neg\emptyset, ?\} \\ \{\emptyset, \neg\emptyset, ?\} & \{\emptyset, \neg\emptyset, ?\} & \{\emptyset, \neg\emptyset, ?\} \end{pmatrix} \tag{2}$$

Scenarios 2-3: Observation on the Boundary/Boundaries

Suppose the observation is conducted only on A's boundary, because the observer's access is not allowed except A's boundary (imagine, for instance, the edge of a volcanic crater or the surface of a planet). By this observation, we can detect the presence or absence of $\partial A \cap B^{\circ}$, $\partial A \cap \partial B$, and $\partial A \cap B^{-}$, but not others. Accordingly, the 3-9i matrix may take the pattern in Eq. 3.

$$\mathrm{M}^{\mathrm{III}} = \begin{pmatrix} ? & ? & ? \\ \{\emptyset, \neg\emptyset\} & \{\emptyset, \neg\emptyset\} & \{\emptyset, \neg\emptyset\} \\ ? & ? & ? \end{pmatrix} \tag{3}$$

Similarly, if the observation is conducted only on the boundaries of both objects, the 3-9i matrix may take the pattern in Eq. 4.

$$\mathrm{M}^{\mathrm{III}} = \begin{pmatrix} ? & \{\emptyset, \neg\emptyset\} & ? \\ \{\emptyset, \neg\emptyset\} & \{\emptyset, \neg\emptyset\} & \{\emptyset, \neg\emptyset\} \\ ? & \{\emptyset, \neg\emptyset\} & ? \end{pmatrix} \tag{4}$$

Scenarios 4-5: Observation from the Boundary/Boundaries

Even if the observer's access is not allowed except A's boundary, it is sometime possible to observe the *neighbor* of A's boundary (i.e., a part of A's interior and that of A's exterior) thanks to visibility or the sensor's effective range. If this is the case, it may be possible to detect the presence of $A^\circ \cap B^\circ$, $A^\circ \cap \partial B$, $A^\circ \cap B^-$, $A^- \cap B^\circ$, $A^- \cap \partial B$, and $A^- \cap B^-$, but not their absence because not all part of A's interior/exterior can be observed. Accordingly, the 3-9i matrix may take the pattern in Eq. 5.

$$\mathrm{M}^{\mathrm{III}} = \begin{pmatrix} \{\neg\emptyset, ?\} & \{\neg\emptyset, ?\} & \{\neg\emptyset, ?\} \\ \{\emptyset, \neg\emptyset\} & \{\emptyset, \neg\emptyset\} & \{\emptyset, \neg\emptyset\} \\ \{\neg\emptyset, ?\} & \{\neg\emptyset, ?\} & \{\neg\emptyset, ?\} \end{pmatrix} \tag{5}$$

Similarly, if the observation is conducted from the boundaries of both objects, the 3-9i matrix may take the pattern in Eq. 6.

$$\mathrm{M}^{\mathrm{III}} = \begin{pmatrix} \{\neg\emptyset, ?\} & \{\emptyset, \neg\emptyset\} & \{\neg\emptyset, ?\} \\ \{\emptyset, \neg\emptyset\} & \{\emptyset, \neg\emptyset\} & \{\emptyset, \neg\emptyset\} \\ \{\neg\emptyset, ?\} & \{\emptyset, \neg\emptyset\} & \{\neg\emptyset, ?\} \end{pmatrix} \tag{6}$$

Scenario 6: Observation from the outside

Suppose A and B have the same dimension with the space (e.g., two bodies in $\mathbb{R}^3$) and their arrangement is observed from the *outside* (i.e., somewhere in $A^- \cap B^-$). In this case, we can always detect the presence or absence of $\partial A \cap B^-$ and $A^- \cap \partial B$. In addition, it is possible to detect the presence or absence of $\partial A \cap \partial B$ when A is not a part of B and vice versa. Consequently, the 3-9i matrix may take the pattern in Eq. 7. Note $A^- \cap B^-$, in which the observer is located, must be non-empty.

$$\mathrm{M}^{\mathrm{III}} = \begin{pmatrix} ? & ? & ? \\ ? & \{\emptyset, \neg\emptyset, ?\} & \{\emptyset, \neg\emptyset\} \\ ? & \{\emptyset, \neg\emptyset\} & \neg\emptyset \end{pmatrix} \tag{7}$$

Scenarios 7-8: Position uncertainty of the boundary/boundaries

In this scenario, we cannot precisely determine the spatial position of A's boundary due to blurring, low resolution, inherit vagueness, and so on. As a consequence, we cannot detect the presence or absence of the intersections related to A's boundary. In addition, we cannot determine the absence of the intersections related to A's interior and exterior. Thus, the 3-9i matrix may take the pattern in Eq. 8.

$$\mathrm{M}^{\mathrm{III}} = \begin{pmatrix} \{\neg\emptyset, ?\} & \{\neg\emptyset, ?\} & \{\neg\emptyset, ?\} \\ ? & ? & ? \\ \{\neg\emptyset, ?\} & \{\neg\emptyset, ?\} & \{\neg\emptyset, ?\} \end{pmatrix} \quad (8)$$

Similarly, if we cannot precisely determine the spatial positions of two objects' boundaries, the 3-9i matrix may take the pattern in Eq. 9.

$$\mathrm{M}^{\mathrm{III}} = \begin{pmatrix} \{\neg\emptyset, ?\} & ? & \{\neg\emptyset, ?\} \\ ? & ? & ? \\ \{\neg\emptyset, ?\} & ? & \{\neg\emptyset, ?\} \end{pmatrix} \quad (9)$$

5.1 Modification of Equations

As stated in Section 4.2, the 3-9i matrix may have fixed elements. Accordingly, the 3-9i matrices in Eqs. 2-9 should be modified to accommodate the fixed elements if they exist. For instance, for $\mathcal{TR}_{\mathrm{RR}-\mathbb{R}^2}$, $m^{\mathrm{III}}{}_{33}$ is fixed as $\neg\emptyset$ and, accordingly, the 3-9i matrix in Eq. 2 is modified into Eq. 2'.

$$\mathrm{M}^{\mathrm{III}} = \begin{pmatrix} \{\emptyset, \neg\emptyset, ?\} & \{\emptyset, \neg\emptyset, ?\} & \{\emptyset, \neg\emptyset, ?\} \\ \{\emptyset, \neg\emptyset, ?\} & \{\emptyset, \neg\emptyset, ?\} & \{\emptyset, \neg\emptyset, ?\} \\ \{\emptyset, \neg\emptyset, ?\} & \{\emptyset, \neg\emptyset, ?\} & \neg\emptyset \end{pmatrix} \quad (2')$$

5.2 Assessment of Topological Information Derived in Eight Observation Scenarios

We have seen eight scenarios of incomplete observations where the 3-9i matrix can be used for charactering uncertain spatial arrangements of the objects. An interesting question is how much topological information we can expect to derive from these incomplete observations. Thus, we conducted the following assessment featuring $\mathcal{TR}_{\mathrm{RR}-\mathbb{R}^2}$. First, we identified the sets of 3-9i matrices effective under Scenarios 1-8 (i.e., all valid 3-9i matrices that satisfy the respective patterns in Eqs. 2-9 except $m^{\mathrm{III}}{}_{33}$). Then, we calculated the average number of relations in a candidate relation set (i.e., the average number of relations to which the 3-9i matrices are mapped by the *candidate-listing* operation) for each scenario. Table 3 shows the result. In Scenarios 7-8 the average numbers of relations in a candidate relation set is much larger than those in other scenarios. This indicates that the knowledge about the presence or absence of boundary-related intersections is critical for deriving a crisper set of possible relations. In Scenarios 1-6, on average one candidate relation set contains 1 to

1.45 relations. Considering that $\mathcal{TR}_{\mathrm{RR}-\mathbb{R}^2}$ consists of 8 relations, incomplete observations in Scenarios 1-6 are highly useful for deducing a small number of possible relations. Especially, the result of Scenario 3 indicates that only from the observations on their boundaries of two regions we can uniquely determine their topological relation.

Table 3. Statistics on the mapping from 3-9i matrices to topological relations in $\mathcal{TR}_{\mathrm{RR}-\mathbb{R}^2}$ under eight scenarios of incomplete observations

Scenario	1	2	3	4	5	6	7	8
Number of corresponding 3-9i matrices	1473	6	8	78	34	11	32	8
Average number of relations in a candidate relation set	1.39	1.33	1	1.05	1	1.45	2.56	4.25

6 Assessment of Topological Information Derived under Different Topological Relation Sets

Section 5.2 assessed the topological information derived from incomplete observations featuring $\mathcal{TR}_{\mathrm{RR}-\mathbb{R}^2}$. In this section we extend the target to the 14 sets of topological relations between simple lines, simple regions, and simple bodies in $\mathbb{R}^1$, $\mathbb{R}^2$, $\mathbb{R}^3$, $\mathbb{S}^1$, and $\mathbb{S}^2$ (Table 1). For simplification, here we focus only on Scenario 1 (neutral incomplete observations), where all valid 3-9i matrices are relevant to the analysis.

First, for each relation set $\mathcal{TR}_X$, we calculated the average number of relations in a candidate relation set, ave_X. For comparison, we calculated its standardized index $1 - ave_X/|\mathcal{TR}_X|$, namely the *crispness of mapping*, on the analogy of the *crispness of compositions* in Egenhofer (2005). We also counted valid 3-9i matrices that are mapped to a unique relation in $\mathcal{TR}_X$, as well as its ratio. The result shown in Table 4 indicates that the crispness of mapping is very high in any relation set, while the ratios of uniquely-mapped 3-9i matrices under $\mathcal{TR}_{\mathrm{LR}-\mathbb{R}^2}$, $\mathcal{TR}_{\mathrm{LR}-\mathbb{S}^2}$, $\mathcal{TR}_{\mathrm{LB}-\mathbb{R}^3}$, $\mathcal{TR}_{\mathrm{RB}-\mathbb{R}^3}$, $\mathcal{TR}_{\mathrm{LR}-\mathbb{R}^3}$, $\mathcal{TR}_{\mathrm{LL}-\mathbb{R}^2}$, $\mathcal{TR}_{\mathrm{LL}-\mathbb{R}^3}$, $\mathcal{TR}_{\mathrm{LL}-\mathbb{S}^2}$, and $\mathcal{TR}_{\mathrm{RR}-\mathbb{R}^3}$ are significantly lower than the others. These nine relation sets have a common feature: one or both objects are lower-dimensional than the space. Assume that A's dimension is equal to the dimension of the space. In this case, A's boundary forms a Jordan curve and, accordingly, if B's topological part p_B intersects with A's interior and exterior, then p_B also intersects with both A's boundary as well. This serves as a constraint on the value of $\partial A \cap p_B$. Conversely, if A is lower-dimensional than the space, the value of $\partial A \cap p_B$ has more freedom. As a result, 3-9i matrices become more rarely mapped to a unique relation.

We also counted the valid 3-9i matrices that are mapped to all relations in $\mathcal{TR}_X$. The result was always one matrix, $\begin{pmatrix} ? & ? & ? \\ ? & ? & ? \\ ? & ? & ? \end{pmatrix}$, for any relation set.

Next, we counted the topologically-equivalent 3-9i matrix classes for $\mathcal{TR}_X$. For comparison, the number was standardized by the number of elements in $\mathcal{TR}_X$'s power-set (i.e., $2^{|\mathcal{TR}_X|}$), because each 3-9i matrix class is represented as a subset of $\mathcal{TR}_X$. For instance, there are 43 topologically-equivalent 3-9i matrix classes for $\mathcal{TR}_{\mathrm{RR}-\mathbb{R}^2}$, which are only 16.8% of $\mathcal{TR}_{\mathrm{RR}-\mathbb{R}^2}$'s power-set. The percentage decreases rapidly as $\mathcal{TR}_X$ contains more relations. This indicates that only slight portion of uncertain states may occur in actual incomplete observations, especially when the base relation set is large.

Table 4. Statistics on the mapping from 3-9i matrices to topological relations in 14 sets of topological relations

Relation set	$\mathcal{TR}_{\mathrm{LL}-\mathbb{R}^1}$	$\mathcal{TR}_{\mathrm{RR}-\mathbb{R}^2}$ $\mathcal{TR}_{\mathrm{BB}-\mathbb{R}^3}$	$\mathcal{TR}_{\mathrm{LL}-\mathbb{S}^1}$	$\mathcal{TR}_{\mathrm{RR}-\mathbb{S}^2}$	$\mathcal{TR}_{\mathrm{LR}-\mathbb{R}^2}$ $\mathcal{TR}_{\mathrm{LR}-\mathbb{S}^2}$ $\mathcal{TR}_{\mathrm{LB}-\mathbb{R}^3}$	$\mathcal{TR}_{\mathrm{RB}-\mathbb{R}^3}$	$\mathcal{TR}_{\mathrm{LR}-\mathbb{R}^3}$	$\mathcal{TR}_{\mathrm{LL}-\mathbb{R}^2}$ $\mathcal{TR}_{\mathrm{LL}-\mathbb{R}^3}$ $\mathcal{TR}_{\mathrm{LL}-\mathbb{S}^2}$	$\mathcal{TR}_{\mathrm{RR}-\mathbb{R}^3}$
# of relations	8	8	11	11	19	19	31	33	43
# of valid 3-9i matrices	1473	1473	3953	3953	463	463	562	2836	3016
Average # of relations in a candidate set	1.39	1.39	1.42	1.42	2.63	2.63	3.53	2.98	3.65
Crispness of mapping	.826	.826	.870	.870	.862	.862	.886	.910	.915
# of uniquely-mapped 3-9i matrices	1016	1016	2694	2694	170	170	134	978	838
% of uniquely-mapped 3-9i matrices	69.0%	69.0%	68.2%	68.2%	36.7%	36.7%	23.8%	34.5%	27.8%
# of 3-9i matrix classes	116	43	113	113	144	144	552	424	558
% of 3-9i matrix classes	45.3%	16.8%	5.52%	5.52%	0.03%	0.03%	2.6×10^{-5}%	4.9×10^{-6}%	6.3×10^{-9}%

Note in Table 4 the columns of $\mathcal{TR}_{\mathrm{RR}-\mathbb{R}^2}$ and $\mathcal{TR}_{\mathrm{BB}-\mathbb{R}^3}$ are integrated, because these two relation sets are *equivalent* with respect to the 9-intersection; that is, they consist of the same number of topological relations, which are represented by the same patterns of 2-9i matrices. For the same reason, the columns of $\mathcal{TR}_{\mathrm{LR}-\mathbb{R}^2}$, $\mathcal{TR}_{\mathrm{LR}-\mathbb{S}^2}$, and $\mathcal{TR}_{\mathrm{LB}-\mathbb{R}^3}$, and those of $\mathcal{TR}_{\mathrm{LL}-\mathbb{R}^2}$, $\mathcal{TR}_{\mathrm{LL}-\mathbb{R}^3}$, and $\mathcal{TR}_{\mathrm{LL}-\mathbb{S}^2}$ are integrated. Surprisingly, the result of $\mathcal{TR}_{\mathrm{LL}-\mathbb{R}^1}$ and that of $\mathcal{TR}_{\mathrm{RR}-\mathbb{R}^2}/\mathcal{TR}_{\mathrm{BB}-\mathbb{R}^3}$ become identical, even though they are represented by different sets of 2-9i matrices. Similarly, the results of $\mathcal{TR}_{\mathrm{LL}-\mathbb{S}^1}$ and $\mathcal{TR}_{\mathrm{RR}-\mathbb{S}^2}$, and those of $\mathcal{TR}_{\mathrm{LR}-\mathbb{R}^2}/\mathcal{TR}_{\mathrm{LR}-\mathbb{S}^2}/\mathcal{TR}_{\mathrm{LB}-\mathbb{R}^3}$ and $\mathcal{TR}_{\mathrm{RB}-\mathbb{R}^3}$ become identical.

Finally, we investigated the relation between the number of '?' elements in the 3-9i matrix and the average number of relations in a candidate relation set for 14 sets of topological relations. Interestingly, for every set of topological relations, the result shows very similar pattern (Fig. 7); that is, the average number of possible relations increases exponentially as the 3-9i matrix has more '?' elements. This supports our intuition that the number of '?' elements in the 3-9i matrix serves as a measure of topological ambiguity.

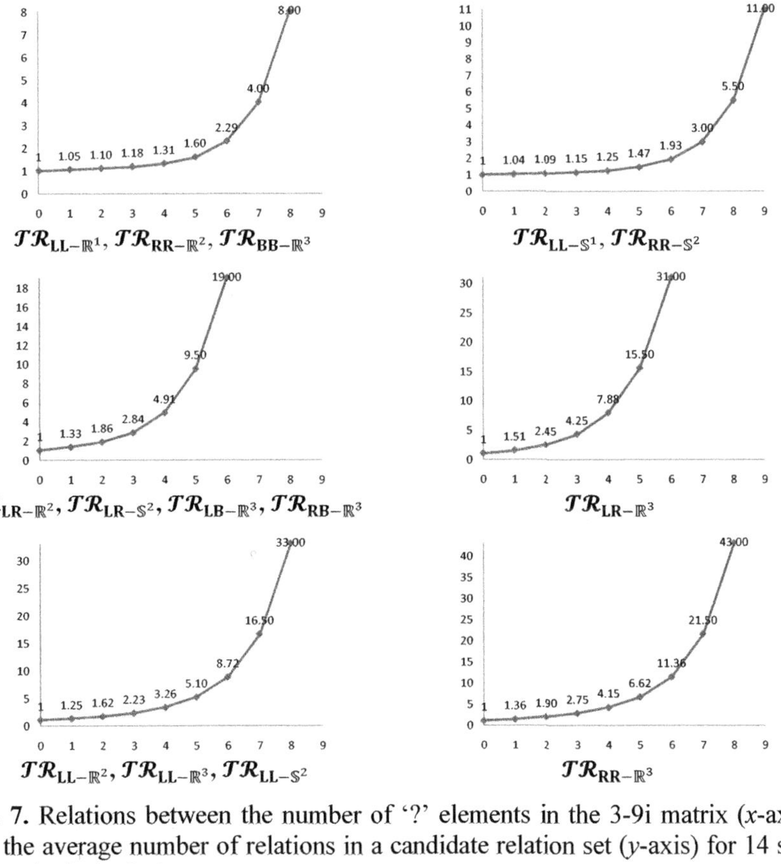

Fig. 7. Relations between the number of '?' elements in the 3-9i matrix (x-axis) and the average number of relations in a candidate relation set (y-axis) for 14 sets of topological relations

7 Conclusions

This paper studied the three-valued 9-intersection matrix by which we can easily represent the spatial arrangement of objects in various scenarios of incomplete observations, and also we can computationally a set of possible topological relations that may hold between the objects. We demonstrated that this candidate set is considerably precise, especially when the 3-9i matrix has fewer '?' elements, for any set of topological relations. This technique for deriving possible topological relations from incomplete observations serves as a complement of qualitative spatial calculi based on the 9-intersection, since rough knowledge about possible relations between objects is prerequisite for conducting qualitative spatial inference. In addition, our assessment shows that only slight portion of uncertain states may occur in actual incomplete observations. This finding is used in qualitative spatial calculi for reducing the search space considerably, as the calculi normally consider the power-set of spatial relations as the representation of all uncertain states of spatial configurations.

The three-valued approach on 4- or 9-intersection can be used also on various extensions of the 9-intersection, such as the 9^{+}-intersection (Kurata and Egenhofer 2007; Kurata 2008). The 9^{+}-intersection can capture a larger variety of topological relations, including those related to directed lines, holed regions, and complex objects. With the *three-valued* 9^{+}*-intersection matrix*, we will be able to process, for instance, uncertain spatial arrangement of motion paths and landmarks or those of temporal intervals under incomplete observations. Thus, the analysis of the three-valued 9^{+}-intersection matrix will be also fruitful for practical applications of qualitative spatial reasoning.

References

Alexandroff, P.: Elementary Concepts of Topology. Dover Publications, Mineola, NY, USA (1961)

Burrough, P.: Natural Objects with Indeterminate Boundaries. In: Bur-rough, P., Frank, A. (eds.): Geographic Objects with Indeterminate Boundaries. Taylor & Francis, London, UK, pp. 3-28 (1996)

Clementini, E.: A Model for Uncertain Lines. Journal of Visual Language and Computing 16(4), 271-288 (2005)

Clementini, E., Felice, P.: An Algebraic Model for Spatial Objects with Indeterminate Boundaries. In: Burrough, P., Frank, A. (eds.): Geographic Objects with Indeterminate Boundaries. Taylor & Francis, London, UK, pp. 155-170 (1996)

Clementini, E., Felice, P.: Topological Invariants for Lines. IEEE Transactions on Knowledge and Data Engineering 10(1), 38-54 (1998)

Cohn, A., Gotts, N.: The 'Egg-Yolk' Representations of Regions with Indeterminate Boundaries. In: Burrough, P., Frank, A. (eds.): Geographic Objects with Indeterminate Boundaries. Taylor & Francis, London, UK, pp. 171-188 (1996)

Du, S., Qin, Q., Wang, Q., Li, B.: Fuzzy Description of Topological Relations I: A Unified Fuzzy 9-Intersection Model. In: Wang, L., Chen, K., Ong, Y. (eds.): ICNC 2005, Lecture Notes in Computer Science, vol. 3612, pp. 1261-1273. Springer, Berlin/Heidelberg, Germany (2005)

Egenhofer, M.: Spherical Topological Relations. Journal on Data Semantics III, 25-49 (2005)

Egenhofer, M., Franzosa, R.: Point-Set Topological Spatial Relations. International Journal of Geographical Information Systems 5(2), 161-174 (1991)

Egenhofer, M., Franzosa, R.: On the Equivalence of Topological Relations. International Journal of Geographical Information Systems 9(2), 133-152 (1995)

Egenhofer, M., Herring, J.: Categorizing Binary Topological Relationships between Regions, Lines and Points in Geographic Databases. In: Egenhofer, M., Herring, J., Smith, T., Park, K. (eds.): NCGIA Technical Reports 91-7. National Center for Geographic Information and Analysis, Santa Barbara, CA, USA (1991)

Erwig, M., Schneider, M.: Vague Regions. In: Scholl, M., Voisard, A. (eds.): 5th International Symposium on Advances in Spatial Databases, Lecture Notes in Computer Science, vol. 1262, pp. 298-320. Springer, Berlin/Heidelberg, Germany (1997)

Kurata, Y.: The 9^+-Intersection: A Universal Framework for Modeling Topological Relations. In: Cova, T., Miller, H., Beard, K., Frank, A., Goodchild, M. (eds.): GIScience 2008, Lecture Notes in Computer Science, vol. 5266, pp. 181-198. Springer, Berlin/Heidelberg, Germany (2008)

Kurata, Y., Egenhofer, M.: The 9^+-Intersection for Topological Relations between a Directed Line Segment and a Region. In: Gottfried, B. (ed.): First International Workshop on Behavioral Monitoring and Interpretation, TZI-Bericht, vol. 42, pp. 62-76. Technogie-Zentrum Informatik, Universität Bremen, Germany (2007)

Lee, B., Flewelling, D.: Spatial Organicism: Relations between a Region and a Spatially Extended Point. In: Egenhofer, M., Freksa, C., Miller, H. (eds.): GIScience 2004, Extended Abstracts and Poster Summaries, pp. 144-147 (2004)

Ligozat, G., Renz, J.: What Is a Qualitative Calculus? A General Framework. In: Zhang, C., Guesgen, H., Yeap, W. (eds.): PRICAI 2004, Lecture Notes in Artificial Intelligence, vol. 3157, pp. 53-64. Springer, Berlin/Heidelberg, Germany (2004)

Liu, K., Shi, W.: Quantitative Fuzzy Topological Relations of Spatial Objects by Induced Fuzzy Topology. International Journal of Applied Earth Observation and Geoinformation 11(1), 38-45 (2009)

Molenaar, M.: Fuzzy Spatial Objects. In: Molenaar, M. (ed.): An Introduction to the Theory of Spatial Object Modelling for GIS. Taylor & Francis, London, UK, pp. 193-224 (1998)

Nedas, K., Egenhofer, M., Wilmsen, D.: Metric Details of Topological Line-Line Relations. International Journal of Geographical Information Science 21(1), 21-48 (2007)

Randell, D., Cui, Z., Cohn, A.: A Spatial Logic Based on Regions and Connection. In: Nebel, B., Rich, C., Swarout, W. (eds.): 3rd International Conference on Knowledge Representation and Reasoning, pp. 165-176. Morgan Kaufmann, San Francisco, CA, USA (1992)

Reis, R., Egenhofer, M., Matos, J.: Topological Relations Using Two Models of Uncertainty for Lines. In: Caetano, M., Painho, M. (eds.): 7th International Symposium on Spatial Accuracy Assessment in Natural Resources and Environmental Sciences, pp. 286-295 (2006)

Schilder, F.: A Hierarchy for Convex Relations. In: 4th International Workshop on Temporal Representation and Reasoning, pp. 86-93. IEEE Computer Society (1997)

Schneider, M.: Uncertainty Management for Spatial Data in Databases: Fuzzy Spatial Data Types. In: Güting, R.H., Papadias, D., Lochovsky, F. (eds.): 6th International Symposium on Advances in Spatial Databases, Lecture Notes in Computer Science, vol. 1651, pp. 330-351. Springer, Berlin/Heidelberg, Germany (1999)

Tang, X., Fang, Y., Kainz, W.: Fuzzy Topological Relations between Fuzzy Spatial Objects. In: Wang, L., Jiao, L., Shi, G., Li, X., Liu, J. (eds.): 3rd International Conference on Fuzzy Systems and Knowledge Discovery, Lecture Notes in Computer Science, vol. 4223, pp. 324-333. Springer, Berlin/Heidelberg, Germany (2006)

Tøssebro, E., Nygård, M.: An Advanced Discrete Model for Uncertain Spatial Data. In: Meng, X., Su, J., Wang, Y. (eds.): 3rd International Conference on Web-Age Information Management, Lecture Notes in Computer Science, vol. 2419, pp. 37-51. Springer, Berlin/Heidelberg, Germany (2002)

Winter, S.: Topological Relations in Hierarchical Partitions. In: Freksa, C., Mark, D. (eds.): COSIT'99, Lecture Notes in Computer Science, vol. 1661, pp. 141-155. Springer, Berlin/Heidelberg, Germany (1999)

Winter, S.: Uncertain Topological Relations between Imprecise Regions. International Journal of Geographical Information Science 14(5), 411-430 (2000)

Winter, S., Bittner, T.: Hierarchical Topological Reasoning with Vague Regions. In: Shi, W., Goodchild, M., Fisher, P. (eds.): International Symposium on Spatial Data Quality '99, pp. 555-565. Hong Kong Polytechnic University, Hong Kong, China (1999)

Worboys, M.: Spatial Reasoning and Uncertainty. In: Worboys, M., Duckham, M. (eds.): GIS: A Computing Perspective. CRC Press, Boca Raton, FL, USA, pp. 323-358 (2004)

Yee, L.: On the Imprecision of Boundaries. Geographical Analysis 19(2), 125-151 (1987)

Zadeh, L.: Fuzzy Sets. Information and Control 8, 338-353 (1965)

Modeling Land Use Change: A GIS Based Modeling Framework to Support Integrated Land Use Planning (NabanFrame)

Karin Berkhoff, Sylvia Herrmann

Institute of Environmental Planning, Leibniz University Hannover,
Herrenhäuser Str. 2, 30519 Hannover,
{berkhoff, herrmann}@umwelt.uni-hannover.de

Abstract. Monoculture rubber plantations replace traditional land use systems in subtropical south west China. This land use change intensified since 1990 and reduced natural diversity. Planning authorities need spatially explicit information for sustainable land use planning.

We developed an integrated modeling cluster to provide decision support for planning authorities. Our definition of an integrated modeling framework was to apply and coordinate agro-economic, ecological and social models which altogether interact with a land allocation model via defined interfaces (no dynamic coupling). Data sources were remote sensing data, data from geographic information systems (GIS), questionnaires, narrative interviews, and ecological field surveys. We conducted GIS analysis (Euclidian distance, Euclidian allocation, focal mean, map algebra) to reference originally non-spatial information to spatial units. The baseline scenario based on the location factors elevation, distance to villages and available labor. We allowed land use change to occur only in regions outside nature protection zones. Villages served as spatial reference (and thus interface) for social information, farm types as spatial reference for agro-economic information.

The result of the modeling framework was a map of land use change for the baseline scenario (2001-2007). Model results showed that rubber covered nearly the whole area that is below the rubber growing limit of 1200 meters in the year 2007. Fields concentrated on the western part of the study area where rubber growing is not possible.

We showed that it was possible to integrate from various sources into a decision support tool. The value of the approach was that all data were referenced to

M. Sester et al. (eds.), *Advances in GIScience*, Lecture Notes in Geoinformation and Cartography,
DOI 10.1007/978-3-642-00318-9_16,

spatial entities. The modeling framework provided land use maps and evaluated the implications of land use change from a social, agro-economic and ecological point of view. Planning authorities can use the results to conduct sustainable land use planning.

1 Introduction

Land use change causes loss of natural vegetation in many parts of the world. Rubber plantations replace tropical rainforests and fallow fields in the case of Yunnan province, south west China. They have an impact on flora, fauna and traditional land use systems. Rubber already covers an area of about 10% of the study area (2007) (Fig. 1). Land use allocation in the study area is mostly driven by economic considerations. Thus, local planning authorities need decision support for land use planning that integrates socio-economic and ecological aspects.

Fig. 1. Typical landscape in the study area with rice paddies (foreground) and terraces prepared for rubber growing (right side) and the nature reserve forests (background)

Due to the need of sustainable land use planning in the study area, our goal was to develop a geographic information system (GIS) based tool that balances socio-economic and ecological needs.

GIS based land use change models exist approximately since the year 1996. Veldkamp & Fresco (1996) present a conceptual model to study the conversion of land use and its effects (CLUE). The model is continuously being enhanced (Claessens et al. 2009; Verburg et al. 2002; Verburg & Veldkamp 2004). Lambin (1997) conclude that social science knowledge needs to be integrated better in land use change models. Lambin et al. (2001) state that the drivers of land use change are often simplified ("poverty", "population"). They state that in the case of tropical deforestation in Southeast Asia the main driver is migration to plantations triggered by government decisions.

Numerous studies focus on urban planning. Pettit & Pullar (2004) use GIS in their modeling approach to disaggregate data, assess land requirements and development constraints, apply land use transition rules and model urban expansion based upon accessibility to services. They base their study on the spatial scenario planning approach developed by Stillwell et al. (1999).

Sante-Riveira et al. (2008) particularly address rural planning with their planning support system. They include the steps of evaluation of land suitability, optimization of land use areas and spatial allocation in their tool. They do not incorporate an evaluation of the socio-economic and ecological impacts of land use change.

Social structures and ownership relationships influence landscape pattern. Sklenicka & Salek (2008) show that land ownership is an important characteristic for land use allocation.

Mottet et al. (2006) present one of the few studies that base their land use modeling on empirical data. They interview farmers to learn about historic land use change on farm level. They integrate several socio-economic driving factors of land use change (land use organization, type of access, remoteness from the farmstead, land ownership) in a European study area. We choose an analogous approach to Mottet et al. (2006) that is characterized by the particular situation in the Chinese study area. It is a rural, remote region with low data availability. Steep slopes make it difficult to interpret remote sensing data because shadows distort the reflection. Several working groups of the LILAC ("Living Landscapes China") project[1] conduct interviews and field studies to derive input data for the socio-

[1] supported by the German Federal Ministry of Education and Research under promotional reference 0330797A

economic and ecological models. The interview results comprise amongst others leasehold relations and innovation level of villages.

We apply an agro-economic, ecological and social model which altogether interact with a land allocation model via defined interfaces (no dynamic coupling). We name this approach the NabanFrame modeling framework. NabanFrame delivers land use maps for scenarios (in this case: current state, business as usual and sustainable land use). In addition, it evaluates the land use maps with regard to their implications from a social, agro-economic and ecological point of view. We visualize model results of NabanFrame by means of a GIS (ArcGIS 9.3).

Our aim is to develop a GIS based modeling framework that incorporates socio-economic and ecological knowledge. It should be able to integrate data from disciplinary models, interviews and field studies. In consequence, it should be suitable for application in rural, remote areas where data scarcity is a problem. A modeling framework that is designed according to these requirements provides decision support for sustainable land use planning in rural areas.

2 The NabanFrame Modeling Framework

The NabanFrame modeling framework formed the umbrella for the application of the sociology, agro-economy, ecology and the land use change model. The current state scenario integrated several parameters (Fig. 2).

The period specified for modeling the current state scenario was 2001 to 2007. NabanFrame proceeded in three steps: pre-processing phase, land allocation and post-processing (evaluation of impact). The main focus of this paper was on the land allocation phase (where the $CLUE_{Naban}$ model was applied) because it brought together the information from the disciplinary models.

2.1 Data

Data sources within NabanFrame were highly heterogeneous with regard to content and type of data (Table 1). We integrated data into the land allocation model (section 0) either within the module location factors, spatial restrictions or conversion settings.

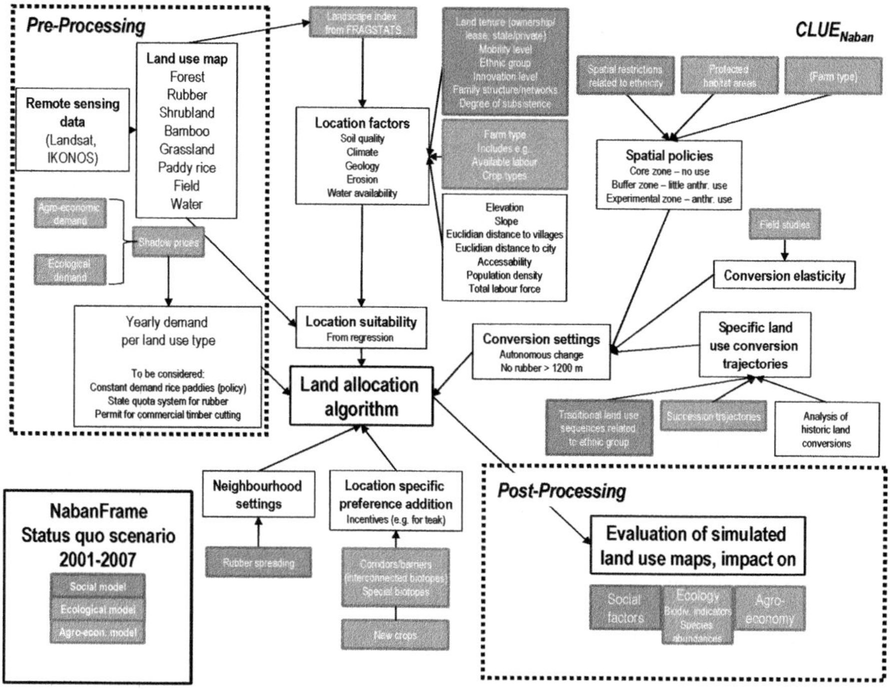

Fig. 2. The NabanFrame modeling framework, operationalized for the current state scenario. Colors refer to model inputs (blue: social model, green: ecological model, orange: agro-economic model). White boxes are in the responsibility of the land allocation model

Table 1. Data sources of NabanFrame

Data Source	Content	Data type
Remote sensing data	Landsat ETM+	Grid
	IKONOS	Grid
GIS data	Land use	Grid
	Infrastructure	Line feature
	Rivers	Line feature
	Elevation	Grid
	Soil	Polygons
	Climate	Point information
Questionnaires	Interviews on social topics (house-holds)	Text
	Agro-economic questionnaires	Text
Narrative interviews	Interviews on innovation diffusion (individuals)	Text
Ecological field surveys	Fauna	Field survey
	Flora	Biodiversity indices

2.2 Pre-Processing and Demand Negotiation Phase

The pre-processing phase served the purpose of preparing data and also defined the land use demands for the land allocation model.
Landsat 7 ETM+ (CNES, 2001) was the basis for classifying land use of the year 2001. We distinguished between 7 land use classes: paddy rice, field, forest, rubber, bamboo, grassland/shrubland and other.

The demand for the different land use types was an external driver for the land allocation model. In NabanFrame, it is usually estimated by balancing the agro-economic and the ecological demand. Social demand will be considered in a next version of NabanFrame. We also considered institutional restrictions of land use demand: The amount of rice paddies was almost stable because it was protected by law, there was a state quota system for rubber, and permits were required for commercial timber cutting (Xu 2006).

2.3 $CLUE_{Naban}$

The second step within NabanFrame was land allocation which was operated by the land allocation model ($CLUE_{Naban}$). $CLUE_{Naban}$ based on the CLUE-S model (Verburg et al. 2006; Verburg et al. 2002). CLUE-S was designed to simulate land use change using empirically quantified relations (regression analysis) between land use and its driving factors combined with the modeling of competition between land use types (dependent on location suitability, neighborhood setting, conversion elasticity and a demand-related iteration variable). It was assumed that locations were assigned to the land use type with the highest total probability. $CLUE_{Naban}$ had the advantage that it was possible to integrate data from the social, agro-economic and ecologic model. It consisted of four modules (Fig. 3).

Only one of the modules, the definition of the 'land use requirements', was non-spatial. It was defined externally of $CLUE_{Naban}$ for the study area as a whole. We described the procedure how land use requirements were defined in NabanFrame in section 2.1.

In the 'location characteristics' module location suitability was derived from location factors by regression analysis and a logit model (Verburg et al. 2002). Location factors in $CLUE_{Naban}$ were delivered from all disciplinary models. E.g. ecological landscape indices were considered, further agro-economic farm-types (describing available labor, farm structure, crop types, and other) and social characteristics influencing land allocation. These are e.g. land tenure, population, mobility level, ethnic group, etc.

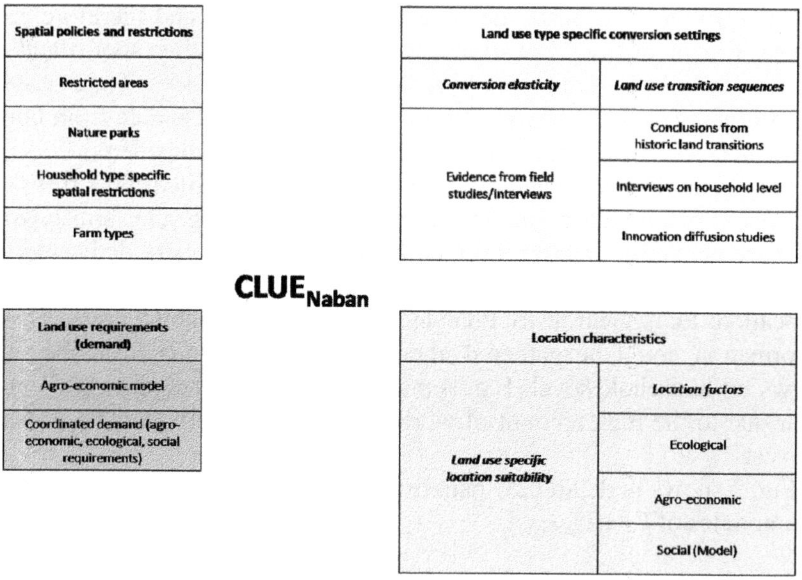

Fig. 3. Modules of the $\text{CLUE}_{\text{Naban}}$ model

Several physical characteristics of the study area like e.g. elevation, soil quality, climate and erosion were also taken as location factors. In section 3 it we describe how we prepare data in ArcGIS 9.3 to meet the requirements of the land allocation model.

The 'land use type specific conversion settings' module of $\text{CLUE}_{\text{Naban}}$ defined the conditions how changes from one land use type to another occurred. This module had two elements: conversion elasticity and land use transition sequences. The conversion elasticity described within a range from 0 to 1 for every land use type if a land use change can be converted back easily (value 0) or if it is irreversible (value 1). In the study area, e.g. fields could be converted back easily and therefore got a conversion elasticity of 0.1. In contrast, paddy rice and rubber both required high initial investments, thus reversion was difficult for farmers (conversions elasticity values of 0.9 and 0.8).

A conversion matrix defined land use transition sequences for every land use type. It described if a conversion to another land use type was possible or not. Natural changes (succession) as well as human induced changes were considered. It was also possible to define differing rules for certain areas; e.g. in the study area only natural succession was allowed in the core zone of the nature reserve. Outside the core zone also human induced changes occurred. A major conversion restriction in the study area was that rubber was easily damaged by cold weather and therefore could not be grown in high elevations. Additional information about land use change could be derived from historic transition sequences. Time series of Landsat imagery were analyzed for this purpose, but also data from household interviews and innovation diffusion studies were considered.

The 'Spatial policies and restrictions' module benefited most from GIS analysis. Spatial restrictions in the study area were e.g. the farm types of the agro-economic model. Regions of the study area were dedicated to a certain farm type which entails several rules and restrictions for land use allocation. Ecological restrictions included protected habitat areas. Further information could be gathered about spatial restrictions from the interviews on household level. E.g. some households planted rice only once a year due to the high amount of work that was necessary to have a second harvest.

Fig. 2 shows in detail how data from the other models is incorporated in the modules of $CLUE_{Naban}$.

2.4 Impact Assessment

The final step within NabanFrame was to evaluate land use change impacts in the post-processing phase. $CLUE_{Naban}$ delivered land use maps for each of the scenarios as input data for the impact assessment (Fig. 2). The three disciplinary models then evaluated the maps with regard to their impact on social, ecological and agro-economic factors. E.g. the social model gave information about the influence of the changed land use on family structures, whereas the ecological model concentrated on changes in biodiversity. The agro-economic model investigated changes in land use with regard to farm income. Finally, at the end of the post-processing phase three evaluation maps were available for decision support in land use planning; all three categories of sustainability (social welfare, ecology, economy) were considered.

2.5 Scenarios

The current state scenario was modeled for the years 2001 to 2007. Landsat imagery was used to classify land use in the starting year of the simulation. Seven land use types were classified, including paddy rice, field, forest, rubber, bamboo, grassland/bushland and other. Since the agro-economic model was not yet implemented we derived land use requirements for each land use type from literature and expert knowledge. We chose elevation, distance to villages/market, available labor, and population as location factors. They were described in detail in section 0. We assumed that land use changes to paddy rice and rubber could not be reconverted easily, whereas field, bamboo and grassland/bushland could be re-converted easily. Only natural succession was allowed in the core zone of the study area. In the other zones also anthropogenic land conversion was possible.

The current state scenario served as calibration model run, because the modeled land use map 2007 could be compared to a land use classification derived from IKONOS 2 imagery (EUSI, 2007). The images were georeferenced and mosaicked and then a supervised land use classification was created. The IKONOS land use map was used to calibrate the current state scenario run. It was important that the main drivers of land use change were implemented correctly into $\text{CLUE}_{\text{Naban}}$.

The NabanFrame modeling framework will be applied for two further scenarios; the modeled period will be 2007 to 2020. One scenario will be business as usual where current land use practices and motivations for land use change will be extrapolated into the future. Changes in external factors (global change, state infrastructure planning etc.) will be considered. For this purpose the scenarios of the Intergovernmental Panel on Climate Change (IPCC) (McCarthy et al. 2007; Watson et al. 2000) will be regionalized for the study area and planning documents will be analyzed. The other scenario will be sustainable land use. For this scenario we will reflect on alternative cash crops and alternative land management, referring to the highly diverse land management the ethnic minorities practiced in former times (Fu et al. 2006; Shiro et al. 2007; Wu et al. 2001; Xu 2006; Xu et al. 2005).

2.6 Visualizing Model Results for Decision Makers

Model results were visualized for the decision makers by GIS (ArcGIS 9.3). The model output of NabanFrame was twofold:

1. Land use maps for every scenario
2. Evaluation of the impacts of land use change by the social, agro-economic and ecological model

The land use maps were important for the land use planning process because they gave spatially explicit information about land use changes in the study area. Maps were created with a spatial resolution of 25 meters, and they included 7 land use types: paddy rice, field, forest, rubber, bamboo, grassland/bushland and other. The maps should help decision makers to identify where changes were likely to occur, and how fractions of land use types might change. Since maps were understood intuitively by people, we considered them suitable as discussion basis in planning processes (Janssen et al. 2006). Data processing and the modeling procedure which finally generated the land use map must be communicated appropriately to secure the decision makers' confidence (Borowski & Hare 2007). An example of a land use map for the current state scenario can be found in section 4.

Planning processes needed additional information about the impacts of modeled changes. Since the objective of the LILAC project was to support sustainable land use planning, the impacts of land use changes were evaluated with regard to social implications, biodiversity and agro-economy. The results of the impact assessment were presented to the decision makers in the form of maps (biodiversity indices), diagrams and tables.

Eventually, another tool will be applied to communicate modeling results to the decision makers. An ArcGIS Server was running for the LILAC project where maps of the study area were published. The advantage of ArcGIS Server was that no desktop GIS had to be available and that it is suitable for interactive operations. E.g. people could zoom to regions they were interested in, deactivate layers that were not needed and perform basic GIS operations. Thus, it could be used in planning processes to communicate geodata to decision makers.

3 Preparing Data in GIS

We explained how data from the disciplinary models were integrated in the land allocation model, $CLUE_{Naban}$ (section 2). In the following we illustrate in detail how these data were prepared in ArcGIS 9.3 to meet the requirements of $CLUE_{Naban}$. Special emphasis was put on the preparation of socio-economic data (distance to next village, available labor, population).

$CLUE_{Naban}$ demanded exactly the same projection, grid size and spatial extent for all data. Universal Transverse Mercator (UTM) projection, zone 47N, 25 meter grid size and the lower left corner of the study area mask were decided on as reference parameters.

3.1 Location Factors

3.1.1 Elevation

Elevation was the most important physical characteristic that served as location factor for rubber growing. The height limit for rubber growing varied from 800 meters to 1350 meters depending on the study (Li et al. 2007; Wu et al. 2001; Xu et al. 1990); in the study area rubber growing could be observed up to a height of 1200 meters. Elevation data with 90 meter spatial resolution (Jarvis et al. 2006) were resampled to 25 grid size. Elevations in the study area ranged from 540 to 2291 meters (Fig. 4).

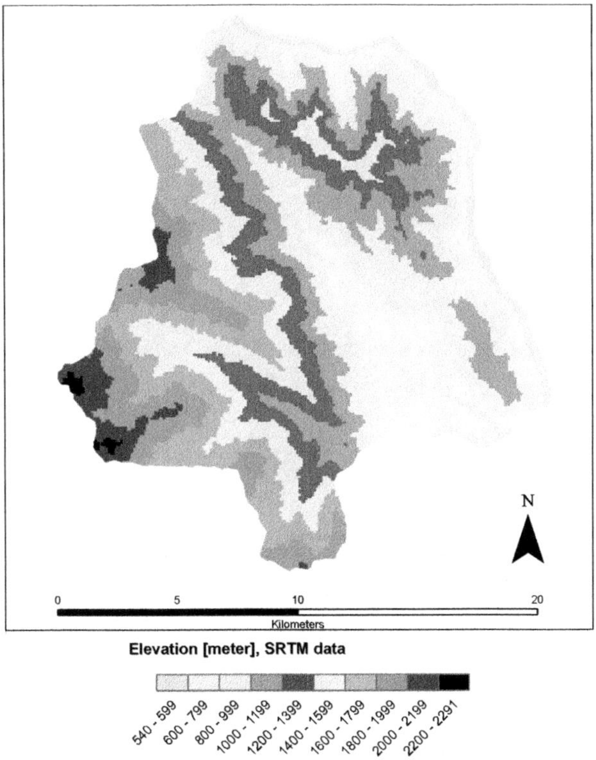

Fig. 4. Elevations in the study area, data from the Shuttle Radar Topography Mission (SRTM, Jarvis et al. 2006)

3.1.2 Distance to Villages

The distance to the next village influenced land use allocation for the land use types paddy rice, field and forest. Rice paddies and fields were labor intensive and thus had to be in close vicinity to villages. Forests were often in vicinity to villages because the inhabitants collect firewood or herbs there. Further, the ethnic minority of the Dai worshiped nature in the form of holy hills covered with forests (Xu 2006).

We digitized villages from the mosaicked and georeferenced IKONOS image and converted them to points. Then, we calculated the Euclidian distance of each grid cell in the study area to the next village using the Spatial Analyst. The distance to the next village varied between 0 and 7.8 kilometers (Fig. 5).

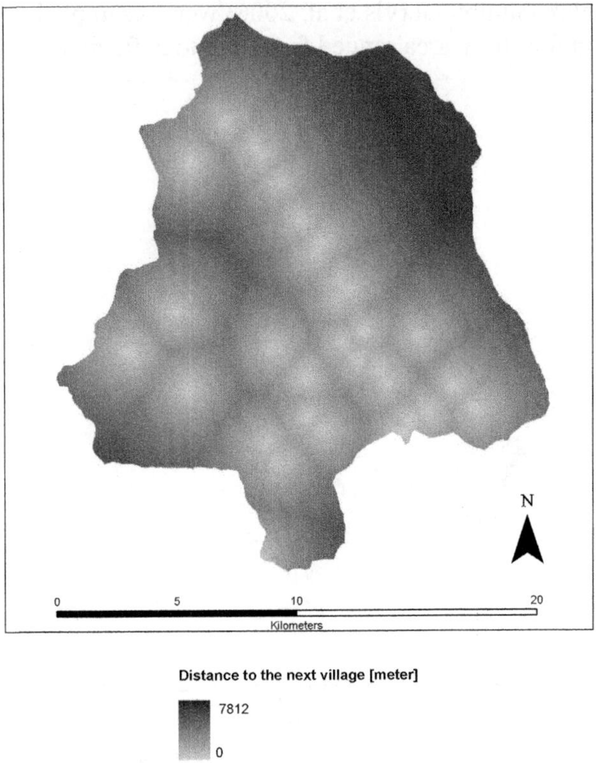

Fig. 5. Distance to the next village

3.1.3 Available Labor

In the regression analysis available labor turned out to be the most important location characteristic for the land use types paddy rice, field and in particular rubber. All three land use types were labor intensive, paddy rice needed the transplantation of seedlings, irrigation and harvesting, fields had to be tilled and also harvested and rubber must be tapped daily.

Information about manpower resources was available for the study area on village level. The administrative boundaries of the villages were not available. Nevertheless, we needed data with spatial reference for $CLUE_{Naban}$. We used the Euclidian allocation function in the Spatial Analyst to overcome this lack of data. Each grid cell in the study area was dedicated to the closest village, up to a maximum distance of 3 kilometers (Xu et al. 1990). As long as data of the real boundaries were not available the artificial boundaries (Fig. 6) served as a proxy and are used in $CLUE_{Naban}$.

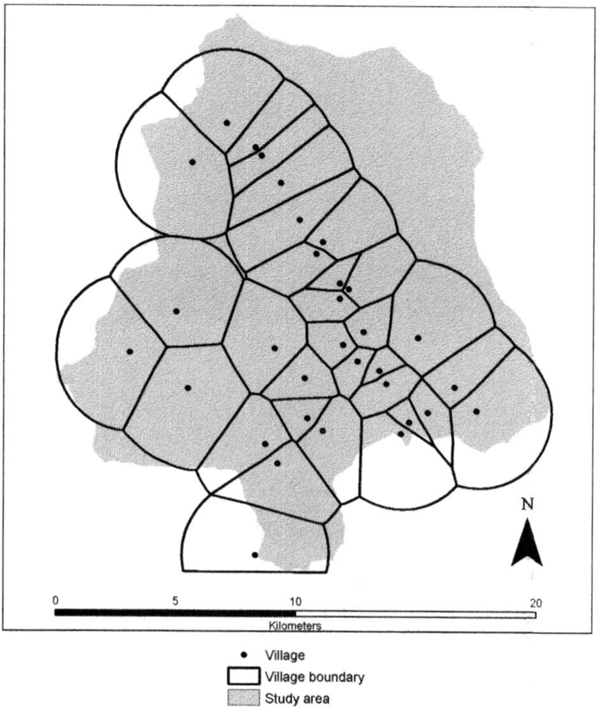

Fig. 6. Hypothetical village boundaries gained through Euclidian allocation

We took the spatial extent of the villages to normalize manpower resources. Available labor could be spatially distributed using the focal mean function (Fig. 7). The same procedure was applied for population; also this information was available on village level.

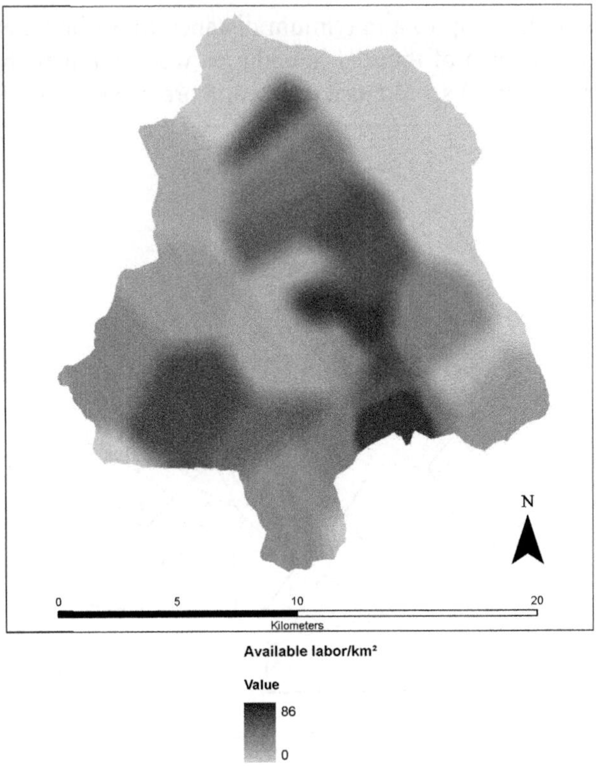

Fig. 7. Available labor per square kilometer in the study area

3.2 Spatial Restrictions

The study area was a nature reserve that consisted of three protection zones. In the core zone, no kind of anthropogenic influence was allowed, land use change only occurred through natural succession. In the buffer zone and also in the experimental zone, anthropogenic land use was possible. We created a mask of the core zone for $\text{CLUE}_{\text{Naban}}$ and used it as a spatial restriction for anthropogenic land use. Another mask contained the grid cells that have higher elevation than the height limit for rubber growing allowed. Since rubber growing can be observed in the study area up to a height of 1200 meters, this is fixed was height limit for rubber growing (Fig. 8).

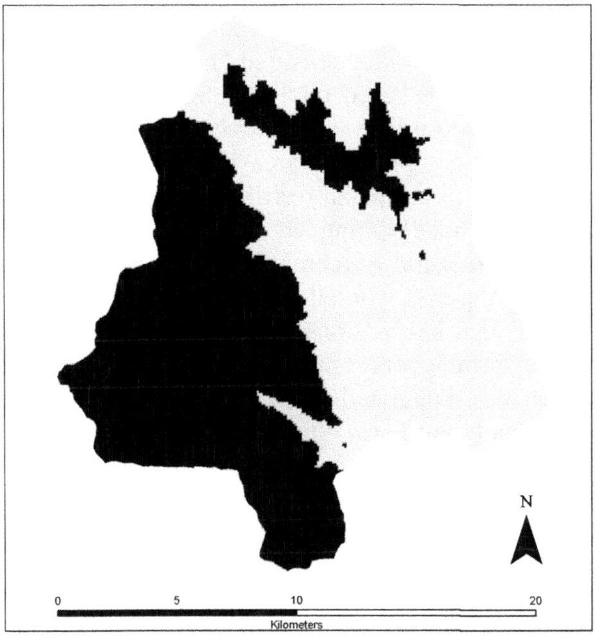

Fig. 8. Areas not suitable for rubber growing (black color) since they exceed the height limit of 1200 meters

3.3 Conversion Settings

The mask of the core zone was also used for the conversion matrix. Basically, in the conversion matrix 0 was used when a change was not possible and 1 if it was possible. The conversion matrix can be extended through linking it to GIS maps (Verburg et al. 2002). The GIS maps were coded according to the syntax in the conversion matrix. They allowed that certain land use changes only referred to special areas and not to the whole area. For $CLUE_{Naban}$ the map of the core zone was reclassified and integrated in the conversion matrix.

3.4 Interfaces

We did not couple dynamically the models within NabanFrame. The results of the social, ecological and agro-economic model were used as input data for the land allocation model. The results of the social model related to the villages in the study area. Through this spatial reference it was possible to integrate them into $CLUE_{Naban}$. E.g. spatial restrictions could be derived that differ from village to village due to ethnicity, special tenure rules, etc. The ecological model delivered landscape indices in grid format for the whole study area, based on FRAGSTATS (McGarigal & Marks 1995). Thus, it is easy to utilize them as location factors in $CLUE_{Naban}$. Interconnected biotopes and protected areas from the ecological model were considered in $CLUE_{Naban}$ as spatial restrictions for anthropogenic land use. The agro-economic model (Fang & Nuppenau 2004) in NabanFrame which was based on General Algebraic Modeling System (GAMS, GAMS Development Corporation[2]) worked on the spatial unit of farm types. The study area was divided into regions that were dedicated to a certain farm type, e.g. rubber farming or cash food crops. The results of the agro-economic model were integrated into $CLUE_{Naban}$ via the spatial reference of, in this case, farm types. Farm type characteristics were relevant as location factors, but also as spatial restrictions for certain land management practices. The agro-economic model gave another input into NabanFrame: it delivered (after negotiation with the ecological model) the land use requirements for $CLUE_{Naban}$, which had to be defined externally.

4 Results

In former times the study area was broadly covered with undisturbed mountainous rain forest. Only in the bottoms of the valleys paddy rice, vegetables, etc. have been planted. Rubber developed particularly in the lower eastern part of the study area (Fig. 9). It moved into the study area from the south (along the road to the city). A rubber spot near the northern boundary of the study area increased since 2003. Finally, rubber covered nearly the whole area that was below the rubber growing limit of 1200 meters (cf. section 0). Fields concentrated on the western part of the study area where rubber growing was not possible. Shifting cultivation is practiced there (Xu et al. 1990). Rubber replaced fields in the lower areas of the study area. The spatial restrictions in the core zone (north-eastern part of the study area) resulted in continuous forest vegetation cover there.

[2] http://www.gams.com/

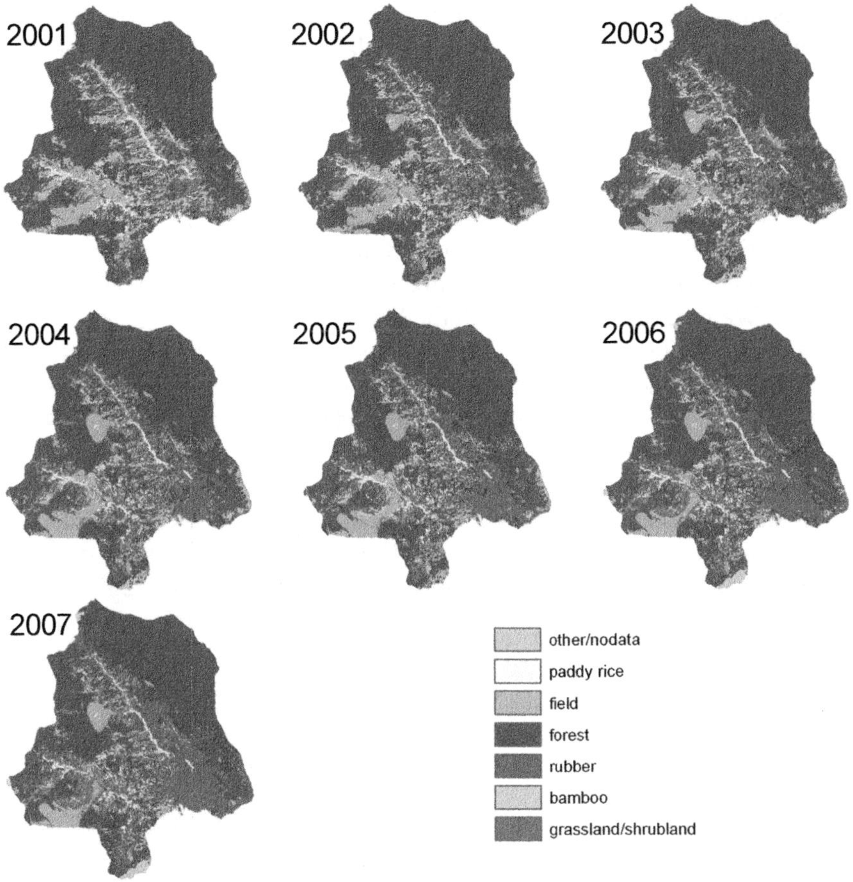

Fig. 9. $CLUE_{Naban}$ simulation results (current state scenario), years 2001-2007

5 Discussion

We provide a GIS based modeling framework (NabanFrame) that integrates socio-economic and ecological information. We found that in the case of a rural study area in China villages and farm types are suitable spatial entities to incorporate this knowledge into the modeling framework. We show that the NabanFrame modeling framework can be applied in a data scarce region.

We cannot yet confirm the conclusion of Lambin et al. (2001) that the main driver of tropical deforestation is migration to plantations that is triggered by government decisions. The social model within NabanFrame evaluates institutional drivers of land use change like these government decisions. This is also true for ownership relations which are considered important for land use change by Sklenicka & Salek (2008).

GIS analysis provides only approximations of reality in the current state of modeling and needs to be refined. We currently use Euclidian distance as an accessibility indicator, but we will replace it by cost-weighted distance as soon as additional information is available.

We find this iterative procedure of data integration useful for study areas where only few data is available at the beginning of the modeling exercise. We proceed similarly with the location factors.

The evaluation phase of NabanFrame is in a conceptual status at the current state of modeling. For this reason, we cannot show its results here, but we regard the evaluation phase as an important and integral part of NabanFrame that also sets it apart from other modeling approaches.

NabanFrame can be applied for sustainable land use planning because it provides land use maps and socio-economic and ecological evaluation of the impacts of land use change. The GIS analyses that we use to overcome data scarcity can be helpful also in other case studies. The regional planning authority and other agencies that are involved in environmental planning can apply NabanFrame in planning procedures.

References

Borowski I. & Hare M. (2007) Exploring the gap between water managers and researchers: Difficulties of model-based tools to support practical water management. *Water Resources Management* 21: 1049-1074.

Centre national d'études spatiales (CNES) (2001) Landsat 7 Enhanced Thematic Mapper Plus (ETM+), acquisition date 12.01.2001.

Claessens L., Schoorl J. M., Verburg P. H., Geraedts L. & Veldkamp A. (2009) Modelling interactions and feedback mechanisms between land use change and landscape processes. *Agriculture Ecosystems & Environment* 129: 157-170.

European Space Imaging (EUSI) (2007) IKONOS Geo, acquisition dates 16.11.2007/02.12.2007

Fang L. & Nuppenau E.-A. (2004) A Spatial Model (SWAM) for Water Efficiency and Irrigation Technology Choices using GAMS - A Case Study from Northwestern China. In: *7th Annual Conference on Global Economic Analysis*, The World Bank, Washington, 07.-09.06.2004, https://www.gtap.agecon.purdue.edu/resources/download/1790.pdf.

Fu Y. N., Guo H. J., Chen A. G. & Cui J. Y. (2006) Household differentiation and on-farm conservation of biodiversity by indigenous households in Xishuangbanna, China. *Biodiversity and Conservation* 15: 2687-2703.

Janssen M. A., Goosen H. & Omtzigt N. (2006) A simple mediation and negotiation support tool for water management in the Netherlands. *Landscape and Urban Planning* 78: 71-84.

Jarvis A., Reuter H. I., Nelson A. & Guevara E. (2006) Hole-filled seamless SRTM data V3. International Centre for Tropical Agriculture (CIAT).

Lambin E. F. (1997) Modelling and monitoring land-cover change processes in tropical regions. *Progress in Physical Geography* 21: 375-393.

Lambin E. F., Turner B. L., Geist H. J., Agbola S. B., Angelsen A., Bruce J. W., Coomes O. T., Dirzo R., Fischer G., Folke C., George P. S., Homewood K., Imbernon J., Leemans R., Li X. B., Moran E. F., Mortimore M., Ramakrishnan P. S., Richards J. F., Skanes H., Steffen W., Stone G. D., Svedin U., Veldkamp T. A., Vogel C. & Xu J. C. (2001) The causes of land-use and land-cover change: moving beyond the myths. *Global Environmental Change-Human and Policy Dimensions* 11: 261-269.

Li H. M., Aide T. M., Ma Y. X., Liu W. J. & Cao M. (2007) Demand for rubber is causing the loss of high diversity rain forest in SW China. *Biodiversity and Conservation* 16: 1731-1745.

McCarthy J. J., Canziani O. F., Leary N. A., Dokken D. J. & White K. S. eds. (2007) *Climate Change 2007 – Impacts, Adaptation and Vulnerability (Contribution of Working Group II to the Fourth Assessment Report of the IPCC)*. Cambridge University Press, also available online: http://www.ipcc.ch/ipccreports/ar4-wg2.htm.

McGarigal K. & Marks B. J. (1995) *Fragstats: spatial pattern analysis program for quantifying landscape structure*, Portland, OR, USA.

Mottet A., Ladet S., Coque N. & Gibon A. (2006) Agricultural land-use change and its drivers in mountain landscapes: A case study in the Pyrenees. *Agriculture Ecosystems & Environment* 114: 296-310.

Myers N., Mittermeier R. A., Mittermeier C. G., da Fonseca G. A. B. & Kent J. (2000) Biodiversity hotspots for conservation priorities. *Nature* 403: 853-858.

Pettit C. & Pullar D. (2004) A way forward for land-use planning to chieve policy goals by using spatial modeling scenarios. *Environment and Planning B: Planning and Design* 31: 213-233.

Sante-Riveira I., Crecente-Maseda R. & Miranda-Barros D. (2008) GIS-based planning support system for rural land-use allocation. *Computers and Electronics in Agriculture* 63: 257-273.

Shiro C., Furtad J. I., Shen L. X. & Yan M. (2007) Coping with pressures of modernization by traditional farmers: A strategy for sustainable rural development in Yunnan, China. *Journal of Mountain Science* 4: 57-70.

Sklenicka P. & Salek M. (2008) Ownership and soil quality as sources of agricultural land fragmentation in highly fragmented ownership patterns. *Landscape Ecology* 23: 299-311.

Stillwell J., Geertman S. & Openshaw S. (1999) Developments in geographical information and planning. In: *Geographical Information and Planning* (eds. J. Stillwell, S. Geertman & S. Openshaw) pp. 3-23. Springer, Berlin.

Veldkamp A. & Fresco L. O. (1996) CLUE: A conceptual model to study the conversion of land use and its effects. *Ecological Modelling* 85: 253-270.

Verburg P. H., Schulp C. J. E., Witte N. & Veldkamp A. (2006) Downscaling of land use change scenarios to assess the dynamics of European landscapes. *Agriculture Ecosystems & Environment* 114: 39-56.

Verburg P. H., Soepboer W., Veldkamp A., Limpiada R., Espaldon V. & Mastura S. S. A. (2002) Modeling the spatial dynamics of regional land use: The CLUE-S model. *Environmental Management* 30: 391-405.

Verburg P. H. & Veldkamp A. (2004) Projecting land use transitions at forest fringes in the Philippines at two spatial scales. *Landscape Ecology* 19: 77-98.

Watson R. T., Noble I. R., Bolin B., Ravindranath N. H., Verardo D. J. & (Eds.) D. J. D. eds. (2000) *IPCC Special Report on land use, land-use change and forestry*. Cambridge University Press, also available online: http://www.ipcc.ch/ipccreports/sres/land_use/index.htm.

Wu Z.-L., Liu H.-M. & Liu L.-Y. (2001) Rubber cultivation and sustainable development in Xishuangbanna, China. *The international journal of sustainable development and world ecology* 8: 337-345.

Xu J. C. (2006) The political, social, and ecological transformation of a landscape - The case of rubber in Xishuangbanna, China. *Mountain Research and Development* 26: 254-262.

Xu J. C., Ma E. T., Tashi D., Fu Y. S., Lu Z. & Melick D. (2005) Integrating sacred knowledge for conservation: Cultures and landscapes in southwest China. *Ecology and Society* 10: -.

Xu W., Shirasaka S. & Ichikawa T. (1990) Farming system and settlements in Xishuangbanna, Yunnan province, China. *Geographical Review of Japan* 62 (Series B): 104-165.

Monitoring System for Assessment of Vegetation Sensitivity to El-Niño over Africa

Pavel Propastin

Department of GIS and Remote Sensing, Institute of Geography, Georg-August-University Göttingen, Goldschmidtstr. 5, 37077 Göttingen, ppropas@uni-goettingen.de

Abstract. The study investigated vulnerability of vegetation to El-Niño Southern Oscillation (ENSO) over Africa by correlating Normalized Difference Vegetation Index (NDVI) data from the Advanced Very High Resolution Radiometer (AVHRR) and two ENSO indices, namely Multivariate ENSO Index (MEI) and Southern Oscillation Index (SOI). The study developed a new monitoring approach (ENSO vulnerability assessment system) that examined and quantified associations between monthly maximum NDVI anomalies and month-to-month correlations with the ENSO indices over the vegetated land areas of Africa throughout the period from 1982 to 2006 at the pixel scale. This system was engaged for an assessment of the long-time vegetation sensitivity to ENSO warm events occurred during the study period. A map of vegetation vulnerability to ENSO was produced. Areas with various vulnerability degrees were measured within main vegetation cover classes. The results suggested that the vulnerability of vegetated tropical land surfaces to climate extremes like EL Nino depends considerably on vegetation type. In particular, it could be shown that equatorial forest areas are more reliable to drought stress than other wooded and non-wooded vegetation categories.

1 Introduction

Global climate anomalies linked to El Niño-Southern Oscillation (ENSO) leading to a redistribution of precipitation and temperature patterns on the Earth surface are one of the major topics of scientific investigation (IPCC, 2007; Chattopadhyay and Bhatla, 1993; Fuller and Murphy, 2006). It is a coupled atmospheric and oceanic mechanism responsible for changes in

M. Sester et al. (eds.), *Advances in GIScience*, Lecture Notes in Geoinformation and Cartography,
DOI 10.1007/978-3-642-00318-9_17,

the Walker circulation system that in turn affects the global atmospheric circulation and, therefore, the weather and climate in other parts of the world (Hoerling and Kumar, 1997; Fox, 2000). The recurrence of this phenomenon with a periodicity of 2 to 7 years has enormous social, economic and ecological impacts worldwide (Glantz, 1996). Africa belongs to the most affected areas. It is well known that the warm event of ENSO or El Niño causes unfavourable climate conditions in broad areas of Africa (Hoerling and Kumar, 2000). Therefore, understanding of the impact of ENSO warm events on ecosystems at all scales is an important component of Earth system science research.

Remote sensing research has also been directed toward the investigation of ENSO impacts in different geographical regions. The Normalized Difference Vegetation Index (NDVI) has been the most used satellite product for these investigations. NDVI has been shown to be correlated with a number of measures of the relative abundance of green biomass, including leaf area index, intercepted fraction of photosynthetically active radiation and density of chlorophyll in plants (Asrar et al., 1984; Sellers et al., 1997; Heinsch et al., 2006). These useful properties of the NDVI led to its adoption as an operational indicator for climate dynamics (Yang et al., 1998; Tateishi and Ebata, 2004; Li et al., 2002; Wang et al., 2003). Time series of satellite-derived NDVI data have been proofed to contain the ENSO signal in various regions and vegetation types (Gutman et al., 2000; Gurgel and Ferreira, 2003; Nagai et al., 2007; Prasad et al., 2007). Furthermore, negative anomalies of NDVI in Africa, indicative of drought, have been shown to be associated with ENSO warm events (Verdin et al., 1999; Anyamba and Estman, 1996; Anyamba et al., 2001).

The goal of this study is to investigate and quantify the vulnerability to ENSO warm events for the vegetated area of Africa throughout the period of 1982-2006. This study developed an effective assessment system for ENSO vulnerability by means of moving window correlation analysis (MWCA) between time series of NDVI and ENSO indices (MEI and SOI) at the pixel scale. Employing this assessment system to vegetated areas of Africa, spatial patterns of land surface response to El-Niño over the period 1982 – 2006 were mapped and discussed. The degree and extent of the vegetation sensitivity to ENSO warm events was also measured within each land cover type.

2 Data

2.1 Remote Sensing Data

In this study we used the NOAA AVHRR NDVI data set compiled by the Global Inventory Monitoring and Modelling Studies (GIMMS) research group. The data cover the period from July 1981 to December 2006 and have a spatial resolution of 8 km. The data are originally processed as 15-day composites using the maximum value procedure to minimize effects of cloud contamination (Holben, 1986) and corrected for sensor drift and sensor degradation (Pinzon et al., 2004). Clouds obscure the land surface and are especially a problem over areas with frequent rainfall and associated dense vegetation. Clouds have low negative NDVI values, whereas vegetated surfaces have NDVI values over zero. The maximum value compositing, e.g. selection of the maximum NDVI over a period of 10 days, 15 days of a month, reduces the effects of clouds by selecting data from clear, cloud-free days with high NDVI values. For the purpose of this research, we created monthly maximum composites from the 15-day composites in any given month to further minimize the effects of clouds on the vegetation and to match the monthly temporal resolution of the associated ENSO indices. However, over tropical forests, very few cloud-free data were available. In such cases the composing period could be expanded (e.g. to two months), or spatial filters could be used that select the maximum NDVI over a window of 3*3 or 5*5 pixels. We removed noisy pixel areas characterized by exceptionally low NDVI values relatively to their pixel neighbourhood by replacing them by a mean value calculated from their spatial neighbourhoods.

In addition, calibration based on the method described by Los (1993) has been applied to the data to minimize the effects of sensor degradation. The NDVI time-series were calibrated against two time invariant desert targets located in the Big Arabian Desert and Nubian Desert. The vegetation-free surface of these desert targets considered to be stable throughout the analyzed time-period and should exhibit NDVI with value of near zero. Any temporal deviations of the NDVI value from zero have to be attributed to a non-vegetation noise and are to be corrected. This method removes effects of sensor degradation remaining in the original GIMMS data and corrects drift between different sensor systems.

For the purpose of this study, the 8-km pixels of the original data set were aggregated to 24-km pixels. The aggregating of the original data set to a coarser resolution aimed: (a) an additional reduction of cloud contamination in the extreme cases over tropical regions, and (b) to relieve the further data analysis through a reduction of the data set quantity. In this study, the data analysis was performed with SAGA-GIS software (http://www.saga-gis.org) by incorporating an additional module specially programmed for the MWCA procedure. This MWCA module worked very slowly if the data set was too large. Taking into account the processing ability of the MWCA module, the size of the original NDVI data set (over 10 Gigabytes) composed of 306 layers (306 months from July 1981 to December 2006) was significantly reduced through the aggregation of the 8-km pixels to 24-km pixels. This reduction enabled the further effective processing with the MWCA module.

2.2 ENSO Indices

Method to describe the state of ENSO system or intensity of a particular ENSO event is to implement proxies - so called indices which are built e.g. on basis of anomalies of climatic variables measured over the various parts of the Pacific Ocean. The dataset of different ENSO indices is freely available from the web-site of the Climate Prediction Centre of National Oceanic and Atmospheric Agency (NOAA) of the USA (http://www.cpc.ncep.noaa.gov/). The ENSO indices used in the analysis include the following:

1. The Southern Oscillation Index (SOI), which represents the basin-wide oscillations in patterns of atmospheric pressure between the eastern and western Pacific, measured as normalized difference in sea level pressure between Papeete (Tahiti) and Darwin (Australia).
2. The Multivariate ENSO Index (MEI), which evaluates the six observed atmospheric and maritime parameters to present the air-sea interaction (Wolter and Timlin, 1998).

Both SOI and MEI have recently been subjects to several studies and showed their high suitability for remote sensing-based analyses of ENSO-vegetation relationships in various geographical regions and vegetation zones (Anyamba et al., 2001; Mennis, 2001; Gutman et al., 2000; Nagai et al., 2007).

2.3 Land Cover Data

The land cover information in this study is based on the AVHRR global land cover product for the reference year 1994 at 8 km spatial resolution (DeFries et al., 1998). The initial land cover map included 14 land cover classes that was spatially resampled to 24-km spatial resolution for the purpose of this study. Two classes, – water and bare ground, - were excluded from the analyses.

3 Methods

First, for each month of the study period images of standardized NDVI anomaly were computed as differences from monthly long-term means divided by the standard deviation. Second, the nature of the general teleconnections between ENSO and patterns in variability in NDVI over Africa was explored using correlation analysis between NDVI anomalies at the per-pixel scale and ENSO indices. Third, areas affected by ENSO were determined for each month at the per-pixel scale. Finally, an ENSO vulnerability map was produced and discussed.

3.1 Correlation Analysis

To explore relationship between ENSO indices and variability in NDVI we implemented a moving window correlation analysis (MWCA) approach. It is a statistical technique which is used to investigate variations in relationships or non-stationarity in regression parameters. The concept of MWCA uses a window with a definitive size that is moving across the data set to be analysed and a separate local correlation coefficient is calculated for each regression point. The MWCA is commonly used to investigate spatial variations when dealing with spatially distributed data (Fotheringham et al., 1996; Lloyd, 2005; Nicolau et al., 2002). The MWCA has been successfully employed to investigation spatiotemporal variations of relationships between satellite-derived NDVI and ENSO over Indonesian archipelago (Erasmi et al., 2009). In the present study, the MWCA was employed in temporal dimension (Figure 1). When applied in temporal dimension, this technique produces a smoother time-series of parameter estimates and represents time-continuity of the analysed process. The equation for calculation of the local correlation coefficient using MWCA is like that for the common correlation analysis but is expanded and rewritten as:

$$r_{in} = \frac{\frac{1}{n}\sum_{j=1}^{n}(x_i - \mu_{xn})(y_i - \mu_{yn})}{\sigma_{xn}\sigma_{yn}} \qquad (1)$$

where r_{in} is the correlation coefficient calculated for month *i; n* is the amount of month included in the local correlation analysis, it refers to the window size; x_i and y_i are values of the dependent and independent variables *x* and *y* in month *i*; μ_{xn} and μ_{yn} are mean values of *x* and *y* calculated within the moving window of a size *n*; σ_{xn} and σ_{yn} are corresponding standard deviations for *x* and *y*.

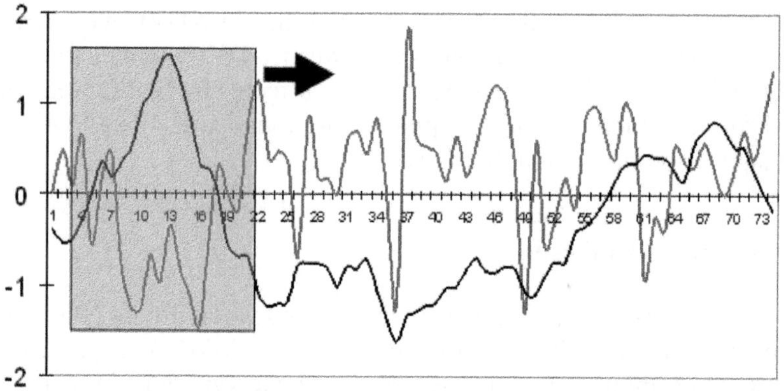

Fig. 1. Example of the MWCA employed in temporal dimension. A regression window of a definitive size (in this case 20 months) is moving across the data set and a correlation coefficient is calculated for the respective point located in the centre of the regression window. Green line is NDVI time series and black line is MEI time series.

The MWCA is a local technique that disaggregates global statistics and calculates the relationship between NDVI and its predicting variables for a certain time-point (month in this study) for each pixel in the area to be analysed. In this research, the MWCA was used to analyse relationships between time-series of monthly NDVI and ENSO indices. In general, the proposed technique disaggregates the entire data set consisting of more than 300 months into separate local correlation models (in temporal dimension) and produces correlation estimates (in the case of the present study - a value of correlation coefficient between NDVI and ENSO indices) for each point (month) of the time series. This way, looking at the resulted time-continuity of the NDVI-ENSO correlation, we can trace the evolution of the response of vegetation to ENSO over the whole study period (see Figure 1 and 3). The use of the MWCA approach helps to detect "hot spots" of ENSO impacts on vegetation both in space and time. Obviously, the results of a MWCA application depend on the size of the moving window. We tested different window widths (from 12 months to 48 months) in order to find the most significant correlation between the NDVI anomalies and the ENSO proxies. The calculations were done both at the concurrent basis and by imposing time lags into the correlation analysis (from 0 to 4 months).

The MWCA applied in this study is a good example for spatiotemporal modelling. Being employed in temporal dimension, the MWCA produces spatial pattern of temporal correlation for each of the layers (months). It means that, as results, the MWCA produced 306 maps of temporal correlation between NDVI and the ENSO indices. These correlation maps were used for further assessment of long time vulnerability of vegetation to El Niño.

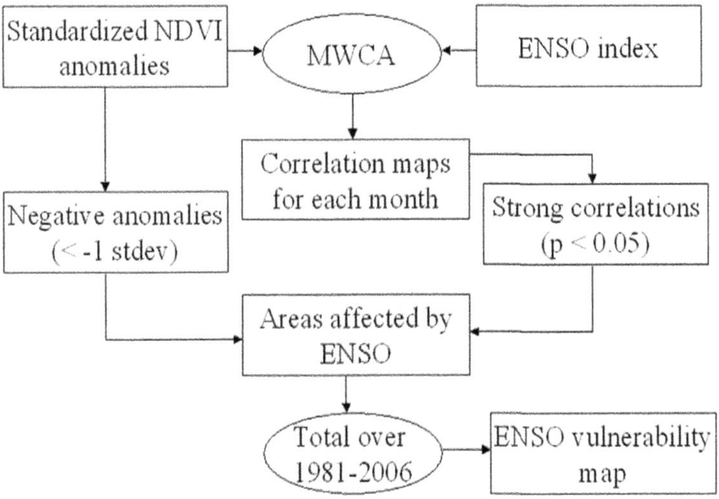

Fig.2. Monitoring system for vegetation vulnerability to ENSO warm events based on the MWCA approach.

3.2 Assessment of Vegetation Vulnerability to El-Niño

A flowchart of work steps used to evaluate vegetation vulnerability to El Niño is shown in Figure 1. First, standardized anomalies of monthly NDVI were calculated for every month at the pixel scale. Second, employing the MWCA technique, correlations between ENSO indices and NDVI anomalies were computed for every month throughout the study period (July 1981 – December 2006) and every pixel in the African mainland. Third, pixels with extensive negative NDVI anomalies (< -1 standard deviation of

NDVI) within all the ENSO warm events (and two months after a certain event) were compared with the results of the correlation analysis. We considered that any monthly NDVI anomaly supported by a strong correlation with MEI or SOI would indicate a statistically significant temporary change of vegetation activity driven by ENSO dynamics. These pixels, which showed both NDVI anomaly < -1 standard deviation and strong correlation with one of the ENSO indices, were detected and extracted. Finally, we calculated the occurrence of these pixels throughout the period from July 1981 to December 2006 for each point of the African area. The magnitude of this occurrence would indicate the vulnerability of vegetation to El Niño. The general mean of the evaluation system is the following: the higher the recurrence of vegetation changes caused by El Niño is, the more vulnerable is the area.

Table 1. Warm episodes of ENSO for the 1982-2006 period based on a threshold of –10 for SOI and +1.0 for MEI and corresponding period of vegetation response to warm episodes ($p < 0.05$)

SOI based warm episodes	MEI based warm episodes	NDVI anomalies < -1 stdev	NDVI/SOI response	NDVI/MEI response
06/82 – 03/83	07/82 – 08/83	01/83 – 04/83	01/83 – 04/83	12/82 – 04/83
11/86 - 10/87	12/86 – 01/88	12/87 – 03/88	none	none
03/91 - 05/91	06/91 – 07/92	01/92 – 02/92	12/91 – 04/92	01/92 – 03/92
09/91-10/91	04/93 – 10/93	12/93 – 03/94	11/93 – 05/94	06/93 – 04/94
12/91 – 01/92	10/94 – 01/95	01/95 – 02/95		02/95
03/92 – 04/92				
04/93 – 10/93				
03/94 – 12/94				
04/97 - 04/98	05/97 - 06/98	12/97 – 04/98	01/98 – 04/98	01/98 – 07/98
11/99 – 04/00	11/99 – 02/00	01/00 – 03/00	none	none
12/02	12/02 – 01/03	01/03 – 02/03	none	none
02/05 – 05/05	none	02/05	01/05 – 02/05	02/05
10/06 – 11/06	09/06 – 12/06	none	none	none

4 Results

4.1 General Dynamics of NDVI and ENSO Indices

Strong negative values of SOI and positive values MEI indicate thresholds for warm event i.e. El Niño. In this study, El Niño threshold values for indices where defined as SOI < -10 and MEI > 1.0 respectively. Based on the thresholds for SST anomaly and MEI time series data, El-Niño events were lined out for the period from 1982 to 2006 (see table 1, left part). The most severe of them are associated with the years 1982-83 and 1997-98. The 1991-94 episode was characterized by relatively weak magnitudes of the ENSO indices and could be treated as a number of single episodes based on the conventions for SOI and MEI thresholds used in this study. The episodes of 2000, 2002, 2005 and 2006 were relative short and had rather weak magnitudes. All the El-Nino events outlined in table 1, with exception of the 2006, affected vegetation activity on the way that the NDVI of areas responded to ENSO should show generally negative anomalies during El-Nino months. A sample site (3*3 pixels) located in southwest Africa centred at 10°N and 6°10`E was selected for use as an example. The NDVI anomalies and NDVI response to the ENSO indices (see table 1, middle and right parts) were compounded from this sample site. Obviously, the response of vegetation to unfavourable climate conditions caused by El-Nino differs between land cover types and individual pixels. This sample site was used only to demonstrate how the monitoring system designed in this study works. The NDVI dynamics of the sample site showed strong negative anomalies (below –1 standard deviation) during the El-Nino events except that of 2006 (table 1, middle row). These NDVI anomalies occurred either within or immediately after a certain El-Nino event.

The MWCA technique enables monitoring of temporal variations of the NDVI-ENSO relation and to trace the evolution of the ENSO influence on vegetation over the study period. Best results in terms of statistical significance (p-value) were achieved using MWCA window widths of 9 months before and after the regression point (19 month) and an assumed time lag between ENSO proxy variability and vegetation response of 0 to 1 month. The results of the MWCA analysis for the sample site based on these parameter settings are presented in Figure 3. The figure shows the non-stationary nature of the relationship between NDVIA and ENSO throughout the time and reveals clear temporal patterns in the response of the vegetation cover to ENSO dynamics. NDVI anomalies basically shows strong positive correlations with SOI and strong negative correlations with MEI during ENSO warm events (e.g. 1982-83, 1991-92, 1994-95, 1997-98 and 2005). It means that vegetation activities over the selected area tend to be influenced by climate changes in El-Nino years. The strongest correlations for NDVI-SOI and NDVI-MEI relationships are observed during the ENSO events of 1982-83 and 1997-98. We extracted from the graphs in figure 2 months with high values of the correlation coefficient ($p < 0.05$) and compared them with the months showing strong negative NDVI anomalies (table 1, right part). Temporal patterns in strong NDVI anomalies (< -1 standard deviation) showed particularly strong association with that of the correlation coefficient between NDVI and both ENSO indices.

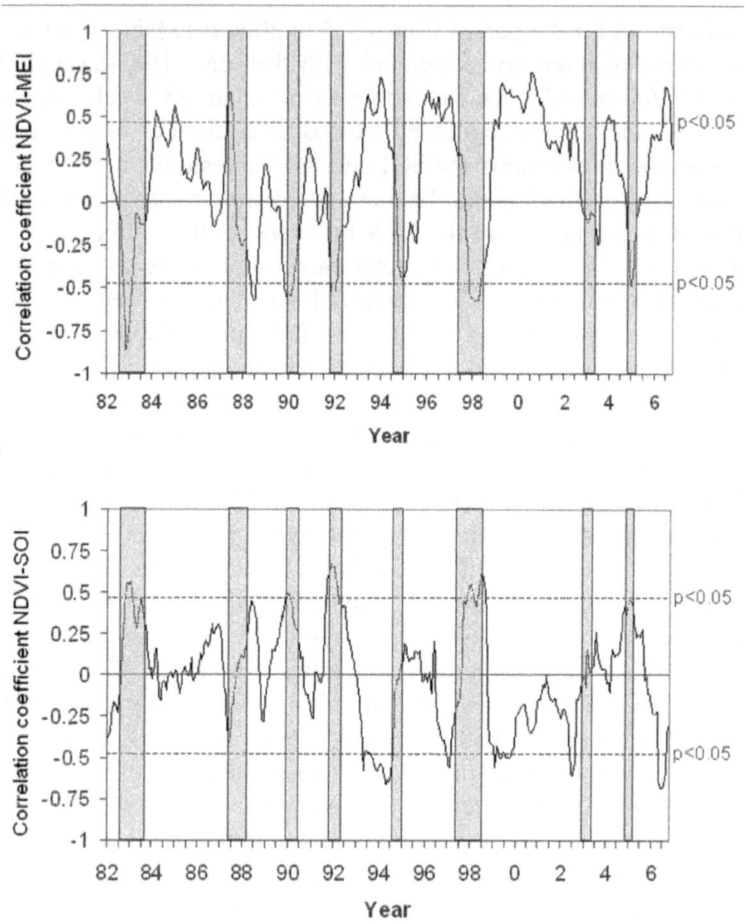

(a)

(b)

Fig. 3. Correlation coefficients between NDVI anomalies and MEI (a) and between NDVI anomalies and SOI (b) during the period from 1982 to 2006, calculated for the sample pixel (10°N and 6°10`E) with a 19-month moving window and with time-lags of 0 months. Duration and location of ENSO warm events within the study period is represented by blue stripes.

We summed up the amount of months which showed both the strong negative NDVI anomalies and statistically significant correlation ($p < 0.05$) with one of the ENSO indices. The results were 15 for the NDVI-SOI and 16 for the NDVI-MEI relationships. It means that, for the sample site, of the 25 cases with strong negative anomalies of monthly NDVI registered during all the El-Nino events over the period 1982 – 2006 15 and 16 cases can be explained (with a statistical probability of 95%) by the dynamics of SOI and MEI, respectively. This quantity can be called as the El-Nino vulnerability index or the index of vegetation sensitivity to ENSO warm events. It should be, however, emphasized that we examined only negative impacts of ENSO warm events on vegetation, which caused a decrease of vegetation activity. Theoretically, over the territory of Africa the index may range from 0 to 65 (the total amount of El-Nino months over the period 1982 - 2006). A value of 0 means no sensitivity of vegetation to El-Nino or the lowest vulnerability of vegetated land areas to El-Nino, whereas a value of 65 means the highest possible sensitivity of vegetation.

4.2 Spatial Patterns of Vegetation Vulnerability to El-Nino

The above described algorithm was employed to each vegetated pixel in African mainland. First, we found out months with strong negative NDVI anomalies for each pixel. After that, we calculated correlation coefficients between NDVI anomalies and the ENSO indices using MWCA technique with different window widths and time lags. Further, at the per-pixel level, we determined months which showed strong negative NDVI anomalies and statistically significant correlation coefficients. These months were summed up and maps of vegetation vulnerability to ENSO warm events were produced. Figure 4 presents the maps resulted from NDVI-MEI and NDVI-SOI relationships. Both used indices produced almost similar patterns. The maps indicate that the most area of the African mainland is not sensitive or only low sensitive to El-Nino. The vulnerability of vegetation to El-Nino impact shows a certain spatial pattern. The maps show a discontinuous broad band of the medium and high vulnerable vegetated areas across the Sahel, the Sudano-Sahelian zone into parts of West Africa. Large clusters of high vulnerability areas occur in the middle part of southern Africa and in the south-eastern coast of the African mainland (most parts of the South-African highland, the Dragon mountains, the Cap mountains).

Recent studies on vegetation dynamics suggested different response of vegetation types to both inter-annual and intra-annual precipitation and temperature variability (Richard and Poccard, 1998; Li et al., 2002; Kowabata et al., 2001; Wang et al., 2003). In order to find discrepancies in the vulnerability of vegetation to El-Nino drought impacts with regard to vegetation types, the areas of different vulnerability degree (low, medium, high) were measured within each vegetation type. The results indicate remarkable variations in the percentage of the ENSO vulnerable areas depending on land cover type (Table 2). In general, evergreen broadleaved forest areas were less affected during the study period 1982 – 2006 than other land cover types. More than 80% and 18% of the total EBF area showed no response or only low response of vegetation to El-Nino events, respectively. Deciduous broadleaved forest was characterized by a much higher percentage of the low vulnerable area. Nonetheless, these both land cover types were characterized by a total absence of high vulnerable areas.

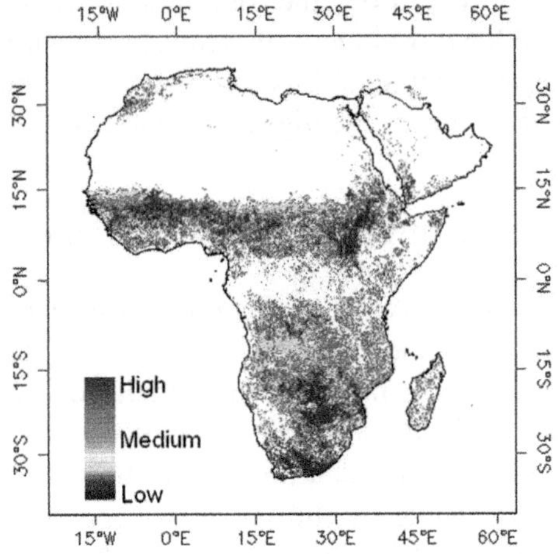

(a)

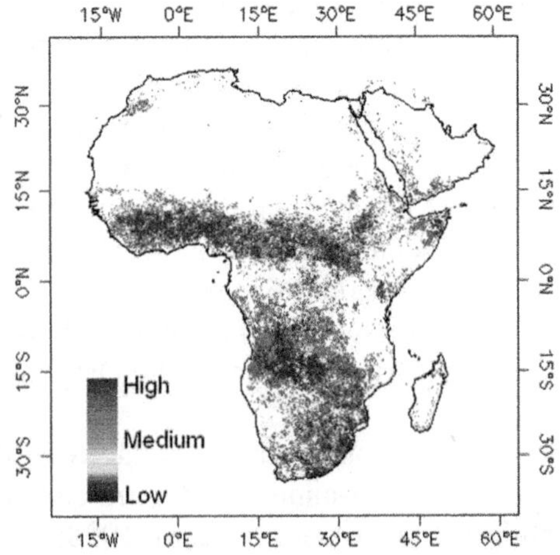

(b)

Fig. 4. Sensitivity of vegetated areas to ENSO warm events: maps show spatial pattern in vegetation response to ENSO warm events throughout the period 1982-2006 based on NDVI-MEI relationship (a), NDVI-SOI relationship (a).

Table 2. Percentage (%) of the land cover types (vegetated land surfaces) in Africa that are significantly affected by ENSO warm events (El-Nino).

Land cover	Area		Vulnerability to El-Nino, % area			
	Pixels	Km²	No	Low	Middle	High
Evergreen Broadleaved Forest	4181	2408256	80.39	18.87	0.74	0.00
Deciduous Broadleaved Forest	1209	696384	53.93	40.52	5.54	0.00
Woodland	9256	5331456	43.22	49.53	6.98	0.26
Wooded grassland	7414	4270464	25.56	62.73	10.41	1.27
Closed shrubland	2790	1607040	47.53	41.68	8.81	1.93
Open shrubland	5854	3371904	61.80	31.53	5.62	1.04
Grassland	2439	1404864	62.03	32.39	5.16	0.37
Cropland	2069	1191744	59.45	36.07	3.33	1.10
Total	35212	20282112	49.97	43.87	5.44	0.69

A decrease of the total tree cover in the vegetation cover leads to a general increase of the vulnerability to El-Nino. Thus, the results exposed that the percentage of low vulnerable area increased consecutively from 18.87% for EBF to 40.52% for DBF, to 49.53% for woodland, and to 62.73% for wooded grassland, while the percentage of middle vulnerable area was increasing in the same direction, too. Altogether, the results support the suggestion of weaker dependence of forested and wooded land cover types on climatic conditions (Li et al., 2002; Wang et al., 2003). The underlying study showed that this suggestion quite shares the climatic anomalies caused by ENSO warm events.

Of the total vegetated area, more than 93% showed no or low sensitivity to El-Nino. About 5% of the total area was characterized by middle vulnerability of vegetation to El-Nino, whereas only 0.7% showed high vulnerability. These results suggest that rather a very small part of the African vegetation cover experiences any repeated damage from devastating impacts of El-Nino. Obviously, each El-Nino event from the period 1982 – 2006 affected vegetation activity within broad areas what is represented in the percentage of the low vulnerability area, 43.87%. However, this percentage reveals only distinct spatial patterns of affected areas that are associated with every ENSO warm event. The areas of middle and high vulnerability that were affected much oftener from El-Nino impacts (considering to the used model, during more than 30-50% of all the delineated El-Nino events) are very small, only 5.44 and 0.69% from the total vegetated area, respectively.

5 Conclusions

In this study, the NDVI data from the AVHRR sensor were combined with ENSO indices to evaluate vegetation susceptibility to unfavourable climatic conditions caused by ENSO warm events during the period of 1982-2006. A new evaluation system for ENSO vulnerability has been presented in this study. This system used the correlation between NDVI anomalies and ENSO indices as major indicator for response of vegetation cover to ENSO impacts. The MWCA technique appears to be a simple but powerful method to investigate temporal dynamics of the vegetation response to El-Niño and to detect local hot spots of this response both in the spatial and temporal dimension. Extensive negative anomalies of monthly NDVI sustained by strong correlations with one of the ENSO indices were considered to indicate the statistically significant ($p < 0.05$) changes of vegetation activity driven by ENSO dynamics. A total quantity of these changes throughout the period 1982 – 2006 calculated for each individual pixel was suggested to be a representative indicator for long-time vulnerability of vegetation to El-Niño events.

The study defines spatial patterns of ENSO vulnerability and key areas of ENSO impact for Africa. These areas show medium and high response during the most of El-Niño events associated with the period 1982-2006. However, these key areas take only about 6 % of the entire African vegetated territory. The explanation for this is that, generally, the spatial distribution of affected areas varies over time and could be dependent on other explanatory factors like e.g. human impact actual at the time of a certain ENSO warm event. Among the studied vegetation types, evergreen broadleaved forest showed the highest resistance against the El-Niño caused drought conditions, whereas woodland, wooded grassland and closed shrubland were the land cover types most suffered by El-Niño events during the period 1982 – 2006. The results of this study may be helpful for forecasting the El-Niño impact on vegetation cover in Africa and can provide basic knowledge and data for planning of protection activities in El-Niño periods.

References

Anyamba A, Tucker CJ and Eastman JR (2001) NDVI anomaly patterns over Africa during the 1997/98 ENSO warm event. Int J Remote Sensing 22: 1847-1859.

Anyamba A, Estman JR (1996) Interannual variability of NDVI over Africa and its relation to El Niño/Southern Oscillation. Int J Remote Sensing 13: 2533-2548.

Asrar GM, Fuchs M., Kanemasu ET, Hatfield JL (1984) Estimating absorbed photosynthetically active radiation and leaf area index from spectral reflectance in wheat. Agronomy Journal 87: 300-306.

DeFries R, Hansen M, Townshend JRG, Sohlberg R (1998) Global land cover classifications at 8 km spatial resolution: The use of training data derived from Landsat imagery in decision tree classifiers. Int J of Remote Sensing 19: 3141-3168.

Erasmi, S., Propastin, P., Kappas, M. and Panferov, O. (2009) Spatial patterns of NDVI variation over Indonesia and their relationship to ENSO warm events during the period 1982-2006. Journal of Climate (accepted).

Fuller DO, Murphy K (2006) The ENSO-fire dynamic in insular Southeast Asia. Climatic Change 74 (4): 435-455

Glantz MH (1996) Currents of Change: El-Niño's impact on climate and society. Cambrige, Cambrige University Press.

Gutman G, Csiszar I, Romanov P (2000) Using NOAA/AVHRR products to monitor El Niño impacts: focus on Indonesia in 1997-98. Bulletin of the American Meteorological Society 81: 1188-1205.

Heinsch FA, Zhao SW, Running JS, Kimball RR, Nemani KJ, Davis PV, Bolstad BD, Cook AR, Desai DM, Ricciuto BE, Law WC, Oechel HK, Luo SC, Wofsy AL, Dunn JW, Munger DD, Baldocchi L, Xu DY, Hollinger AD, Richardson PC, Stoy MBS, Siqueira RK, Monson S, and Flanagan. LB (2006) Evaluation of remote sensing based terrestrial productivity from MODIS using AmeriFlux tower eddy flux network observations. IEEE Transactions on Geoscience and Remote Sensing 44:1908-1925.

Hoerling MP, Kumar A (1997) Origins of extreme climate states during the 1982-83 ENSO winter. J of Climate 10: 2859-2870.

Hoerling MP, Kumar A, (2000) Understanding and predicting extratropical teleconnections related to ENSO. In: Diaz HF, Markgraf V (Eds.) El Niño and the Southern Oscillation: Multi-scale Variations and Global and Regional Impacts. Cambridge University Press, 57-88.

Holben BN (1986) Characteristics of maximum-value composite images from temporal AVHRR data. Int J of Remote Sensing 7:1417-1434.

IPCC (2007) IPCC WGI Fourth Assessment Report, Summary for policy makers. Paris, February 2007.

Li B, Tao S, Dawson RW, (2002) Relation between AVHRR NDVI and ecoclimatic parameters in China. Int J of Remote Sensing 23: 989-999.

Lloyd CD (2005) Assessing the effect of integrating elevation data into the estimation of monthly precipitation in Great Britain. J of Hydrology 308: 128-150.

Los S O (1993) Calibration Adjustment of the NOAA AVHRR Normalized Difference Vegetation Index Without Resource to Component Channel 1 and 2 Data. Int J Remote Sensing 14: 1907-1917.

Mennis J (2001) Exploring relationship between ENSO and vegetation vigour in the southeast USA using AVHRR data. Int J Remote Sensing 22: 3077-3092.

Nagai S, Ichii K, Morimoto H (2007) Inter-annual variations in vegetation activities and climate variability caused by ENSO in tropical rainforest. Int J Remote Sensing 28: 1285-1297.

Nicolau R, Ribeiro L, Rodriges RR, Pereira HG, Camara AS (2002) Mapping the spatial distribution of rainfall in Portugal. In: Kleingeld WJ, Krige DG (Eds.) Geostatistics 2000. Cape Town, pages 548-558. Geostatistical Association of Southern Africa, South Africa.

Pinzon JE, Brown ME, Tucker CJ (2004) Global Inventory Modeling and Mapping Studies (GIMMS) AVHRR 8-km Normalized Difference Vegetation Index (NDVI) dataset. Product Guide. http://glcf.umiacs.umd.edu/data/gimms/

Prasad AK, Sarkar S, Singh RP, Kafatos M (2007) Inter-annual variability of vegetation cover and rainfall over India. Advances in Space Research 39: 79-87.

Sellers P, Randall DA, Betts AH, Hall FG, Berry J., Collatz GJ, Denning AS, Mooney HA, Nobre CA, Sato N, Field CB, Henderson-Sellers A (1997) Modelling the exchanges of energy water and carbon between continents and atmosphere. Science 275: 502-509.

Tateishi R, Ebata M (2004) Analysis of phenological change patterns using 1982-2000 Advanced Very High Resolution Radiometer (AVHRR) data. Int J Remote Sensing 25: 2287-2300.

Verdin J, Funk C, Klaver R, Robert D (1999) Exploring the correlation between Southern Africa NDVI and Pacific sea surface temperatures: results for the 1998 maize growing season. Int J Remote Sensing 20: 2117-2124.

Wang J, Rich PM, Price KP (2003) Temporal responses of NDVI to precipitation and temperature in the central Great Plains, USA. Int J of Remote Sensing 24: 2345-2364.

Wolter K, Timlin MS (1998) Measuring the strength of ENSO – how does 1997/98 rank? Weather 53: 315-324.

Yang L, Wylie B, Tieszen LL, Reed BC (1998) An analysis of relationships among climate forcing and time-integrated NDVI of grasslands over the U.S. Northern and Central Great Plains. Remote Sensing Environment 65: 25–37.

A Storage and Transfer Efficient Data Structure for Variable Scale Vector Data

Martijn Meijers, Peter van Oosterom, Wilko Quak

Delft University of Technology,
OTB Research Institute for Housing, Urban and Mobility Studies
{b.m.meijers, p.j.m.van.oosterom, c.w.quak}@tudelft.nl

Abstract. This paper deals with efficient data handling of variable scale vector data. Instead of pre-building a collection of data sets on different scales, we create an index structure on the base data set (largest scale data) that enables us to extract a map at exactly the right scale the moment we need it. We present both the classic version of the tGAP (topological Generalized Area Partitioning) data structure for storing our variable scale map, as well as an ameliorated version, both based on topological concepts. We prove that the classic structure needs in a worst case scenario $O(e^2)$ edges (with e the number of edges at largest scale). In practice we observed up to a factor 15 more edges in the variable scale data structure. The tGAP structure has been optimized to reduce geometric redundancy, but the explosion of additional edges is due to the changing topological references. Our main achievement finds its roots in the reduction of the number of edge rows to be stored for the 'lean' version (by removing the topological referential redundancy of the classic tGAP), which is beneficial both for storage and transfer. We show that storage space for the data set, plus the index, is less than twice the size of the original data set. The 'lean' tGAP, as the classic tGAP, offers true variable scale access to the data and has also improved performance, mainly due to less data communication between server and client.

1 Introduction

There is a growing tendency to focus data management of spatial datasets on the highest level of detail and manage the other levels of detail as data that is automatically derived from this base data set (e.g. Bobzien et al., 2006; Ellsiepen, 2007; Stoter et al., 2008).

M. Sester et al. (eds.), *Advances in GIScience*, Lecture Notes in Geoinformation and Cartography,
DOI 10.1007/978-3-642-00318-9_18,

Basically there are two methods of managing a data set on different levels of detail (cf. Cecconi and Galanda, 2002): The multi-scale approach and the variable scale approach. The multi-scale approach works by creating several smaller scale versions of the map. Every time a map is needed on a specific scale the most appropriate scale from the pre-defined collection is chosen and displayed. Instead of pre-building a collection of maps on different scales the variable scale approach creates an index structure on the base map that enables you to extract a map at exactly the right scale the moment you need it. This means that when you want a map on a specific scale for a specific region it is constructed for you on the fly. Advantages of the variable-scale approach are that only one dataset needs to be managed and that data can be displayed at any scale.

This paper proposes a new data structure for the management of a variable-scale map product and is an improvement on the tGAP data structure as described in Van Oosterom, 2005. The idea of the tGAP data structure is to run automatic map generalization on the base data set and instead of storing the result of the generalization on different scales the whole process of the generalization is stored in a tree structure where every node of the tree corresponds to the application of a cartographic generalization operator. Each generalization operator is performed at a specific level of detail (or scale). If a map is needed at a given level of detail the generalization tree structure is used to get the right data at the right level of detail. The implementation of the tGAP structure maintains a valid topological structure on all levels of detail by tracking which nodes, edges and faces are visible on each level of detail. The structure stores the node and face data very efficiently. However there is a lot of redundancy in the way the edges are stored in the model (for details, see Section 3). It turns out that in the worst cases $O(e^2)$ edges (with e the number of edges in the largest scale map; see Section 4.4) have to be stored in the tGAP structure (and in practice we observed up to a factor 15 of edges to be stored; see Section 5). This paper describes how this redundancy can be removed without loss of functionality. The new data structure resolved the redundancy for edges so that every edge is stored only once. Saving storage space also implies saving data transfer times as one of the main application areas will be a variable-scale server in a web-based environment.

The classic tGAP structure is offering non-redundant geometric data storage for arbitrary levels of detail. Technically, the problem with the data structure is that too much data storage is needed. Analysis shows that this is due to the high number of changing references in the data structure causing new versions of edge representations to be stored, resulting in an unreasonable growth of edge data in the scale dimension. For 2D geographic information the scale dimension is considered to be the third dimension within the tGAP structure. A 3D spatial index is used to efficiently retrieve a spatial selection at a specific scale.

In this paper we give an overview of some design alternatives we considered to solve the problem of the growing number of redundant edges and present our final solution. The rest of the paper is structured as follows: In section 2 we give an overview of previous work that is done in this area. In section 3 we describe the structure that we wish to improve in more detail and in section 4 we give a few of the alternatives for improvement. Experiments done on these alternatives and our resulting structure are described in 5. Finally in 6 we conclude with a discussion and summarize the most important contributions of this paper and present a number of open problems to be addressed for further improvements.

2 Previous Works

Most research on the management of variable scale datasets is done on the multi-scale approach where a fixed set of layers is managed. This might stem from the paper map production process where it is very expensive to produce products at different scales. In this digital era it could be possible to manage vector data at arbitrary levels of detail (and disseminate this data via web services). Few solutions are known for this kind of variable scale data access.

Buttenfield and Wolf (2007) have a pyramid structure (called MRVN) that is able to represent a data set at a multiple scales while maintaining topology. To achieve this, all the topological nodes of the original dataset cannot be removed. If the scale range is very big the resulting number of nodes can still be very large making this a method that works for a limited scale range. Xinlin and Xinyana (2008) presented the Zoom quad-tree is. In the Zoom quad-tree all objects of the original dataset are stored in nodes of the tree dependent on the size of the object. The initially sparse tree is filled with generalized versions of the original features. As described in the paper it is not clear whether the structure can maintain a polygonal partition on the different levels of detail. In the original GAP-tree data structure (Van Oosterom, 1995) a scale-less structure was described that could manage a polygonal partition. Disadvantage of the structure is that full polygons are stored at various levels of detail making it a very redundant structure.

The first attempt at a fully topological hierarchical structure was done by Vermeij et al. (2003). The structure worked by extending all tables of a standard topological model with two extra attributes (a minimum and maximum scale). Now a generalization algorithm is run on the dataset. The algorithm works by replacing nodes, edges and faces with other nodes edges and faces at a lower level of detail. Instead of deleting the old nodes, edges and faces their max-scale is set meaning that the object is not needed anymore from that scale. By retrieving all nodes, edges and faces that are needed for a specific level of detail a complete topology at that level can be reconstructed. The big disadvantage of this structure is that it produces a lot of redundant data. An ameliorated version (the tGAP structure) was therefore proposed by Van Oosterom (2005), for which the first implementation results were described by Van Oosterom et al. (2006). This structure is described in more detail in the next section.

3 Classic tGAP Structure

This section first summarizes the classic tGAP structure as it will be the basis for the improved version described in the next section. The datasets that are currently supported within the tGAP structure have to be modeled as a two dimensional polygonal map, i.e. it is a partition of the plane in a geometric sense, without gaps and overlaps. The physical storage of the data takes place in a database management system (DBMS) in an extended topological, node-edge-face data structure. The exact table definitions are given in Figure 1.

Each polygon of the map is represented by a topological face (this is a one-to-one relation). The level of detail (LoD) can be regarded as third dimension and represented by the concept of 'importance'. The importance of objects is based on their size and feature classification. E.g. a large forest area can have lower importance than a small city area. A functional spatial index on a 3D bounding box (bbox) is used to efficiently access the 2D spatial data extended by the third dimension: the importance (or scale) range for which a certain representation is valid.

3.1 Filling the Face Table

As we want to reduce the LoD for display at smaller scales, we have to generalize our original data. A generalization process reduces the number of polygonal objects, based on the importance. The object that has the least importance is removed first. Plain removal of the object is not allowed, because a gap would exist after this operation. Therefore we let the most compatible neighbor take the space of the object to be removed. Based on the shared boundary length and the feature class compatibility this neighbor is chosen. The merging operation creates a new object. This new object then has a new identity and is given the feature class of the most compatible neighbor. The importance of this object is recomputed (and several different options have been tested for this: e.g. taking the sum of the importance of the two merged objects). This process continues until only one object is left.

```
CREATE TABLE tgap_faces
(
face_id integer,
parent_face_id integer,
imp_low numeric,
imp_high numeric,
imp_own numeric,
feature_class_id integer,
area numeric,
bbox geometry
);
```

(a) Face table

```
CREATE TABLE tgap_edges
(
edge_id integer,
imp_low numeric,
imp_high numeric,
start_node_id integer,
end_node_id integer,
left_face_id integer,
right_face_id integer,
geometry geometry
);
```

(b) Edge table

```
CREATE TABLE tgap_nodes
(
node_id integer,
imp_low numeric,
imp_high numeric,
geometry geometry
);
```

(c) Node table

Fig. 1. Table definitions for the classic tGAP structure.

During this merging process the importance range for all objects is also created and stored. This range is intimately related to the importance assigned to all faces present at the largest scale. The importance is stored with all the faces as the 'imp own' attribute that clearly defines the ordering of the generalization process. The importance range (stored with an 'imp_low' and an 'imp_high' attribute for each face) defines the lifespan of objects in the LoD dimension and allows selection of the right objects at an arbitrary LoD (using an importance level for selection, 'imp_sel').

The importance range for the objects is created as follows: The objects used as starting point will be assigned an imp_low value of 0. The example in figure 2(a) and table 1 shows that the imp_low value of all original faces (1-6) is indeed 0. Then, in each generalization step, the lifespan of two objects will be ended and a new one will be created. In our example, face 1 is the least important face, and is merged with its most compatible neighbor (face 5); a new object (face 7) is formed. Both ended objects are assigned the importance own value of the least important object, named 'imp_remove', as their importance high attribute (face 1 has an imp own of 150, this is assigned to both face 1 and 5 as imp_high value). The new object that is formed in the generalization step will be assigned the sum of the own importance of the two old objects and the imp_high as the imp_low value. The resultant of this process is that the sum of all own importance of the original objects is equal to the importance high value of the last remaining object. This means that the sum of importance for all objects valid at any given scale (LoD) in the complete map does not change.

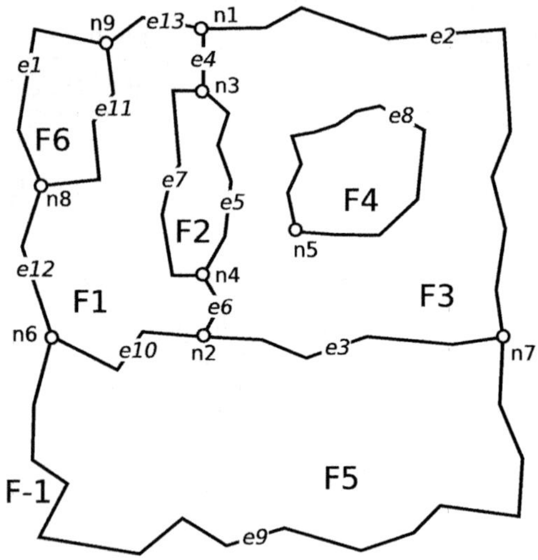

(a) The map of the initial configuration, with imp_sel = 0 (note that the nodes, edges, and faces are labeled with their identity)

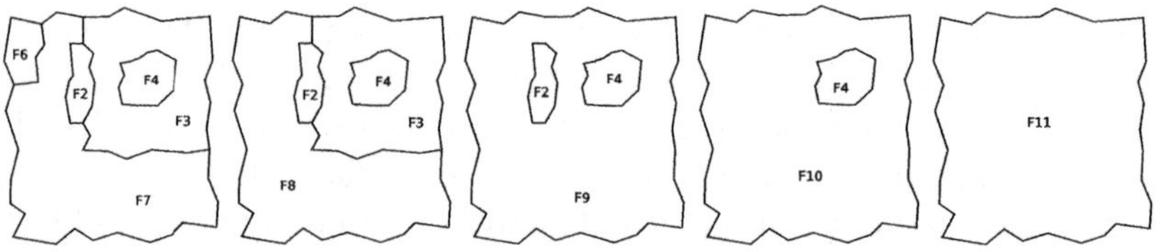

(b) 150 (c) 325 (d) 395 (e) 505 (f) 610

Fig. 2. Example map with 6 polygonal regions. Subfigures (b) – (f) show the map at the imp_sel value mentioned in their caption.

3.2 Filling the Node and Edge Tables

When merging two faces, the life of the edges between the two old faces is ended by setting their imp_high value to the imp own value of the face that is removed (imp_remove). The remaining edges are now adjacent to the newly created object, so also these edge versions are terminated (their importance high value is set to imp_remove) and new, updated versions for those edges are created (with imp_remove as their importance low value). These updated versions get the same identity as before, but with a different left or right face pointer and a new importance low value). In our example edge 10 is removed in the first face merge step, when face 1 is merged to face 5 (the imp_high value of edge 10 is set to 150, see table 3). Edge 11 is an example of an edge that is changed due to the change of the neighboring face. This edge was adjacent to face 1, but is after the merge adjacent to face 7. So, a new version of this edge is created.

Furthermore, the nodes that are having only a relationship with two edges after the merge, are as well ended and the incident edges are merged; see the node information from Table 2. A new version for those incident edges is created, with merged geometry based on the geometry of the two old edges. This is shown in our example for the edges 9 and 12 (forming a new edge 14) and the edges 3 and 6 (forming the newly created edge 15). In the classic tGAP structure the edge geometry is represented by a Binary Line Generalization (BLG) tree. For leaf edges this is a directly stored version. For non-leaf edges this is a BLG-tree with a new top and references to the two BLG-trees of the child-edges. So no redundancy in the storage of geometry, but the result can be having to trace a lot of references during usage of the structure. An alternative therefore is to create a new (redundant) geometric representation of the merged edge (a non-BLG-tree representation). For this new geometry there are two options: 1. keep all original vertices or 2. keep half of the original vertices (after applying line simplification). Both solutions introduce (controlled) geometric redundancy, but will be easier to use.

face_id	parent_face_id	imp_low	imp_high	imp_own	feature_class
1	7	0	150	150	corn
5	7	0	150	750	grass
6	8	0	325	325	grass
3	9	0	395	395	forest
2	10	0	505	505	lake
4	11	0	610	610	town
7	8	150	325	900	grass
8	9	325	395	1225	grass
9	10	395	505	1620	grass
10	11	505	610	2125	grass
11	-1	610	2735	2735	grass

Table 1. The tGAP face table for the sample data set, which is graphically depicted in Figure 2(a) (note there is a bbox and an area value stored, but this is not shown).

node_id	imp_low	imp_high
1	0	2735
2	0	150
3	0	395
4	0	505
5	0	610
6	0	150
7	0	395
8	0	325
9	0	325

Table 2. The tGAP node table. Note that each node has a point geometry, but this is not shown.

edge_id	imp_low	imp_high	left_face	right_face	start_node	end_node
1	0	325	-1	6	8	9
2	0	395	3	-1	7	1
3	0	150	3	5	2	7
4	0	150	3	1	1	3
4	150	325	3	7	1	3
4	325	395	3	8	1	3
5	0	395	3	2	3	4
6	0	150	1	3	2	4
7	0	150	1	2	4	3
7	150	325	7	2	4	3
7	325	395	8	2	4	3
8	0	395	4	3	5	5
8	395	505	4	9	5	5
8	505	610	4	10	5	5
9	0	150	5	-1	6	7
10	0	150	5	1	2	6
11	0	150	6	1	8	9
11	150	325	6	7	8	9
12	0	150	-1	1	6	8
13	0	150	-1	1	9	1
13	150	325	-1	7	9	1
14	150	325	7	3	7	4
14	325	395	8	3	7	4
15	150	325	7	-1	8	7
16	325	395	-1	8	7	1
17	395	505	9	-1	1	1
17	505	610	10	-1	1	1
17	610	2735	11	-1	1	1
18	395	505	9	2	4	4

Table 3. The classic tGAP edge table with the example content (Note: a. the repeated versions of edges, due to the left/right reference changes, b. The geometry of the edges is stored, but this is again not shown).

edge_id	imp_low	imp_high	left_face	right_face	start_node	end_node
1	0	325	-1	6	8	9
2	0	395	3	-1	7	1
3	0	150	3	5	2	7
4	0	395	3	1	1	3
5	0	395	3	2	3	4
6	0	150	1	3	2	4
7	0	395	1	2	4	3
8	0	610	4	3	5	5
9	0	150	5	-1	6	7
10	0	150	5	1	2	6
11	0	325	6	1	8	9
12	0	150	-1	1	6	8
13	0	325	-1	1	9	1
14	150	395	7	3	7	4
15	150	325	7	-1	8	7
16	325	395	-1	8	7	1
17	395	2735	9	-1	1	1
18	395	505	9	2	4	4

Table 4. The lean tGAP edge table with the example content (note the geometry/line is not displayed but present in the structure).

3.3 Using the Structure Dynamically

The structure is used dynamically by providing a spatial extent (for the view port) and an importance value (for the LoD). The importance value can be derived from a given extent: A smaller extent means more detail to show and finally a lower importance value for querying the data structures with (imagine a user zooming in, more detail can be shown for all objects). Contrary, if a larger extent needs to be shown, due to a user zooming out, a higher importance value needs to be used for selecting less objects. The mapping between importance and spatial extent is currently done in such a way that it honors the rule of 'a fixed number of objects' to be retrieved and shown on the screen. For this mapping the Radical law could have been applied, by which the best number of objects for a certain scale can be calculated (cf. Töpfer and Pillewizer, 1966).

Note that the topological data structures used give more degrees of freedom for modeling what information to store and thus allow us to take more different design decisions than when we would have used plain geometry (e.g. simple feature polygons). It must be noted that this paper focuses on saving storage space. For large data sets this also implies saving time as a more compact storage structure requires less disk pages to be read and less communication between server and client (assuming that the structure itself still supports the most important actions). These considerations are the subject of the next section.

4 Design Alternatives for a Lean tGAP Structure

During the design of a more data storage (and transfer) efficient version of the tGAP structure a number of different alternatives were explored, they were labeled with the following symbolic names: no_lr, abox, use_tree. In Van Oosterom (2005) it was already mentioned that less columns in the table structure directly implies less storage (a column less to store), but also indirectly implies less storage – if scale changes are reflected only in a column that is removed then there is no need for a new row with the new value. This was explained by showing how the tGAP edge table which has four edge-to-edge references could be reduced in size by removing two edge-to-edge references and only keeping two edge references (edge_lr and edge_fl). In the example data sets this resulted in both less columns and less rows. In the implementation reported in (Van Oosterom et al., 2006) all edge-to-edge references were removed, but even in that case the tGAP edge table for a realistic data set still did have up to 15 times more rows than the original edge table (and the theoretic worst case is even $O(e^2)$ with e the number of edges at the largest scale). This was mainly due to the changing references to the left and right faces after merging two neighbor faces (and not so much due to merging to existing edges into one new edge). One of the approaches followed was splitting the edge table into two parts: one part with attributes that do not changes in the tGAP structure (e.g. geometry, and references to start and end node, if stored) and attributes that do change for different scales/importance values (e.g. left and right face references). However, for the changing part of the edges the number of rows is still the same factor higher, only the fixed part is not repeated, saving some storage space. So the aim is to further reduce the required storage, but without loosing performance during the most relevant operations. The most important operation is selecting and visualizing a part of the data set at a certain scale. Another important operation is selecting refinement differences between two scales (for a given part of the map). Further, in the future update operations should be supported (at the most detailed level and then propagated upwards, but this is outside the scope of the current paper).

The selection and visualization of a part of the map at a certain scale, called imp_sel, functions as follows: select all faces and edges that overlap the selection rectangle and that have their imp_low-imp_high range containing imp_sel. Note that these are efficient queries (assuming proper 3D spatial clustering and indexing) and this is all the interaction needed with the database server. Then at the client side some topology processing is done: for every face the relevant edges are selected (based on the left/right face references they contain) and rings are created (and if needed inner-rings are properly included in the outer ring). Due to the fact that the edges are selected based on their bbox overlap, not all edges needed to complete the rings of faces partly included in the search rectangle may be present. This is solved by first clipping the selected edges against the selection rectangle (and also splitting the selection rectangle at the intersection points and creating temporary edges). Together, the clipped edges and the temporary edges created from the selection rectangle are sufficient for forming closed loops, which together cover the whole selected area. For sure every ring contains at least a part of an original edge. The left/right information of such an edge provides a reference to the face which can then be colored according to its classification. This is the setting of the use of the tGAP structure and it is clear that the left/right information is needed (for classifying and coloring the faces) despite the fact that it is storage expensive; the 'row explosion of edges'. Now we are going to discuss our three alternatives, no_lr, abox and use_tree, to make the structure more storage efficient.

4.1 Alternative I: no_lr

We started out with a very lean topology data structure: no left/right references (as these caused most of the storage overhead), only edge geometry and a point inside a face region ('spaghetti with meatballs'-approach); Tables: Nodes (id, location, imp_low, imp_high), Edges (id, geometry, imp_low, imp_high), Faces (id, mbr, point_on_surface, imp_low, imp_high). The rings are formed based on topology processing without left/right information. There are three steps: 1. creating rings, 2. assigning island rings to their parent and 3. association of the right identifier with the area (outer ring). Step 1: The procedure starts with an arbitrary edge and then starts forming rings by finding all edges incident with the end (node) coordinates (using the geometry of edges), sorting all incident edges based on angle and then takes the first edge left (for counter-clockwise orientation), this process is repeated until the start edge is reached again and the ring is closed. This procedure is then repeated with the next unused edge and a new ring is formed. The ring production terminates when all edges are used twice (once in forward and once in backward direction). Step 2: some of the rings do not have the expected counter-clockwise orientation, and these correspond to islands in the face. The parent outer-ring can be found by a point-in-polygon test (use arbitrary point from inner-ring and finding the smallest outer ring that contains this point). Step 3: Now all faces with holes are created and have to be assigned an identifier. This is done again with a point-in-polygon test (the point now being the point on surface from the Faces table). For both step 2 and 3 the use of an R-tree (or other type of spatial index) will speed up the point-in-polygon test, building the R-tree once takes $O(n \log n)$ time and then the repeated searches take $O(\log n)$ time.

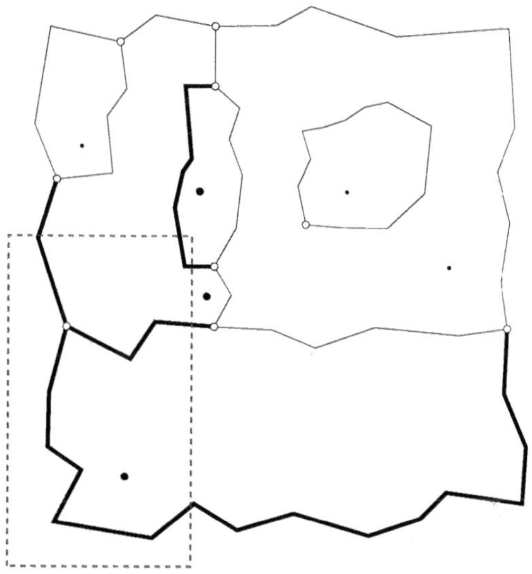

Fig. 3. The 'spaghetti with meatballs' approach. The retrieved edges (overlapping with the selection rectangle in dashed lines) are given with the thickest lines. After clipping, 3 rings are formed, but the two rings at the top of the selection rectangle cannot be labeled with the correct face information as the point on surface for these faces is outside the formed ring.

This approach does work for having a complete extent of area partition within the view port while visualizing. It does not work well when clipping the data: areas cannot be reconstructed any more, without having a complete set of edges. An option is to clip the selected edges again (as described above). The result is that now areas can be created covering the selection rectangle. However, faces crossing the boundary might have their point on surface outside the rectangle (and therefore the area can not be identified. There might be some solution to go back to the database server for each unidentified area, but this is both a non-trivial query and time expensive as it has to be repeated for every unidentified area.

4.2 Alternative II: abox

In an attempt to solve the identification of the clipped areas, the adjacency box (Van Oosterom and Vijlbrief, 1994), or abox for short, instead of the bbox of edges was proposed for selection. The abox of an edge is the union of the bbox of the faces left and right of the edge. The result is that more edges are selected based on the abox, but for sure these are enough to completely reconstruct all faces in the selected rectangle. However, in order to have the aboxes available in the edge table they have to be maintained (stored). Due to merging of faces in the tGAP structure also the aboxes have to be updated. Actually this is then exactly the same increase in rows as what would be obtained by maintaining the left and right face references. So, no real storage reduction, rather the opposite as the abox will take more storage space that the left and right reference. The advantage of the abox solution is that it allows easier reconstruction of faces at the client side resulting in full unclipped areas. In theory the explicit storage of aboxes might be avoided by introducing them in a view (which uses a function to compute the abox). But again this is non-trivial without the left and right references. Therefore we concluded that this was also not the ideal solution and continued investigating another alternative with fewer drawbacks.

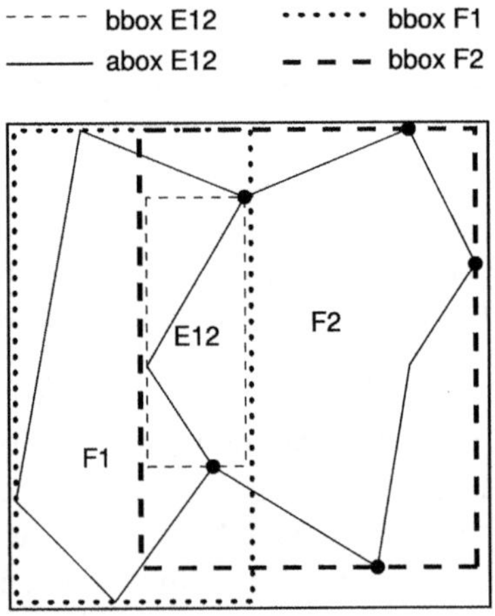

Fig. 4. Adjacency box (abox).

4.3 Alternative III: use_tree

Looking at edges that are changing due to changes in the left and right side information (and not in the edge geometry); we considered merging the rows related to the same edge in one row. This results in no change for the geometry, start and end nodes, and id attributes. The imp_low and imp_high attributes contain the union of all imp ranges of the edge (which are per definition adjacent ranges). The next question is what to do with the differences in left and right references? Store the left/right reference corresponding to the lowest imp range or to the highest imp range? Take for example edge 4 in Table 3 and 4: storing the right face reference corresponding to the lowest imp range [0 - 150) would imply a reference to face 1, and storing it related to the highest imp range [325 - 395) would result in a reference to face 8. It was decided to store the left and right face references related to the lowest imp-range, for reasons that will be explained below when assigning the proper identity to the created areas. Anyhow, just storing only rows for edges that are really new (because these edges are merged) safes a lot of storage (rows) as will be explained in section 4.4 (the number of rows is for sure always below a factor 2 as edges are merged pair wise). The left/right information and the tGAP face-tree can then be exploited to properly identify the areas at a certain importance level (scale). With this solution we have combined both the requirement to be storage efficient (as the factor 15 of records in the edge table is solved), while still having an efficient solution for the most relevant operation.

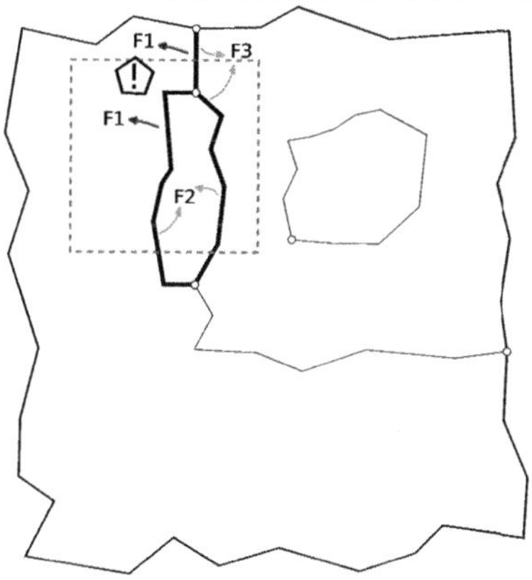

Fig. 5. Rewriting of face-id's with the use_tree variant. Edges are retrieved, at imp_sel = 330, based on their bounding box and the selection rectangle (dashed). After clipping, only the thickest lines are used for forming rings. The most-left ring is formed based on edge 4, 7 and temporary edges stemming from the selection rectangle. Both edges 4 and 7 do not point to the correct neighboring face and rewriting has to take place (face 1 is rewritten using the tGAP face tree as face 8).

The identification of areas in a given search rectangle of a specified importance level imp_sel proceeds as follows. All edges are retrieved a. based on a selection rectangle and b. having an imp range that includes imp_sel. The faces are also selected based on these two criteria. Then the clipping is applied to the edges and rings are created as described above and inner-rings are again assigned to outer-rings. During the creation of rings the left/right information is used to find the identity of the face. As the edges carry the left/right information of the lowest imp-range (which may be below the requested imp_sel) not all edges directly have a pointer to the correct face (that is at the requested imp_sel level). In many cases however there will be at least one edge with the proper (w.r.t. imp_sel) left/right information and this is then indeed the identity of the area. In some cases this information is not present (1. when this edge is outside the selection rectangle, 2. when an island is not yet merged with its parent). In these cases the referred face (and the corresponding edge) with the highest imp_low level is used as start in the tGAP face-tree and the tree is traversed upwards until the face identifier at the right imp level is found. The final layout of data structure is (again) based on topology and has the following tables: Nodes (id, geometry, imp_low, imp_high), Edges (id, start node, end node, left face, right face, geometry, imp_low, imp_high), Faces (id, parent face id, feature class, bounding box, imp_low, imp_high, imp own); see Table 1, 2 and 4 for the sample data set in Figure 2(a), the sample map, and Figure 6, a visual representation of the tGAP face-tree.

The drawback of using the tGAP face-tree is that this tree is not present at the client side (after the two face and edge selection queries). An efficient solution is to send one third query to the server requesting the 're-writing' of the face-id's which correspond to a too low imp level and get back face-id's that correspond with imp_sel. An easier solution is not to draw these faces at all: the drawback is of course that white spots will occur on the map (most often near the boundaries of the selection rectangle).

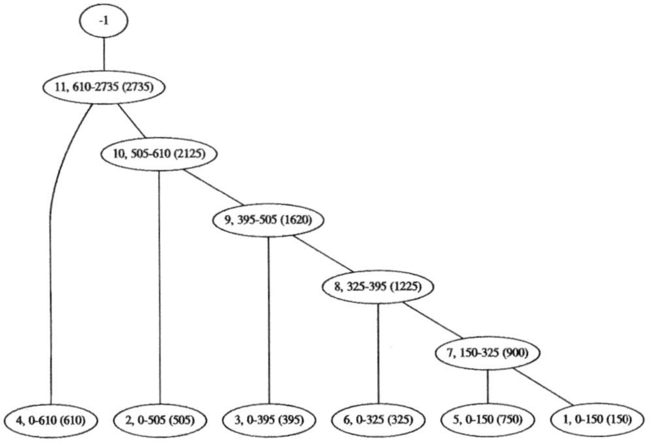

Fig. 6. The tGAP face-tree, corresponding to the data set of Figure 2(a).

4.4 Theoretical Numbers for Faces and Edges

In the previous section we sketched a more optimal solution for storing data in the edge table. Here, we continue our investigations by finding the theoretical upper bounds after filling the data structures for both the classic and the lean variant. These bounds are expressed in numbers of edges (e) and faces[1] (f) present in the original dataset.

Lemma 4.1. *The number of total faces stored in the tGAP structure is, after the generalization process, equal to:*

$$2 \cdot f - 1$$

Proof. The generalization process starts with f original faces. Merging can be executed until we have only one face left. This means we can merge u times, with $u = f - 1$. Each time we merge two faces, we add 1 new face to *f*. In total we add u times a face to f. The total number of faces will thus be $u + f$, or, expressed differently:

$$2 \cdot f - 1$$

---

[1] Numbers for faces here do *not* include the concept of a universal face

Lemma 4.2. *The total number of edges in the classic tGAP structure, that is, filled with the original method (generating all intermediate edge versions), is at most:*

$$\sum_{i=0}^{f-1}(e-i)$$

Proof. Faces are merged in $f-1$ steps. Faces that are neighbors are adjacent in, at least, one edge (due to the planar map criterion). With each merge step thus at least one edge will disappear. The worst case is that in every generalization step all remaining edges will be duplicated due to new left/right references. These observations lead to Lemma 4.2.

Corollary 4.3. *The total number of edges in the classic tGAP structure, that is, filled with the original method (generating all intermediate edge versions), can be quadratic:*

$$O(e^2)$$

Proof. Assume a configuration (similar to the one shown in Figure 7) with one big face (described by one big edge) containing many small islands (small faces, each one described by one edge). Then in the summation of Lemma 4.2 it is clear that $f=e$ and this results in a total of $e\cdot(e+1)/2=O(e^2)$ edges.

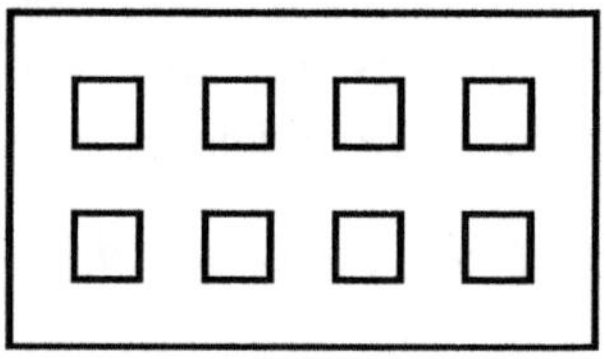

Fig. 7. A worst case initial configuration.

Our new, lean approach performs significantly better in this respect:

Lemma 4.4. *The total number of edges stored in the tGAP structure, filled with the new 'use_tree' method, is dependent on the number of original edges and faces and is at most:*

$$2\cdot e-f$$

Proof. All original edges will be present once in the output. The merging of edges is what brings new edge versions. Suppose this edge merging is performed with all start edges as input, as follows: two edges will be merged at a time, until 1 edge is left. The resultant of this process is then one large polyline with self-intersections. The total number of edges in the output will then be at most two times the original number of edges minus 1 (cf. Lemma 4.1). However, in each generalization step, to merge two faces, at least one edge has to be removed, i.e. the number of edges to be removed is the number of faces minus 1 (as that is the amount of merges that will take place). Taking both steps into account, results in a number of edges that is equal to:

$$(2 \cdot e - 1) - (f - 1) = 2 \cdot e - f$$

This is a worst case estimate, as in each merge step more than one edge might be removed.

5 Experiment and Results

To judge whether our theoretical investigations described above would yield valid results in practice, we implemented both variants (classic and lean) of filling the tGAP structures using PostgreSQL[2] extended with PostGIS[3] (for the geometrical attributes) as DBMS. For filling the tGAP structures in the DBMS with our generalization procedure of merging faces and for retrieving and visualizing the data from the DBMS, we wrote some scripts using the Python[4] programming language. Table 5 highlights the number of faces and edges for the original data, the amount of data after using the classic variant and for the lean variant of filling the structures.

To verify the lemmas from section 4.4, we started by creating two artificial test data sets (1 and 2). It is clear that the number of faces follows Lemma 4.1 in all cases, independently from which filling variant is used. Further, it is also clear that our concerns with respect to the duplication of edge rows are valid: To see whether the upper bound for the number of edges could exist in practice, we created a data set (set 2) consisting of one polygon containing 2500 islands polygons. Each polygon was described with one line, resulting in 2501 faces and 2501 edges. In practice, this data set can occur when an archipelago is mapped and in which all islands are merged to the surrounding ocean. The factor for the classic variant of filling is an abominable result (on average each edge is duplicated 1251 times, that is indeed $O(e^2)$), especially compared to the lean version (in which only the original edge versions are present once).

---

[2] www.postgresql.org
[3] www.postgis.org
[4] www.python.org

Besides artificial data sets we also used some data sets containing real world data. That the factors are higher for the sets 5 and 6 compared to the factors for 3 and 4, is explainable by the fact that the last two sets do not contain any island polygons, while set 5 and 6 do contain some polygons with a few hundred islands. Filling the structures in the classic way leads then to even more duplicated edge rows. Although the theoretical upper bounds are, by far, not met by these data sets, the factors of the classic filling variant are still high (and we suspect that this will even be worse for larger data sets), while our new variant significantly performs better.

Data set	Original	FacestGAP faces	Original Edges	Edges Classic (increase factor)	Edges Lean (increase factor)
Set 1	6	11	13	29 (2.2)	18 (1.4)
Set 2	2501	5001	2501	3128751 (1251)	2501 (1.0)
Set 3	525	1049	1984	11091 (5.6)	2975 (1.5)
Set 4	5537	11073	16592	77585 (4.7)	26787 (1.6)
Set 5	50238	100475	178815	2663338 (15)	264950 (1.5)
Set 6	173187	346373	426917	3544232 (8.3)	630944 (1.5)

Table 5. Number of faces and edges for the different test data sets. Numbers are shown for the original data, the data after using the classic variant of filling (i.e. edge version duplication) and for the lean variant (only each first edge version is stored). Both data set 1 and 2 were created artificially. The data sets 3 – 6 contain real world data. Data set 3 and 4 both contain land cover data. Set 5 contains cadastral parcels and data set 6 contains topographical data.

Fig. 8. Data set 3, visualized with different imp_sel values.

6 Conclusion and Discussion

With our design and implementation exercise, we learnt the following lessons: First, our 'use_tree' alternative performs a lot better when looking at the storage part, compared to our initial solution, not only in theory, but certainly also in practice. Reducing storage is not the only achievement here; The reduction in the number of edge rows will be very beneficial for the case when the tGAP structure will be used in a web service environment and data is sent (progressively, using increments) to a client. Second, we designed a structure that now has a better trade off between storage and calculation-when-needed than before (much less data is to be stored and transferred, but with our lean alternative sometimes it is necessary to perform a lookup operation of the correct neighboring face). Third, we support – with less than twice the original dataset size – all intermediate scales for visualization at an arbitrary scale.

Irrespective of these accomplishments, we also realize that our main contribution is currently based on one generalization operation (the merging of objects) and based on the heuristics this operation brings (geometry is always removed and gradually becomes less). The field of map generalization however offers more operations, like line simplification, collapse/splitting of area objects, displacement and typification of groups of objects, to name a few. Some of these operations, by definition, introduce new geometry (e.g. splitting of objects will introduce new boundaries). An optimal solution for a data structure, in terms of data storage, has thus to take into account the heuristics of these generalization operations. Therefore our investigations will continue and topics we would like to focus on in the (near) future include:

- Topologically correct line simplification (taking into account the neighborhood, while performing line simplification and preventing (self)-intersections that cause a change in topology, similar to what is described in Saalfeld (1999) and Bertolotto and Zhou (2007)). Creating and using data structures for variable scale access to this geometry (e.g. finding an alternative algorithm for filling the BLG-tree structure, currently Douglas-Peucker is used) is another topic that deserves attention.
- More operations for the (currently simplified) generalization process; As a first step, we would like to, instead of merging, allow splitting, probably based on a triangulation (cf. Bader and Weibel, 1997), and experiment what happens when using such a split operator with weights for all neighbors, until no further splitting is possible.
- Inclusion of more and different semantics in order to take better decisions which generalization operator to choose (instead of the current 'one fits all' approach); e.g. apply a different generalization operator for infrastructure type of objects than for other terrain objects.
- Making the structure dynamic: perform updates at the largest scale (and propagate these upwards in the tGAP structure).

References

Bader, M. and Weibel, R. (1997). Detecting and Resolving Size and Proximity Conflicts in the Generalization of Polygonal Maps. In Proceedings of the 18th International Cartographic Conference., pages 1525–1532, Stockholm.

Bertolotto, M. and Zhou, M. (2007). Efficient and consistent line simplification for web mapping. International Journal of Web Engineering and Technology, 3(2):139–156.

Bobzien, M., Burghardt, D., Petzold, I., Neun, M., and Weibel, R. (2006). Multi-Representation Databases with Explicitly Modelled Intra-Resolution, Inter-Resolution and Update Relations. In Proceedings Auto-Carto 2006, Vancouver.

Buttenfield, B. and Wolf, E. (2007). "The road and the river should cross at the bridge" problem: Establishing internal and relative topology in an MRDB. In Proceedings of the 10th ICA Workshop on Generalization and Multiple Representation 2-3 August 2007, Moscow, Russia.

Cecconi, A. and Galanda, M. (2002). Adaptive Zooming inWeb Cartography. In Computer Graphics Forum, volume 21, pages 787–799. Blackwell Synergy.

Ellsiepen, M. (2007). Partial regeneralization and its requirements on data structure and generalization functions. In Kremers, H., editor, Proceedings 2nd ISGI 2007: International CODATA symposium on Generalization of Information, Lecture Notes in Information Sciences, pages 72 – 84, Germany. CODATA.

Saalfeld, A. (1999). Topologically Consistent Line Simplification with the Douglas-Peucker Algorithm. Cartography and Geographic Information Science, 26(1):7–18.

Stoter, J., Morales, J., Lemmens, R., Meijers, M., Van Oosterom, P., Quak, W., Uitermark, H., and van den Brink, L. (2008). A data model for multi-scale topographical data. In Headway in Spatial Data Handling 13th International Symposium on Spatial Data Handling, pages 233–254.

Töpfer, F. and Pillewizer, W. (1966). The principles of selection, a means of cartographic generalization. Cartographic Journal, 3(1):10–16.

Van Oosterom, P. (1995). The gap-tree, an approach to "on-the-fly" map generalization of an area partitioning. In Müller, J., Lagrange, J., andWeibel, R., editors, GIS and Generalization, Methodology and Practice, page 120–132. Taylor & Francis.

Van Oosterom, P. (2005). Scaleless topological data structures suitable for progressive transfer: the gap-face tree and gap-edge forest. In Proceedings Auto-Carto 2005, Las Vegas, Nevada. Cartography and Geographic Information Society (CaGIS).

Van Oosterom, P., de Vries, M., and Meijers, M. (2006). Vario-scale data server in a web service context. In Ruas, A. and Mackaness, W., editors, Proceedings of the ICA Commission on Map Generalisation and Multiple Representation, pages 1–14, Paris, France. ICA Commission on Map Generalisation and Multiple Representation.

Van Oosterom, P. and Vijlbrief, T. (1994). Integrating complex spatial analysis functions in an extensible gis. In Proceedings of the 6th International Symposium on Spatial Data Handling, pages 277–296, Edinburgh, Scotland.

Vermeij, M., Van Oosterom, P., Quak, W., and Tijssen, T. (2003). Storing and using scaleless topological data efficiently in a client-server dbms environment. In 7th International Conference on GeoComputation, Southampton.

Xinlin, Q. and Xinyana, Z. (2008). Multi-representation geographic data organization method dedicated for vector-based webgis. In Proceedings of the XXXVI congress of ISPRS, volume Part B4 ommision IV, pages 815–819.

Line Decomposition Based on Critical Points Detection

Eric Guilbert

Department of Land Surveying and GeoInformatics
Hong Kong Polytechnic University
Hung Hom, Kowloon, Hong Kong
lseguil@polyu.edu.hk

Abstract. The problem of line simplification is a recurrent problem in cartography. The purpose is to remove irrelevant details while emphasising the main features of the line. Most of the current techniques belong to the spatial domain (least square method, active contour, point selection). However, some techniques applying to the frequency domain (Fourier transform, wavelets) have also been introduced. These latter methods are mostly employed for simplification and compression purposes where information about line features is rarely taken into account, thus limiting their usefulness for cartographic applications. This paper presents the principle of Empirical Mode Decomposition which belongs to the frequency domain. It is used in signal processing to decompose a signal into its different frequencies. The method for line simplification has been studied, showing that line features can be taken into account by introducing a new decomposition method based on the detection of critical points. Results obtained at different levels of detail are discussed. Finally, future directions for work are presented.

1 Introduction

In cartographic generalisation, line generalisation is a process consisting of the removal of non relevant information on the line while keeping and emphasising details of prime importance for visualisation (Weibel and Dutton, 1999). Line generalisation is often perceived as a line simplification process. The objective is to reduce the amount of data by selecting some characteristic points on the lines or to approximate the overall shape of the line with a new and simpler definition. Simplification techniques can be

M. Sester et al. (eds.), *Advances in GIScience*, Lecture Notes in Geoinformation and Cartography,
DOI 10.1007/978-3-642-00318-9_19,

classified according to two approaches. First is the compression or smoothing approach building up an approximated line within a predefined error tolerance of the original line. Second is the feature point selection approach detecting and selecting feature points on the line corresponding to important shape features. In line generalisation, the cartographer often has to strike a balance between smoothing the line and removing some features on one side and preserving the shape and retaining too many features and details which may hinder legibility on the other side. Among the different techniques, the Douglas and Peucker algorithm (Douglas and Peucker, 1973) is the most popular method as it is considered a good compromise between compression and shape preservation.

Simplification techniques can also be classified within filter theory under the spatial and frequency domains of application (Burghardt 2005). The most commonly used smoothing methods in generalisation are found in the spatial domain. Smoothing is considered as a filtering process applied to the point coordinates of the line. Coordinates are filtered according to a geometrical criterion such as a distance threshold (Douglas and Peucker 1973) or by using optimisation techniques (Bader 2001, Burghardt 2005). Smoothing methods in the spatial domain are also interesting in that they can be combined with other generalisation operations (displacement, enlargement, aggregation…) which are also defined using geometrical constraints (Harrie 1999, Guilbert and Saux 2008).

Smoothing methods in the frequency domain effectively use low-pass filters. Filtering is performed by converting a signal from the spatial domain to the frequency domain and removing the high frequencies to keep only low frequency information corresponding to large variations. For this reason, low-pass filters are mostly used for signal noise reduction or for data compression purposes. In line generalisation, smoothing methods based on Fourier transforms or Fourier series (Fritsch and Lagrange 1995, Lawford 2006), wavelets (Fritsch and Lagrange 1995, Saux 2003) and empirical mode decomposition (Li et al. 2004) have been introduced. These methods are applied either to the line coordinates (Saux 2003, Li et al. 2004, Lawford 2006) or to the line curvature (Fritsch and Lagrange 1995). In the latter case, authors first compute curvature as a function of the curvilinear abscissa of the line, filter it, and by reverse transformation, compute a smoother line. However, the transformation is not unique and the line has to be adjusted by taking some characteristic points and matching them with the original line.

Empirical mode decomposition which has been introduced in signal theory by (Huang et al. 1998) offers more adaptative decompositions as a one dimensional signal can be decomposed in components related to the local extrema of the signal. EMD provides a data driven decomposition approach whereby basis functions are adapted to the signal while other methods perform decomposition based on a predefined set of basis functions.

In cartography, spatial methods are more popular because simplification parameters can be more easily expressed as spatial constraints. However, frequency domain methods present the advantage of performing decomposition at different levels of detail in a straightforward way. Each detail corresponds to a signal component at a given frequency and different signal approximations can be built by combining these different components. This approach is used to represent the signal at different resolutions starting from the original to a coarse one showing the signal trend. However, these methods were originally designed for one-dimensional signals and have been extended to higher dimensions by simply applying the process to each coordinate or by projecting the signal in different directions. Therefore, when processing a two-dimensional line, features are not directly taken into account. An improvement in the context of line generalisation would be to perform decomposition based on critical points at each level of resolution so that each component can be expressed with regard to these points and features can easily be extracted.

For that purpose, EMD seems to be a promising method as it is data driven and has already been applied to one and two dimensional signals. Therefore, this paper describes an investigation into the application of EMD to line simplification and presents a decomposition method based on critical points detection. The paper is organised as follows: the principle of EMD and related works that can apply to line simplification with their limitations are first describe. Secondly, the authors' decomposition method is introduced. Its interesting features and limitations are then discussed before future developments and perspectives of this work for line generalisation in cartography are presented.

2 Empirical Mode Decomposition

2.1 Decomposition of a Real Value Signal

2.1.1 Principle of Empirical Mode Decomposition

Empirical mode decomposition (EMD) was originally defined for real-valued time series (Huang et al. 1998). It decomposes a signal into slow oscillations (low frequency components) and fast oscillations (high frequency components). The main principle of EMD is that the basis of decomposition is derived from the data. The method is therefore adaptative and can handle any kind of data.

Each component corresponding to a different frequency is called an intrinsic mode function (IMF). An IMF should be a zero-mean oscillation where all maxima are positive and all minima are negative. An example of EMD[1] is given in Figure 1 where the curve on top is the original signal. Four IMFs are extracted. The last curve is the signal residual which represents the global trend.

2.1.2 Definition of the IMF

Discrimination between fast and slow oscillations is obtained through an algorithm called the sifting process. Starting from an original signal $x(t)$, all the local maxima and minima on the curve are found. A curve e_{max} joining all the maxima and a curve e_{min} joining all the minima are next defined (Figure 2). The mean $m_1 = \frac{e_{min} + e_{max}}{2}$ of the two envelopes is then computed.

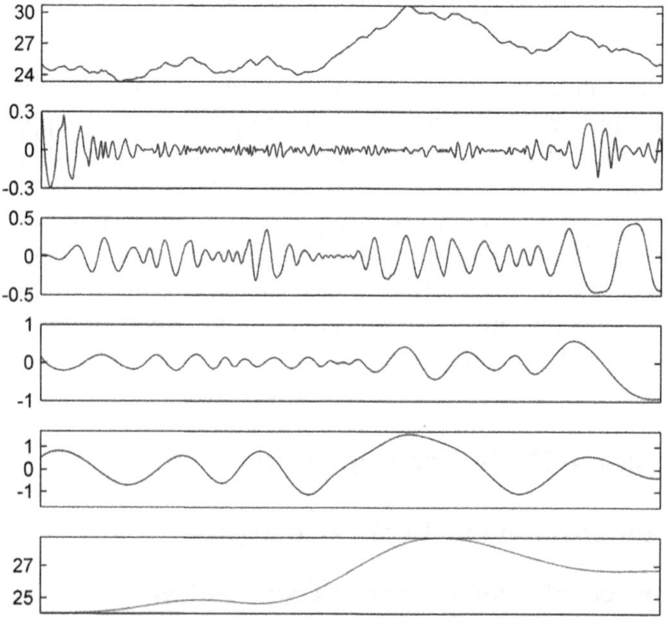

Fig. 1. Empirical Mode Decomposition of a signal. Top is the original signal. Four following signals are four IMF representing signal components at different frequencies. Last curve is the signal residual

[1] All examples in this section have been produced using the Matlab/C codes available at http://perso.ens-lyon.fr/patrick.flandrin/emd.html

The difference $h_1 = x - m_1$ between the mean curve and the original curve is an IMF if it is a zero-mean signal. If it is not, the process has to be reiterated on the difference until a zero-mean signal is obtained: starting with h_1 as the data, the mean m_{11} and the difference h_{11} are computed so that $h_{11} = h_1 - m_{11}$ and the process is repeated (Eq. 1) until a stop criterion is reached. This stop criterion is based on the maximum amplitude of the signal or the standard deviation between two consecutive signals (Huang et al. 1998).

$$h_{1k} = h_{1(k-1)} - m_{1k} \tag{1}$$

The last h_{1k} signal corresponds to the first IMF d_1. The IMF can be subtracted from the original signal and the remainder $r_1 = x - d_1$ can be processed as a new input signal to extract the next IMF. All the components can be extracted from the data as follows:

$$r_i = r_{i-1} - d_i. \tag{2}$$

The process stops when the remainder becomes a monotonic function or when there are not enough extrema to extract another IMF. Based on Eq. 2, the signal x can be decomposed in n IMFs d_i and a residual r_n as follows:

$$x = r_n + \sum_{i=1}^{n} d_i \tag{3}$$

As with multi-resolution analysis methods such as wavelets, the residual can be seen as a function showing the global trend of the signal and the IMF are the detail components which are added together to reconstruct the original signal. Figure 3 shows an example of signal reconstruction where the top signal is the residual to which IMFs are added until the original signal at the bottom is obtained.

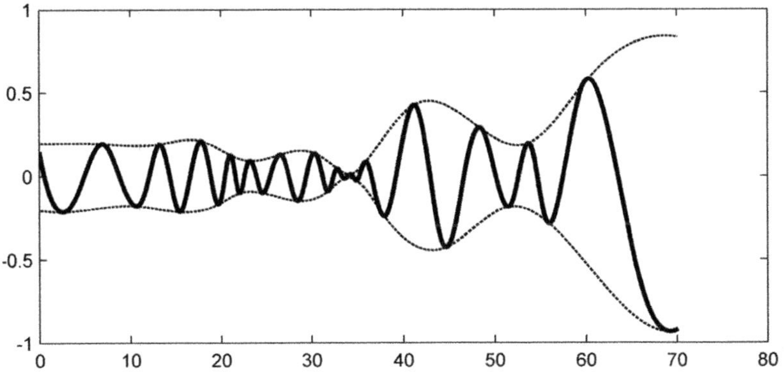

Fig. 2. Definition of the lower and upper envelopes interpolating the local maxima and minima of the signal

2.2 Decomposition of a Bivariate Signal

2.2.1 Limitations of Classical EMD to Two-Dimensional Signals

The previous method defined for real-valued time series in one dimension can be applied to 2D line decomposition by applying the process to each coordinate (Li et al. 2004). The authors showed that a line can be decomposed at different levels of smoothness with results similar to those traditional smoothing methods.

However, this method has limitations as both decompositions are performed independently. Firstly, the two EMDs may have a different number of IMFs, corresponding to different oscillation speeds. Secondly, extrema obtained for IMF construction do not correspond to critical points of the curve. Building up a two-dimensional envelope from two one-dimensional envelopes would have limited interest for shape analysis as the resulting line may cross the curve and connect points which are located on different sides of the curve. Thirdly, the line decomposition depends on the coordinate system. For example, if the system is rotated, different extreme points would yield a different decomposition.

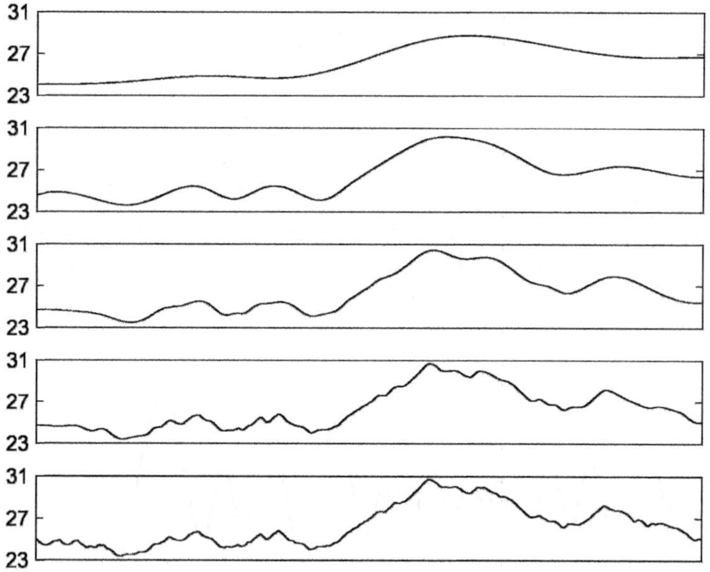

Fig. 3. Reconstruction of the signal by adding the different IMF components. Top signal is the residual. Bottom signal is the original signal. From top to bottom, an IMF is added to the previous signal. Each signal can be seen as a representation of the original signal at different levels of resolution

2.2.2 Bivariate Empirical Mode Decomposition

An extension of EMD to bivariate time series (i.e. complex-valued) has been introduced (Rilling et al. 2007). A two-dimensional line can be processed as a complex-valued signal where the x-coordinate corresponds to the real part and the y-coordinate corresponds to the imaginary part (Figure 4).

Slow oscillations are extracted as the mean of a 3D envelope enclosing the signal. This envelope is computed by considering a given number of directions: the signal is projected in each direction and an envelope is computed for each of these directions. As an illustration, on Figure 4, right, dashed lines correspond to projections along the x and y directions. These projected lines are used to extract extreme points and to compute an upper and a lower 2D envelope in each direction. These envelopes correspond to points on the 3D envelope surface. The mean curve is computed as the mean value of these points at each instant of time. Based on that, each component is extracted through a similar sifting process until a sufficiently regular signal is obtained.

Applied to line decomposition, the method gives better results than the previous one as both coordinates are decomposed together and more IMF components can be identified. However, results depend on the number of directions. When the number of directions increases, the mean curve is computed with a larger number of points and the decomposition is less dependent with respect to rotations of spatial coordinates.

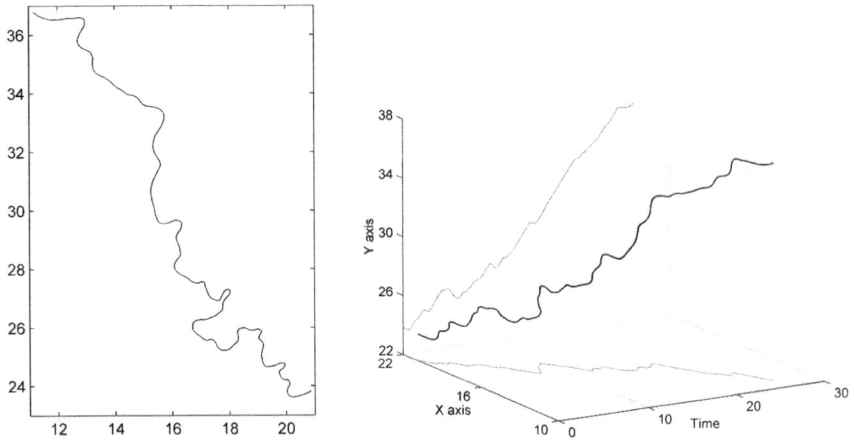

Fig. 4. Left: Two-dimensional curve. Right: Representation of the curve as a complex valued signal. Plain line is the original curve. Dashed lines are curve projections along the x and y axes representing the real and imaginary parts of the signal

Different results obtained with the curve of Figure 4 are presented in figures 5, 6 and 7. By applying classical EMD (figure 5), two IMFs are identified for each coordinate, meaning that the curve can be defined at three levels of detail. By using bivariate EMD with two directions of projection, we have five levels of detail (figure 6). Projections have been made along the x-axis and y-axis directions. With sixteen directions, six levels of detail were obtained (figure 7). Results obtained with more than sixteen directions were very similar. Tests performed on other lines provided similar results. The number of directions to choose, however, may depend on the complexity of the shape.

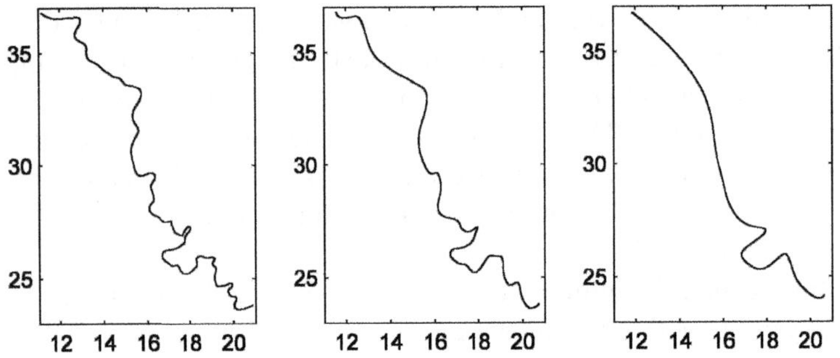

Fig. 5. Line decomposition using classical EMD applied to each coordinate

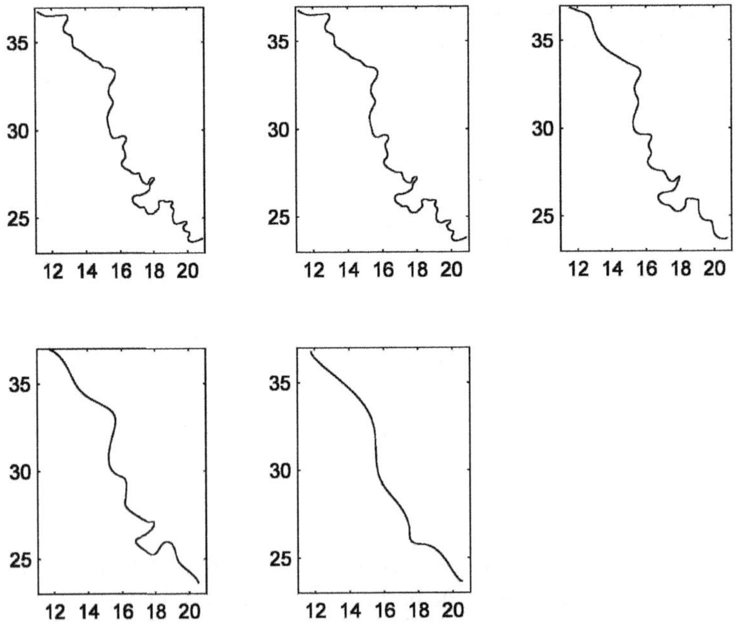

Fig. 6. Bivariate EMD decomposition with two directions

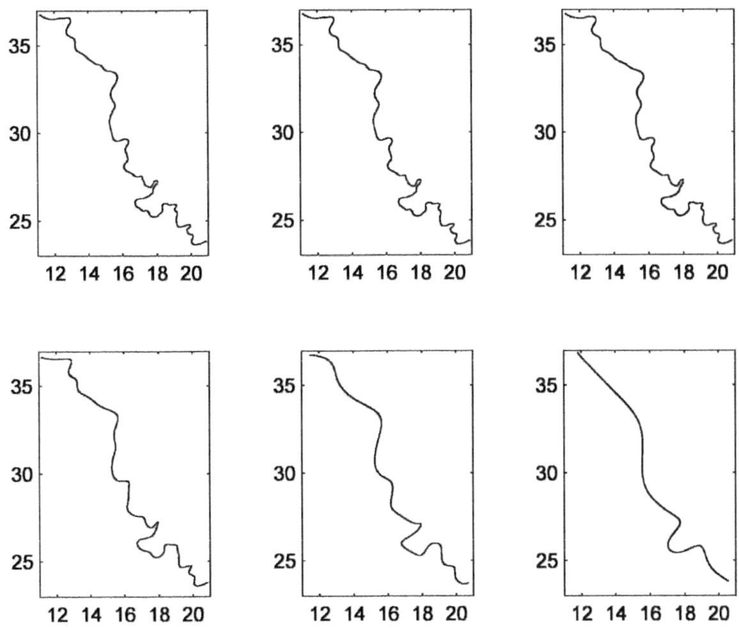

Fig. 7. Bivariate EMD decomposition with sixteen directions

3 Decomposition Based on Critical Points Identification

3.1 Significance of Critical Points

In cartography and in computer visualisation, critical points are those able to describe and to characterise the shape of an object (Attneave 1954). On a line, critical points are end points and curvature extrema corresponding to locations where the line direction changes more rapidly. Identifying critical points on a line is therefore a common process for identifying the main features of a line or to decompose a line in compound segments for generalisation (Lecordix et al. 1997, Plazanet et al, 1998). During the generalisation process, the cartographer can edit a line and still maintain its features and preserve its topological consistency by ensuring that critical points still correspond to the same curvature extrema. In the same way, a feature is removed from the line if all its corresponding critical points are no longer critical. From this point of view, critical points can be used to control the shape of the line by acting on them. Other points can be expressed in a local frame defined on each line feature with the corresponding critical points.

Generalisation also concerns the editing of cartographic data from one scale to a smaller scale. In visualisation, representation of a line at different scales corresponds to its representation at different levels of detail. In the same way as before, critical points at one level can be used to express critical points at a higher level of detail so that modifications made at one level can automatically be transmitted to higher levels. Having a line decomposition method based on these critical points would therefore be of interest as it would provide handles for the user to edit a curve at different levels of detail either independently or by propagating the changes to higher levels. Such an approach is already used with B-spline representation (Elber and Gotsman 1995) where a hierarchy of control points (which are not critical points) is defined. The technique has been applied to line representation with B-spline wavelets in (Saux 2003).

Bivariate EMD is an efficient decomposition technique and is used by its authors to analyse rotating signals and to detect phase shifts or changes in sense of rotation. However, in order to compute the main components of a line, the method identifies a large number of extreme points in each direction which do not necessarily correspond to critical points. Extreme points cannot be used to control the shape as there are too many of them which are not significant for the user. Therefore, although the method can still be used for compression purpose, it would be interesting to develop a technique that takes into account critical points as these points are important for decomposing a line based on its main features and for preserving its topology.

Therefore, a new decomposition method is introduced based on the EMD principle which extracts curvature extrema to define the line components. These points are used because they are critical points and correspond to extreme points through which it is possible to build an upper and lower envelope. The next sections detail the decomposition method. Firstly how the lower and upper envelopes are built up is explained. Secondly, how components are extracted to obtain the whole decomposition is explained.

3.2 Construction of Upper and Uower Envelopes

In order to build the envelopes, extreme points must be detected. As mentioned above, these points correspond to curvature extrema. The higher envelope is built by interpolating points of maximum curvature while the lower envelope is built by interpolating points of minimum curvature.

A line can be described by a parametric function f: $[a, b] \rightarrow \mathbb{R}^2$ where for each parameter $t \in [a,b]$, a point $P = f(t)$ of the plane is associated. A point $P = f(\tau)$ is a local maximum of curvature if for all points $f(t)$ in the vicinity $V(\tau)$ of τ we have:

$$\forall t \in V(\tau), \kappa(\tau) > \kappa(t). \tag{4}$$

In the same way, a local minimum of curvature at a point $f(\tau)$ is characterised by:

$$\forall t \in V(\tau), \kappa(\tau) < \kappa(t). \tag{5}$$

If working with a polygonal line $(P_i)_{i=0}^{n}$, the parameterisation is the curvilinear abscissa and each point P_i is associated with an $f(t_i)$. A maximum of curvature at a point P_i is defined by $\kappa(t_i) > \kappa(t_{i-1})$ and $\kappa(t_i) > \kappa(t_{i+1})$. A minimum of curvature is defined by $\kappa(t_i) < \kappa(t_{i-1})$ and $\kappa(t_i) < \kappa(t_{i+1})$.

In the following, noting that $(Q_j)_{j=0}^{m}$ are the extreme points of the polygonal line and $(\tau_j)_{j=0}^{m}$ their parameters, then $Q_j = f(\tau_j)$. These points and parameters are subsets of $(P_i)_{i=0}^{n}$ and of $(t_i)_{i=0}^{n}$. As two consecutive points Q_j and Q_{j+1} cannot be two of the same kind (i.e. two maxima or two minima), it can be supposed, in order to simplify notations without loss of generality that a point Q_{2j} is a maximum of curvature and a point Q_{2j+1} is a minimum of curvature. The upper envelope $e_{\max}$ is a curve interpolating the points Q_{2j} and the lower envelope $e_{\min}$ is a curve interpolating the points Q_{2j+1}. However, unlike previous methods, envelopes cannot be computed directly by simple interpolation. In parts of the line with important shape variations, the envelopes may intersect so that the higher envelope may cross under the lower envelope. More constraints must be imposed on the interpolation to ensure a correct location of the envelopes. Supplementary constraints are set by imposing points on the envelopes for all parameters τ_j i.e. we impose values of $e_{\max}$ at parameters τ_{2j+1} and values of $e_{\min}$ at parameters τ_{2j}.

Let us consider two consecutive maxima $Q_{2j} = f(\tau_{2j})$ and $Q_{2j+2} = f(\tau_{2j+2})$. C_{2j} is defined as the cubic curve segment joining these two points with derivatives at each extremity equal to the line derivatives.

$$\begin{aligned} C_{2j}(\tau_{2j}) = f(\tau_{2j}), \quad C'_{2j}(\tau_{2j}) = f'(\tau_{2j}) \\ C_{2j}(\tau_{2j+2}) = f(\tau_{2j+2}), \quad C'_{2j}(\tau_{2j+2}) = f'(\tau_{2j+2}) \end{aligned} \tag{6}$$

Such a segment can be defined either as a Bézier curve or through Hermite interpolation (Farin 2001). As Q_{2j} and Q_{2j+2} are maxima of curvature, segment C_{2j} is located above the curve on the interval $[\tau_{2j}, \tau_{2j+2}]$ so $e_{\max}(\tau_{2j+1})$ is set equal to $C_{2j}(\tau_{2j+1})$. By applying this to all segments, location for all parameters $(\tau_j)_{j=0}^{m}$ of $e_{\max}$ can be defined (figure 8). The whole envelope is finally computed by spline interpolation: given the coordinates of the points at parameters $(\tau_j)_{j=0}^{m}$, a cubic spline curve interpolating all the points $e_{\max}(\tau_j)$ is defined. The same process is applied to define the points $e_{\min}(\tau_{2j})$ and compute the lower envelope for all values on interval $[a, b]$.

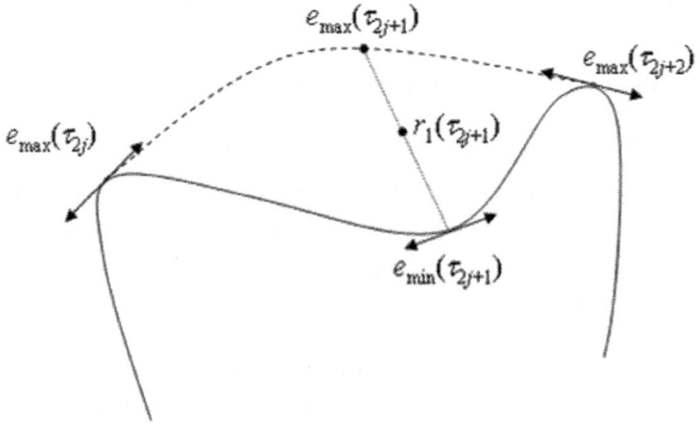

Fig. 8. Definition of the mean curve: computation of $e_{\max}(\tau_{2j+1})$ and $r_1(\tau_{2j+1})$

3.3 Extraction of the Components

Once both envelopes are computed, the mean curve r_1 passing through the middle of the envelopes is computed (figures 8 and 9):

$$r_1 = \frac{e_{\min} + e_{\max}}{2} \tag{7}$$

As with other decomposition methods, a component is defined by the difference between the mean curve and the curve $d_1 = f - r_1$ with values $d_1(t)$ being vectors as the functions are defined in the two-dimensional plan. The same process can be repeated on the new curve in order to extract the next component. The process is stopped when the curve shows a sufficiently small number of extrema. Finally, a similar decomposition to Eq. 3 is obtained.

$$f = r_n + \sum_{i=1}^{n} d_i \tag{8}$$

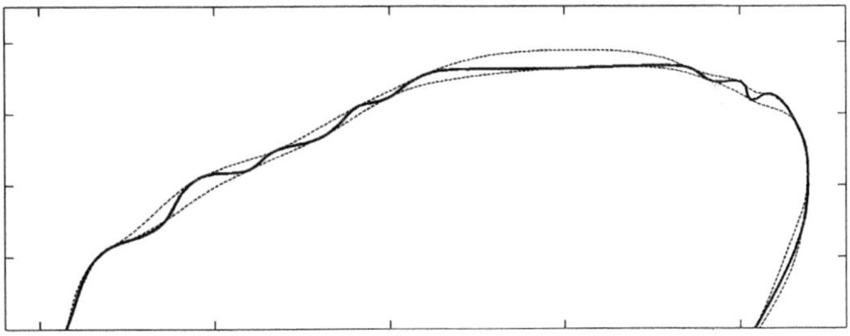

Fig. 9. Curve segment with its lower and upper envelopes

3.4 Results

Results obtained with bivariate EMD and the technique presented in this paper can be compared in figures 7 and 10. As the method still suffers some limitations detailed below in this section, the purpose here is to show that the method yields valid results. Doing a thorough comparison of the compression level achieved between this approach and other line simplification techniques is beyond the scope of this paper due to the limitations mentioned.

In the first decompositions, results are similar for both methods. Both act to smooth the line by removing small details. In the last decompositions, the authors' method keeps more details and the line displays bigger variations but still has the same global shape as produced by the bivariate EMD. An advantage of our method is that this was obtained with fewer parameters (number of directions does not need to be set) and that it is independent of the coordinate system.

However, the method is based on detection of local maxima and minima of curvature. Computing the curvature of points of polygonal lines is an operation which is very sensitive to numerical errors and approximations. It becomes a problem in the last stages of the decomposition as, although we have a line which appears very smooth, the algorithm still detects an artificially large number of extrema. In order to improve the convergence and reliability of the algorithm, components computed during the process for the envelope and the mean curve interpolation are expressed with Bézier or B-spline curves which provide a more reliable form than is the case on a canonic polynomial basis or with direct computation on polygonal line approximations.

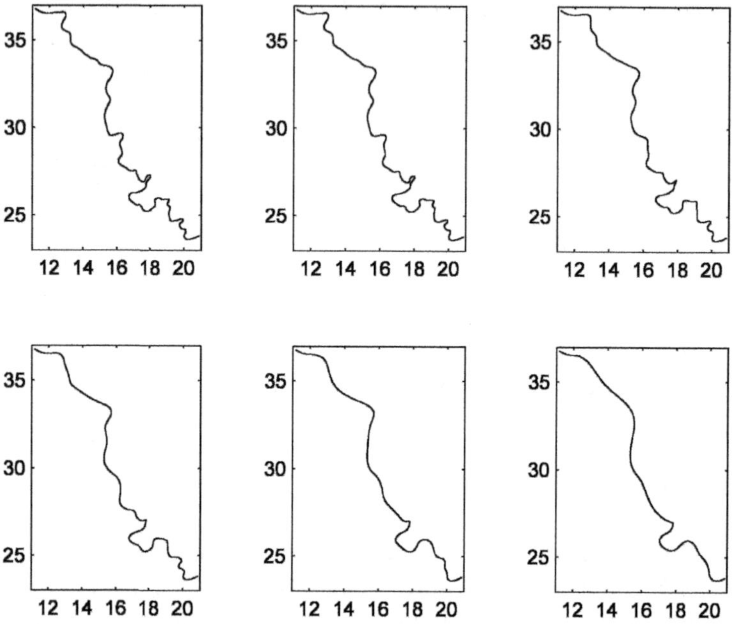

Fig. 10. Line decomposition based on critical points

Finally, the components extracted by the method do not exactly match with the definition of IMFs for empirical mode decomposition. In EMD, IMFs are built through a sifting process to guarantee that they are symmetrical zero mean signals. The purpose of the sifting process is to eliminate riding waves and to smooth uneven amplitudes (Huang et al. 1998). Although EMD may still be performed without sifting, sifting is necessary in signal processing for each IMF to physically represent a meaningful frequency and to avoid large amplitude fluctuations. In the authors' method, no condition requiring symmetry of the components or their means is imposed. This may result in amplifying some local extrema by increasing the absolute curvature value at these points as the mean curve may have large fluctuations between the envelopes. However, it does not prevent the algorithm from converging as the number of extrema between two consecutive iterations is always reduced.

4 Perspectives

This paper is concerned with line simplification for generalisation and focuses on empirical mode decomposition, a method originally designed for signal processing. A common drawback to most methods issued from the frequency domain is that simplification is performed globally with no connection to the line features. The objective here has been to study the principles of EMD and to propose an approach allowing a decomposition of the line based on its critical points.

In the first part, principles of EMD for one-dimensional and two-dimensional signals have been presented. Although existing methods can be applied to two-dimensional line decomposition, they are used mostly for signal processing purposes focusing on the extraction of oscillatory components. In cartography, line decomposition is usually performed for visualisation purposes at different levels of detail concentrating upon the identification of critical points characterising line features at each level. Therefore, a decomposition method based on the EMD principle using points of extreme curvature for construction of the components has been introduced. The results look promising as the method, in addition to presenting similar results as the bivariate EMD in terms of extraction of levels of detail, gives a decomposition based on critical points which can be later used for analysis or editing for generalisation purposes. However, the work presented in this paper is still at a preliminary stage and should be developed further. Two main directions for future work are identified.

As stated in the previous section, the first issue concerns the reliability of the method itself and the current IMF definition. The method still suffers some limitations in the definition of the decomposition process. Firstly, curvature is defined as the inverse radius of the osculating circle at a point. This definition is not reliable enough as it is too sensitive to numerical errors. It is also too local within this context and should take into account the scale or the resolution to limit approximation errors and detection of extrema which are too local. An improvement would be to consider more global definitions of the curvature taking into account a larger neighbourhood, such as the visual curvature as defined in (Liu et al. 2008).

Furthermore, needing investigation is the definition of a sifting method adapted to our approach so that components are symmetrical in order to improve the decomposition and to fully satisfy the definition of IMF from a physical point of view. A major issue also is to compare the method with other existing line simplification techniques including other EMD techniques and the Douglas Peucker algorithm.

A second direction concerns the application of the method to line visualisation and editing at different levels of resolutions. The objective of extracting critical points is to define a line via its critical points so that each feature is expressed in a local frame. A model has to be defined to store and access the line points through these frames. In this way, editing can be performed only by acting on the critical points so that the user retains easier control of the shape. Critical points are also used to define the different IMF components of the line: critical points at a level are expressed through the critical points at the lower level. Such a model allows the editing of several levels at the same time as modifications can be transferred to higher levels. It is also useful for visualisation operations such as continuous zooming.

A major issue when simplifying lines relates to the preservation of topological consistency. In the longer term, this issue should be the main focus. By ensuring that the relative locations of critical points on a line are maintained (for example, a maximum is always higher than a minimum), we can guarantee that the shape is maintained. In a set of lines, topological relationships may also be linked with relationships between critical points. A main objective would be to define generalisation operators which are directly applicable to critical points. However, it will be necessary to establish if any further constraints must be set depending upon the type of line or the context of the map as operations do not always need to be applied to all levels at the same time.

References

Bader M (2001) Energy minimization methods for feature displacement in map generalization. PhD thesis, Department of Geography, University of Zurich, Switzerland

Burghardt D (2005) Controlled line smoothing by snakes. GeoInformatica, 9(3), pp. 237-252

Douglas DH and Peucker TK (1973) Algorithms for the reduction of the number of points required to represent a digitized line or its caricature. The Canadian Cartographer, 10(2), pp. 112-122

Elber G and Gotsman C (1995) Multiresolution control for non uniform b-spline curve editing, In The Third Pacific Graphics Conference on Computer Graphics and Applications, pp. 267-278

Farin G (2001) Curves and Surfaces for CAGD, 5th edition, San Francisco, CA: Morgan-Kaufmann

Fritsch E and Lagrange JP (1995) Spectral representation of linear features for generalisation. In Proceedings of COSIT'95, pp. 157-171, Austria, Springer Verlag

Guilbert E and Saux E (2008) Cartographic generalisation of lines based on a B-spline snake model. International Journal of Geographical Information Science, 22(8), pp. 847-870

Harrie L (1999) The constraint method for solving spatial conflicts in cartographic generalization. Cartography and Geographic Information Science, 26(1), pp. 55-69

Huang NE, Shen Z, Long SR, Wu ML, Shih HH, Zheng Q, Yen NC, Tung CC and Liu HH (1998). *The Empirical Mode Decomposition and Hilbert spectrum for nonlinear and non-stationary time series analysis*. Proc. Roy. Soc. London A, vol. 454, pp. 903-995

Lawford GJ (2006) Fourier series and the cartographic line. International Journal of Geographical Information Science, 20(1), pp. 31-52

Lecordix F, Plazanet C, Lagrange JP (1997) A platform for research in generalization: application to caricature, Geoinformatica, 1(2), pp. 161-182

Li Z, Khoshelham K, Ding X and Zheng D (2004) Empirical mode decomposition (EMD) transform for spatial analysis. In Advances in Spatial Analysis and Decision Making, Z. Li, Q. Zhou and W. Kainz (Eds), pp. 19-29 (Lisse, Netherlands, Balkema)

Liu H, Latecki LJ, Liu W (2008) A unified curvature definition for regular, polygonal and digital planar curves, International Journal of Computer Vision, 80, pp. 104-124

Plazanet C, Bigolin N, Ruas A (1998) Experiments with learning techniques for spatial model enrichment and line generalization, Geoinformatica, 2(4), pp.. 315-333

Rilling G, Flandrin P, Gonçalves P, Lilly JM (2007) Bivariate Empirical Mode Decomposition, IEEE Signal Processing Letters, 14(12), pp. 936-939

Saux E (2003) B-spline functions and wavelets for cartographic line generalization.Cartography and Geographic Information Science, 30(1), pp. 33-50

Weibel R and Dutton G, (1999) Generalising spatial data and dealing with multiple representations. In Geographical Information Systems, P.A. Longley, M.F. Goodchild, D.J. Maguire and D.W. Rhind (Eds), pp. 125-155 (New York: Wiley).

Generalization of 3D Buildings Modelled by CityGML

Hongchao Fan, Liqiu Meng, Mathias Jahnke

Department of Cartography, Technische Universität München
Arcisstr. 21, D-80333, Munich, Germany
{fan, meng, mathias.jahnke}@bv.tum.de

Abstract. CityGML (City Geography Markup Language) not only represents the shape and graphical appearance of 3D buildings but specifically addresses the object semantics and the thematic properties, taxonomies and aggregations. The generalization algorithm presented in this paper takes this advantage of CityGML. That means that our approach considers the semantic information associated with geometrical objects of buildings to be generalized. Experiments show that the approach can reduce about 90% of the storage space of 3D buildings while keeping the information amounts as far as possible.

Keywords: Generalization, 3D building, CityGML, Typification, Simplification

1 Introduction

With the aim to reduce storage space, speed up network transmission and geometric computation, hence improve rendering performance, a number of algorithms have been made available for generalization of 3D buildings. Staufenbiel (1973) presented an early approach for the simplification of building data by defining rules that are successively applied to the polygonal outline of the building. Mayer (2005) and Forberg (2007) apply operators of mathematical morphology e.g. opening, closing, erosion and dilation to simplify a building. Sester (2000, 2005) suggests using the least square adjustment as a technique for the simplification of building ground plans. Kada introduces his approaches for building generalization by defining parts of simplified buildings as intersections of half-planes (2006) and by cell decomposition and primitive instancing (2007). Thiemann (2002)

M. Sester et al. (eds.), *Advances in GIScience*, Lecture Notes in Geoinformation and Cartography,
DOI 10.1007/978-3-642-00318-9_20, © Springer-Verlag Berlin Heidelberg 2009

proposes to decompose a building into basic 3D primitives. Primitives with small volume are eliminated. More considerable research can be found in (Sester 2007) and (Meng & Forberg, 2007).

However, the abovementioned methods do not consider the semantic information associated with geometrical objects of buildings. This could lead to eliminating some features which are important for visual impression or merging features which belong to different objects.

This paper presents a generalization algorithm that takes the semantic information of 3D buildings into account. The 3D buildings are modelled with City Geography Markup Language (CityGML) that defines the classes and relations for the most relevant topographic objects in cities and regional models with respect to their geometrical, topological, semantic and appearance properties. CityGML has five Levels of Detail (LoDs), each for a different purpose. For the ongoing research project "Integrating time-dependent features in 3D building models" we model 3D buildings at LoD3. In order to make the analysis and visualization of the results at different hierarchical geometrical resolutions more efficiently the buildings in LoD3 need to be simplified. The simplification should be able to preserve the semantic information of the building components as much as possible.

In contrast to the input models for the abovementioned methods the buildings in CityGML are modelled with respect to their geometrical, topological, semantic and appearance properties (Gröger et al. 2008). Even small components of a building such as windows, doors etc. are documented with high geometrical and semantic precision, where the semantic information indicates the meanings of the individual object components. Taking semantic information into account in a generalization process can avoid (i) deleting some features which are important for visual impression; (ii) merging polygons which belong to different entities.

The paper is structured as follows: as the input, buildings of our approach are modeled with CityGML which is introduced in Section 2. Section 3 is dedicated to the semantic-driven generalization algorithm: at first, the process to extract the exterior shells of buildings from their LoD3 models is presented; then the ground plan is simplified with regard to the semantics of the corresponding wall and roof part. Subsequently, the surface is simplified and window structures are typified. In Section 4 the implementation and results are shown. The last section summarizes the conducted work.

2 The CityGML for 3D Building Modeling

CityGML stands for the City Geography Markup Language. It is an open data model and XML-based format for the representation and exchange of virtual 3D city models. CityGML not only represents the shape and graphical appearance of city models but specifically addresses the object semantics and the representation of the thematic properties, taxonomies and aggregations (Kolbe, 2008). Since 20th, August, 2008 CityGML (City Geography Markup Language) version 1.0.0 has been adopted by the Open Geospatial Consortium, Inc. (OGC) for modeling 3D urban objects especially 3D buildings.

The thematic information in CityGML goes beyond graphic exchange formats and allows users to employ virtual 3D city models for comprehensive analysis in different application domains such as simulation, urban data mining, facilities management, decision support and thematic inquiries.

The building model is the most detailed thematic concept of CityGML. It allows for the representation of thematic and spatial aspects of buildings, building parts and installations in four levels of detail LoD1 - LoD4 (Figure 1). In LoD1, 3D buildings are represented by block model with flat roofs. In LoD2, buildings have differentiated roof structures and thematically differentiated surfaces. LoD3 denotes architectural models with detailed wall and roof structures, balconies, bays and projections. LoD4 completes a LoD3 model by adding interior structures.

As mentioned above, the facade of a building of LoD3 can be semantically differentiated. Furthermore the openings such as doors and windows are represented as thematic objects. If a building of LoD3 is observed more closely, it would be found that every component e.g. wall is represented as a cubiod (multi surface) instead of a planar surface (see Figure 2). This means that at least six polygons are required in order to model a simple wall without any opening objects in LoD3. As a result, a single building in LoD3 needs a rather large storage space. Therefore, modelling all the buildings in a district or even the whole city leads to enormous storage space.

This paper proposes algorithms of generalization based on LoD3 in CityGML with the objective to reduce the storage space, speed up the network transmission and make rendering more efficient.

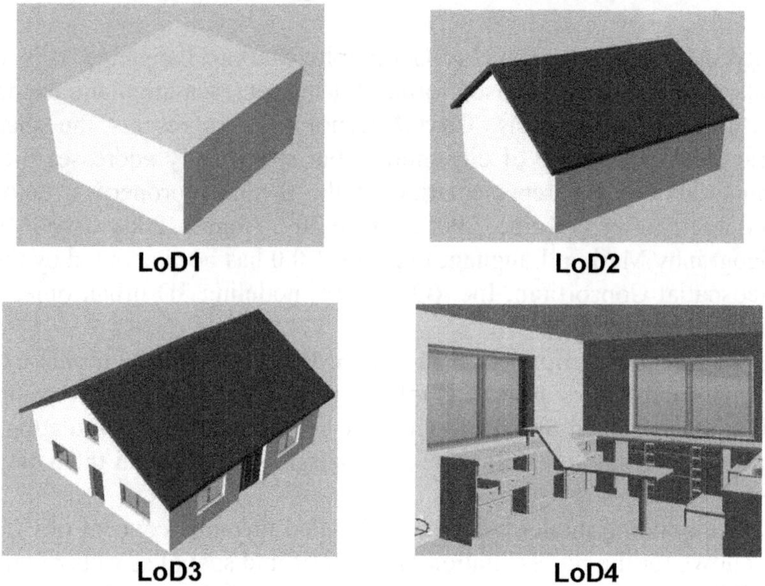

Fig. 1. The four levels of detail (LoD) defined by CityGML for building model (FZK-House modeled by Forschungszentrum Karlsruhe, Institute for Applied Computer Science), visualized in LandXplorer CityGML Viewer

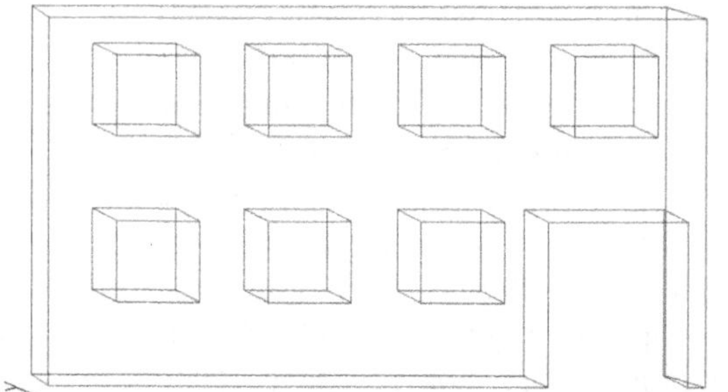

Fig. 2. Wall, windows as well as door in LoD3 are modelled as cuboid

3 The generalization Based on CityGML Modelling

The presented generalization algorithm is designed for 3D buildings modelled in CityGML at LoD3. The approach can be realized in four steps: (i) extraction of exterior shell of building; (ii) simplification of ground plan, (iii) simplification of surface for facade and (iv) typification of windows.

3.1 Extraction of Exterior Shell of Building

The first step of the generalization is to extract the exterior shell of building from its LoD3 model. First of all the walls in CtiyGML are converted into point clouds. For each wall its centroid $M_i = [M_x, M_y, M_z]_i$ would be calculated and an adjusting plane F_i can be found by using the following equation:

$$F_i: \quad A_i x + B_i y + C_i z + D_i = 0 \tag{1}$$

where $\vec{n}_i = [A_i, B_i, C_i]$ is the normal vector of the plane F_i and D_i the closest distance of the plane to the origin of the coordinate system.

The average point $M = [X_m, Y_m, Z_m]$ of all centroids of the walls can be easily calculated which would be treated as the centroid of the building. The point obtained in this way can guarantee that it lies more probably in the middle of the building than the point obtained by averaging all boundary points of walls.

For each polygon which belongs to the same wall a plane F_{ij} would be computed by inputting the coordinates of all its vertices in the equation 1. Thereafter the angle θ_{ij} between the plan F_{ij} and F_i can be derived using:

$$\theta_{ij} = \arccos\left(\frac{A_i A_{ij} + B_i B_{ij} + C_i C_{ij}}{\sqrt{A_i^2 + B_i^2 + C_i^2} \cdot \sqrt{A_{ij}^2 + B_{ij}^2 + C_{ij}^2}}\right) \tag{2}$$

If the angle θ_{ij} is close to 90°, the two planes are assumed to be orthogonal. The corresponding polygon should be deleted. But if the angle θ_{ij} is close to 0° or 180° the two planes can be either coplanar or parallel. Therefore, the corresponding polygon should be preserved. After this process all the remaining polygons which belong to the same wall should be either coplanar or parallel. The distances from the centroid M to the planes of these polygons are calculated by:

$$d_{m,ij} = \frac{\left|A_{ij} \cdot X_m + B_{ij} \cdot Y_m + C_{ij} \cdot Z_m\right|}{\sqrt{A_{ij}^2 + B_{ij}^2 + C_{ij}^2}} \tag{3}$$

The maximum value of the distances indicates the polygons which represent the exterior shell of the wall. The coefficients of the plane for these coplanar polygons $[A_i^{ex}, B_i^{ex}, C_i^{ex}, D_i^{ex}]$ will be used for the subsequent stage. In order to preserve the spatial details of windows and doors within the wall, it is necessary to project them onto the exterior shell according to:

$$\begin{pmatrix} X \\ Y \\ Z \end{pmatrix}_{projection} = \begin{pmatrix} x_{wd} \\ y_{wd} \\ z_{wd} \end{pmatrix} + \left(-D_i^{ex} + \begin{pmatrix} A_i^{ex} & B_i^{ex} & C_i^{ex} \end{pmatrix} \cdot \begin{pmatrix} x_{wd} \\ y_{wd} \\ z_{wd} \end{pmatrix} \right) \cdot \begin{pmatrix} A_i^{ex} \\ B_i^{ex} \\ C_i^{ex} \end{pmatrix} \tag{4}$$

where $\begin{pmatrix} x_{wd} & y_{wd} & z_{wd} \end{pmatrix}^T$ is a point of a window or a door, ex stands for the plane of exterior shell and i for the currently involved wall .

Those polygons that have been projected as lines on the exterior shell should be deleted because they lie perpendicular to the exterior shell. On the other hand if some polygons are identical to one another, only one of them should remain. As a result windows and doors are reduced to planes on the exterior shell.

For the roof of the building there are two cases of modelling in CityGML:

(1) The roof is modelled as several *RoofSurface* and every *RoofSurface* is a 6-sided prism whose four faces are rectangular.
(2) The roof is modelled as one single *RoofSurface*.

In the first case the exterior shell of roof will be extracted using the same method as for the wall. In the second case the situation turns out relatively complex. Firstly, for each roof polygon its plane is computed by using equation (1). If the third coefficient of the plane is close to zero ($C \approx 0$), the polygon is orthogonal to the xy-plane, i.e. the ground plan. This polygon will be deleted. The remaining polygons will be classified into several clusters according to their orientations: the normal vectors of the planes for the polygons in the same cluster are the same and the absolute difference between their D values is smaller than 0.3 meter (since the thickness of roofs is normally less than 0.3m). In each cluster the distances from the centroid M to the planes of the polygons are computed by using the equation (3). The maximum distance indicates then the polygons on the exterior shell of roof.

Now the overall exterior shell of the building has been extracted. Figure 3 shows the original model (left) of a house in LoD3 and its exterior shell model (right). In comparison to the original model the exterior shell preserves almost all the details (features) of the house, but it needs very few polygons (around 1/8 numbers of polygons) for the modelling.

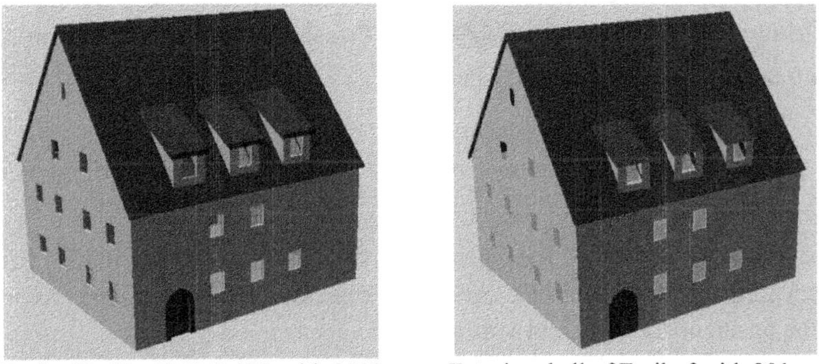

Freihof in CityGML LoD3 with 2429 polygons (from the city model of Et tenheim in Germany)

Exterior shell of Freihof with 301 polygons

Fig. 3. An example house in LoD3 and its exterior shell

As a matter of fact, this kind of optical illusion could also be found in our real world. In order to conduct the renovation work for a building without influencing its exterior appearance, its facade is painted on a planar cloth which hangs in front of the building. The painted facade can be viewed as the exterior shell of the facade. It keeps all the spatial details, thus gives pedestrians a similar visual impression.

Fig. 4. Painted facades give pedestrians similar visual impression (the left building is located on the Neuhauserstr.6 and the right one in Kaufingerstr.26, Munich, Germany. The photos were taken on the 30th, November. 2008)

3.2 Simplification of Ground Plan

Prior to the simplification, the ground plan of a building has to be derived from the exterior shell model by projecting the wall on the ground and connecting the footprint into a closed polygon.

Similar to [Sester et al. 2004] the operations are going to be triggered by a building side S_n that is smaller than a threshold. However, one more important semantic rule must be obeyed:

> *The side should not be removed, if there is an object important for visual impression e.g. a window or a door etc. on its corresponding wall.*

For this reason, before the operations with intrusions, extrusions, offsets and corners the involved side should be checked, whether there is window or door on its corresponding wall or not.

Intrusion and extrusion

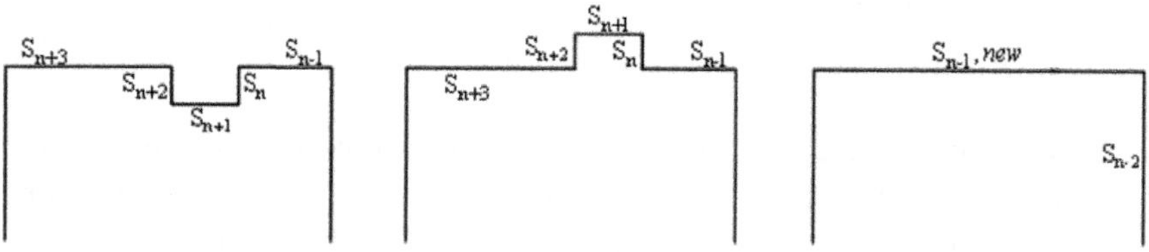

Fig. 5. Removing intrusion and extrusion in the ground plan

Actually, the operation for intrusion and extrusion is same. First of all the three sides of the intrusion or extrusion (see Figure 5) are checked:

(1) If there is window or door on the wall corresponded to S_n or S_{n+2}, the intrusion or extrusion would not be removed.

(2) If there is no window or door on the wall corresponded to S_n or S_{n+2}, the side S_{n-1} will be merged with side S_{n+3} into one side $S_{n-1,new}$ (the right figure in the Figure 5).

(3) On the base of the second case the side S_{n+1} will be checked. If there is window or door on the wall corresponding to S_{n+1}, the window or the door will be projected on the new wall corresponding to $S_{n-1,new}$. At the same time the semantics of the window or door will be indorsed.

Offset

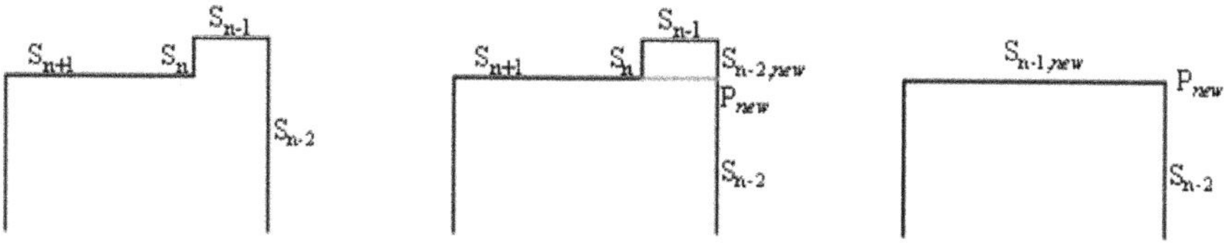

Fig. 6. Removing offset in the ground plan

In case of offset (see Figure 6) the longer side S_{n+1} of the adjacent neighbours of the shortest edge S_n will be extended. This leads to an intersection with the side S_{n-2} and creates a new node P_{new} and a new side $S_{n-2,new}$ (the middle figure of Figure 6). Now the three sides $S_{n-2,new}$, S_{n-1} and S_n will be checked. The situation here is comparable to the intrusion and extrusion described above and the operations are also the same.

Corner

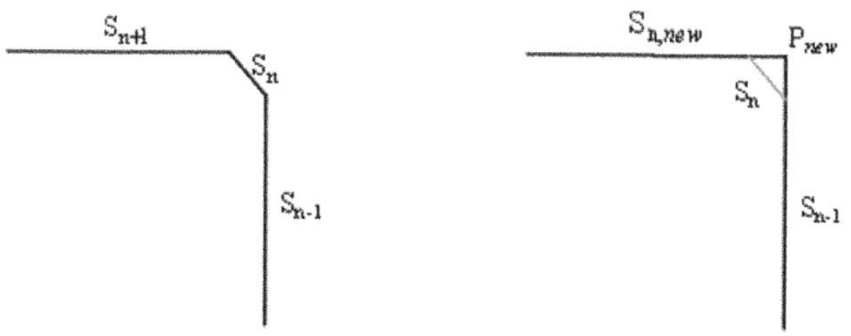

Fig. 7. Removing a corner in the ground plan

If the shortest side of a corner (Figure 7) is smaller than the threshold, this side will be checked at first. If there is window or door on the wall corresponding to this side, the side S_n should not be changed. Otherwise, the two adjacent neighbours of S_n will be extended. A new node will be produced at the intersection point P_{new}. At the same time the side S_n will be removed (right figure in Figure 7).

After the ground plan is simplified the roof structure should be adjusted accordingly. If the roof is flat, it will be simplified in the same way as for the ground plan. Otherwise, the polygons of the roof should be projected onto the ground. Normally, the projection of roof outline and the ground plan are contained within each other. Therefore, they will be processed in the same way. For the new nodes created during the process their corresponding points on the roof can be easily computed:

$$\begin{pmatrix} X_{roof\_new} \\ Y_{roof\_new} \\ Z_{roof\_new} \end{pmatrix} = \begin{pmatrix} X_{new} \\ Y_{new} \\ \dfrac{-D_i - A_i X_{new} - B_i Y_{new}}{C_i} \end{pmatrix} \tag{5}$$

Where $(X_{new}\ Y_{new}\ Z_{new})^T$ stands for the new created point, the A_i, B_i, C_i and D_i are plane coefficients of the involved polygon, and $(X_{roof\_new}\ Y_{roof\_new}\ Z_{roof\_new})^T$ is the new created point on the roof.

3.3 Generalization for Façade

The polygons on and within the facade represent the decorations of the wall, windows as well as doors. The generalization for these polygons is achieved mainly by using simplification at first and aggregation in the second step.

Simplification: this operation can be carried out by eliminating the sides of polygon that are smaller than the given threshold.

Aggregation: using this operation the polygons will be combined to form one larger polygon, if the distances between them are smaller than the given threshold. However, the aggregation is only allowed when the involved polygons belong to the same object e.g. wall element, window or door. In other words, for example, a polygon of window must not be merged with a polygon of the wall in which the window is situated.

The figure 8 above demonstrates the aggregation on a window. This kind of window could be oft found in walls of church in Europe. Normally, this kind of wind needs quite a lot of polygons (98 polygons in the case of the example) for the modeling. After the aggregation the amount of polygon is reduced to 11.

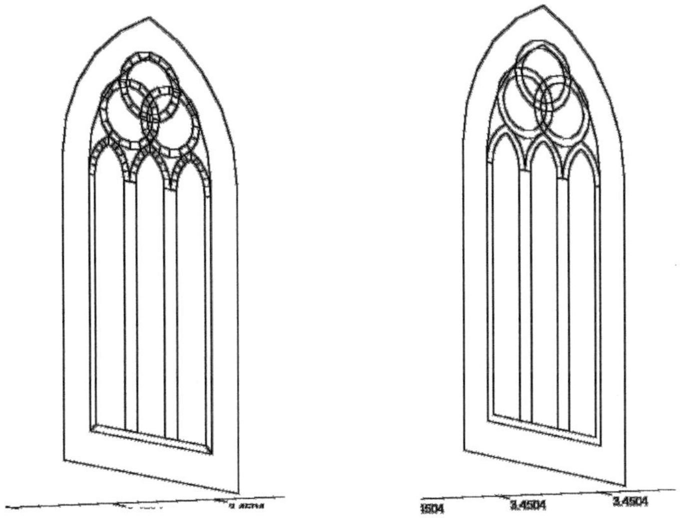

a). original shape of a window (modeled with 98 polygons)

b). after the aggregation the window is modeled with 11 polygons

Fig. 8. Aggregation conducted on a church window (in 3D view)

In fact, the windows of normal buildings are rather simple and their exterior shells are modeled in CityGML by three polygons: two rectangles for the outside frame and one irregular polygon for the inside skeleton (see Figure 9). Since most of windows in CityGML are modeled in this way, it is necessary to develop a method for their generalization:

(i) The original shape (Figure 9.a left) can be viewed as a tree with a trunk and six branches (Figure 9.a right).

(ii) For the first level of aggregation the branches $B1$, $B3$, $B4$ and $B6$ will be removed. At the same time the width of branches $B2$ and $B5$ will be increased two times (Figure 9.b left).

(iii) The shape after the first level of aggregation is treated as a tree with a trunk and two branches (Figure 9.b right).

(iv) For the second level of aggregation the two branches of the tree will be removed (Figure 9.c).

(v) For the third level of aggregation the window is represented only by one polygon (Figure 9.d).

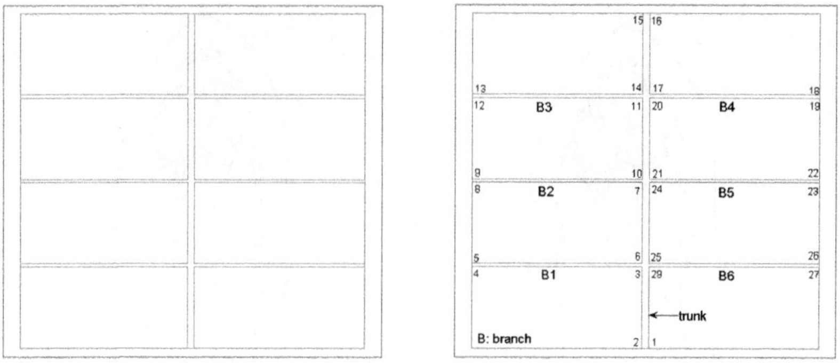

a) shape of a normal window (original)

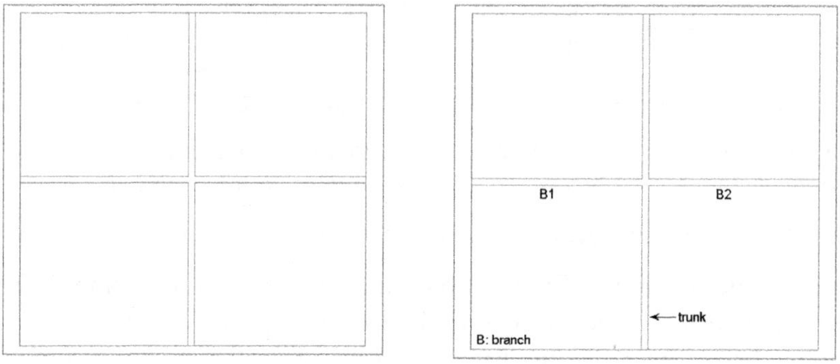

b) after the first aggregation

c) after the second aggregation d) after the third aggregation

Fig. 9. Aggregation on a normal window

3.4 Typification for Windows

For the further decrease in the level of detail the windows on the same facade can be typified. Typification denotes the process of replacing the originally large number of objects by a smaller number of uniform shaped objects. Although this kind of operation is used in many literatures for generalization and various results of typification are presented in (van Kreveld, 2001; Thiemann, 2002; Sester & Brenner 2004, Li, 2004 and etc.), it is not discussed why their results are reasonable. In our project a user test for results of typification was carried out.

3.4.1 User Test for Typification of Windows

For the test, three sets of windows were extracted from three different facades. In Figure 10 one set of windows is shown as an example. For the windows extracted from the image (Figure 10.b) three kinds of typification are possible: (i) the shape of window is conformal and the ratio of the distance between windows in horizontal direction to that in vertical direction is preserved; (ii) the area is conformal and the ratio of the distance between windows in horizontal direction to that in vertical direction is preserved and (iii) typification according to the algorithm in (Sester & Brenner 2004) that creates a new polygon by connecting centers of four neighboring polygons.

The user test was carried out during a lecture at the Technical University of Munich. 21 students had 15 seconds to stare at the extracted windows (b) and its three kinds of typification (c, d & e). Then they have to make a decision: which kind of typification is best associated to the original extracted windows, which is the second and the third choice.

a) original image of facade

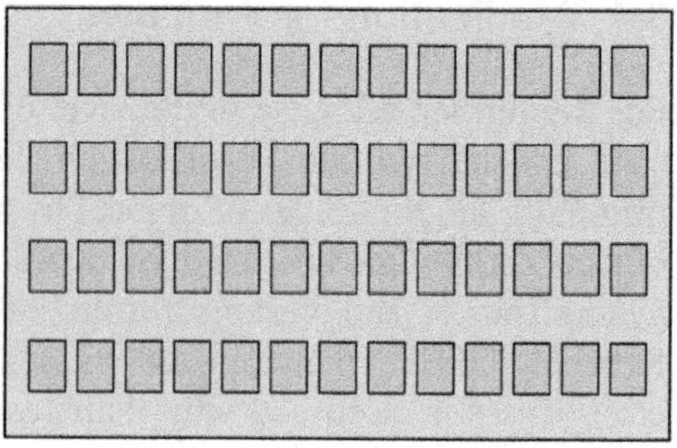

b) windows extracted from the image

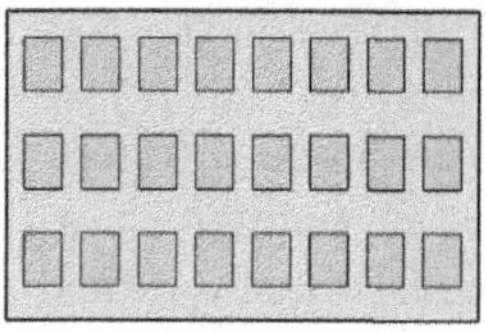

c) Typification based on conformal shape and the ratio of the distance between windows in horizontal direction and that in vertical direction.

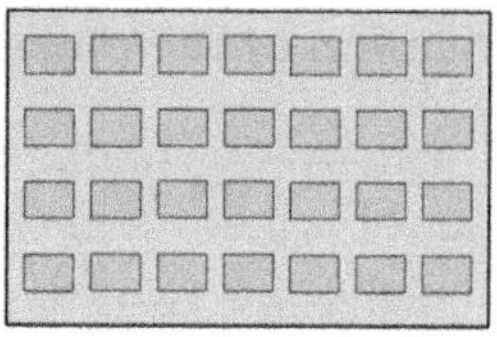

d) Typification based on area conformal area and the ratio of the distance between windows in horizontal direction and that in vertical direction.

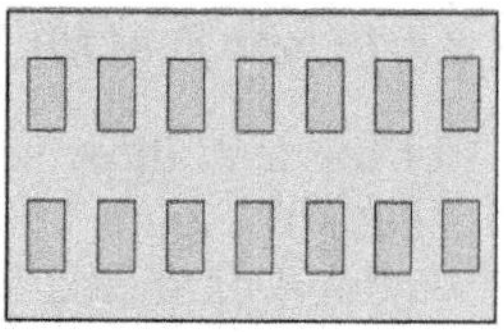

e) Typification according to algorithm in (Sester and Brenner 2004)

Fig. 10. User test for typification

3.4.2 Results of the User Test

The results of the test are summarized in the following table (Table 1).

Table 1. The results of the user test

Typification\Ranking	Best	Middle	Bad
Typification 1	17	3	1
Typification 2	4	16	1
Typification 3	0	4	17

Table 1 indicates that 81 percents of students found the first kind of typification could be best associated to the original façade; the second kind of typification lies in the middle and the third kind of typification is not appropriate for the façade whose windows are distributed similar to the façades of our test. The results of the test for the other two facades are quite the same as the results given in the example. That means the shape and the spatial distribution are most important for representation of the windows on a façade and these should be preserved during the typification.

4 Implementation and Results

The generalization algorithm has been implemented and tested on a number of 3D buildings modelled in LoD3 of CityGML. It shows good results for buildings with many windows or doors on the facades. For a selection of examples see Figure 11. In all cases the complexity as well as the storage space of the building modelled in LoD3 could be substantially reduced without destroying the overall appearance and losing the semantic information of the building.

For the example in Figure 11 the generalization was conducted in six steps as introduced in Section 3. Obviously, our approach is able to keep the information amount represented by the original model. But the storage space has been substantially reduced, in particular from the original model to the exterior shell. Table 2 gives an overview for the decrease of storage space after each step of generalization. The decrease of storage space is intuitively illustrated in Figure 12.

Tabel 2. Decrease of storage space after each step of generalization

step	Name	Storage space in [KB]
0	Original model in LoD3	1051
1	Exterior shell	128
2	Simplification of ground plan	124
3	The 1st generalization of window	76
4	The 2nd generalization of window	71
5	The 3rd generalization of window	49
6	Typification of windows	43

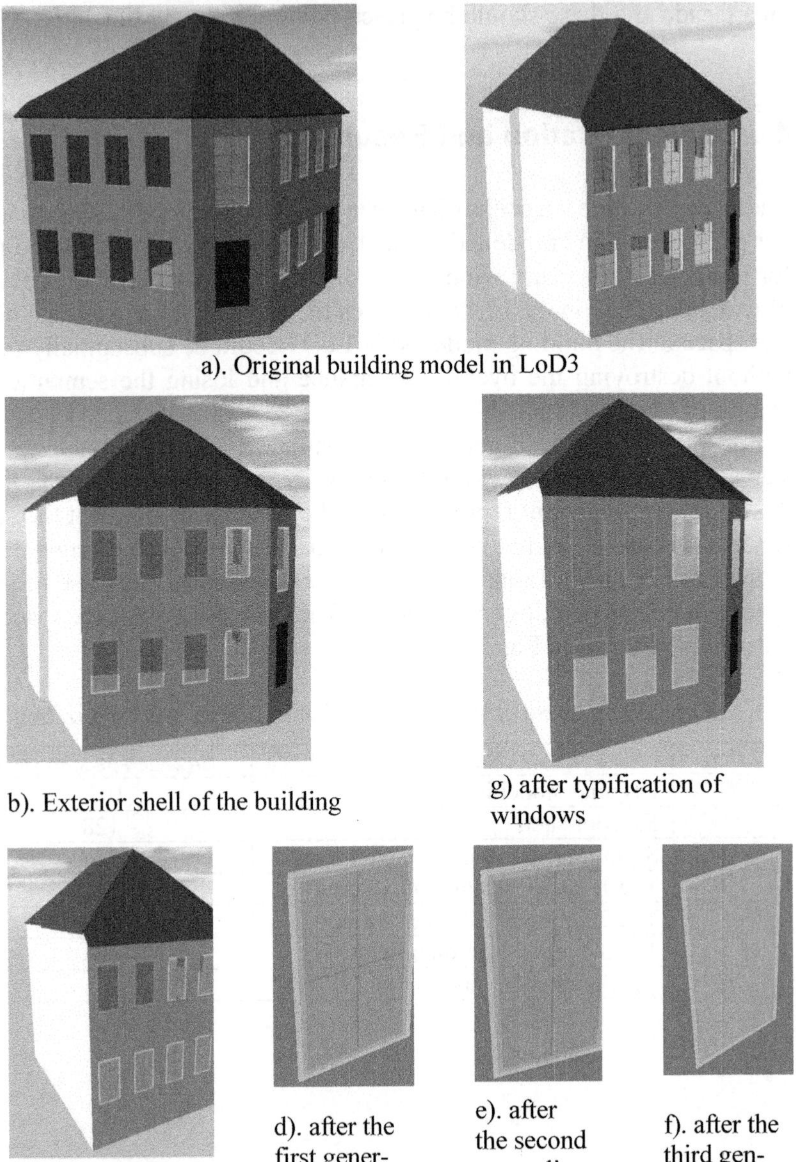

a). Original building model in LoD3

b). Exterior shell of the building

g) after typification of windows

c). after simplification of the ground plan

d). after the first generalization of window

e). after the second generalization of window

f). after the third generalization of window

Fig. 11. Generalization of a 3D building

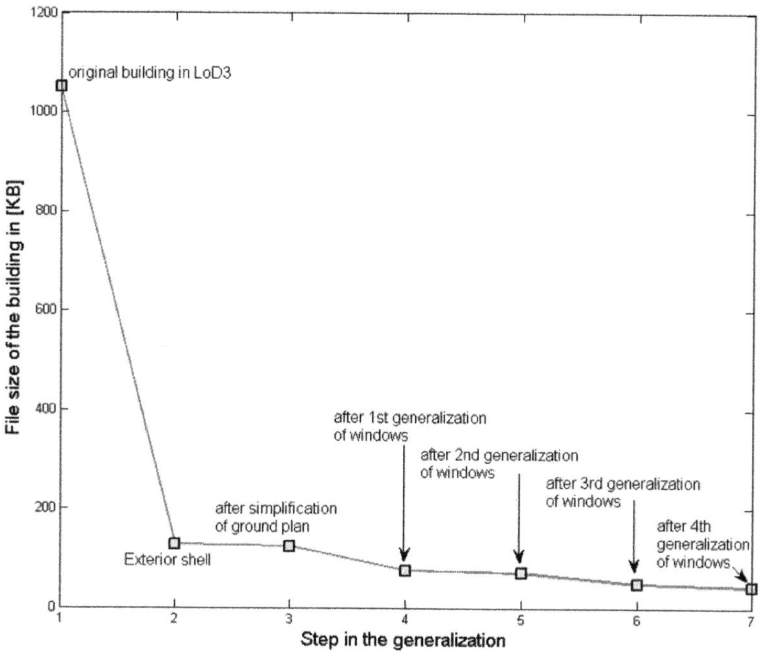

Fig. 12. Storage space of the building is reduced after each step of generalization

5 Conclusion

The paper presented a new approach for the generalization of 3D building models. Unlike other algorithms that have been proposed in the past, the new approach considers the semantic information of the building's components. The input 3D buildings are modelled in LoD3 by CityGML, as CityGML not only represents the shape and graphical appearance of 3D buildings but specifically addresses the object semantics and the representation of the thematic properties, taxonomies and aggregations. The generalization for a complex building can be realized step by step. At first the exterior shell is extracted from the building's LoD3 model. Secondly, the ground plan will be simplified with regard to the components on the walls. At the same time the roof structure will be adjusted according to the change of the ground plan. Then the generalization for façades is carried out by using simplification and aggregation of polygons which belong to the same object. Finally, the windows on a façade will be typified for a further decrease in level of detail. In order to find a reasonable performance for the result of typification a user test was carried out. The test shows that the shape and the spatial distribution are most important for representation of the windows on the façade.

Our approach for generalization has been implemented and tested on a number of 3D buildings modelled in LoD3 of CityGML. It shows good results for buildings with many windows or doors on the facades. In all cases the complexity as well as the storage space of the building modelled in LoD3 could be substantially reduced without destroying the overall appearance and losing the semantic information of the building.

So far our approach except the typification for windows can be automatically applied for generalization of 3D buildings modelled in LoD3 using CityGML. For the time being, our typification operation for windows can only be carried out interactively. The difficulties lie in determining how many windows are appropriate for a good visual impression after the typification. Further experiments will therefore be based on the combination of the findings from visual perception with some empirical cartographic rules such as the Toepfer's radical law (Töpfer and Pillewizer, 1966).

Acknowledgements

The research work presented in this paper is financed by the International Graduate School of Science and Engineering (IGSSE) of the Technische Universität München, Germany.

References

Gröger, G., Kolbe, T.H., Czerwinski, A. & Nagel, C. (2008). OpenGIS® City Geography Markup Language (CityGML) Implementation Specification. http://www.opengeospatial.org/legal/.

Forberg, A. (2007). Generalization of 3D building data based on scale-space approach. In: ISPRS Journal of Photogrammetry and Remote Sensing 62 (2007), pp. 104-111.

Kada, M. (2006). 3D Building Generalization based on Half-Space Modeling. In: Proceedings of the ISPRS Workshop on Multiple Representation and Interoperability of Spatial Data, Hannover.

Kada, M. (2007). Generalisation of 3D Building Models by Cell Decomposition and Primitive Instancing. In: Proceedings of the Joint ISPRS Workshop on "Visualization and Exploration of Geospatial Data", Stuttgart, Germany.

Kolbe, T.H. (2008). Representing and Exchanging 3D City Models with CityGML. In: Lecture Notes in Geoinformation and Cartography, Springer Berlin Heidelberg 2009, ISSN 1863-2246, pp. 15-31.

Mayer, H. (2005). Scale-spaces for generalization of 3D buildings. In: International Journal of Geographical Information Science. Vol. 19, No. 8-9, September-October 2005, pp. 975-997.

Meng L. and Forberg A. (2007). 3D building generalization. In: Mackaness, W., Raus, A. and Sarjakoski, T. (Eds): Generalization of Geographic Information: Cartographic Modelling and Applications, Elsevier, 2007.

Sester, M. (2000). Generalization Based on Least Squares Adjustment. In: International Archives of Photogrammetry and Remote Sensing, Amsterdam, Netherlands, Vol. XXXIII, Part B4, pp. 931-938.

Sester, M. & Brenner, C. (2004). Continuous Generalization for Visualization on Small Mobile Devices. In: Proceeding of Spatial Data Handing 2004, Springer-Verlag, pp. 469-480.

Sester, M. (2005). Optimization approaches for generalization and data abstraction. In: International Journal of Geographical Information Science. Vol. 19, No. 8-9, September-October 2005, pp. 871-897.

Sester, M. (2007). 3D Visualization and Generalization. In: Photogrammetric Week 07, Wichmann, 03.09-07.09.2007. Stuttgart, Germany, pp. 285-295.

Staufenbiel, W. (1973). Zur Automation der Generalisierung topographiser Karten mit besonderer Berücksichtigung großmaßstäbiger Gebäudedarstellungen. PhD thesis (in German), Univeristät Hannover, Germany.

Thiemann, F. (2002). Generalization of 3D building data. In: The International Archives of the Photogrammetry, Remote Sensing and Spatial Information Science 34 (Part 4).

Töpfer, F. and Pillewizer W. 1966. The Principles of Selection. In: the Cartographical Journal, Vol. 3, No. 1, pp. 10-16.

van Kreveld, M. (2001). Smooth Generalization for Continuous Zooming. In: Proceeding of the ICC, Beijing, China, 2001.

Li, Z., Yan, H., Ai, T. & Chen, J. (2004). Automated building generalization based on urban morphology and Gestalt theory. In: International Journal of Geographical Information Science. Vol. 18, No. 5, July-August 2004, pp. 513-534.

http://www.iai.fzk.de/www-extern/index.php?id=1051#c1225

http://www.citygml.org/1539/

3D Wayfinding Choremes: A Cognitively Motivated Representation of Route Junctions in Virtual Environments

Tassilo Glander[1], Denise Peters[2], Matthias Trapp[1], Jürgen Döllner[1]

[1] Hasso-Plattner-Institute, University of Potsdam,
Prof.-Dr.-Helmert-Strasse 2-3, 14482 Potsdam, Germany
{tassilo.glander, matthias.trapp, juergen.doellner}@hpi.uni-potsdam.de
[2] Transregional Collaborative Research Center SFB/TR 8 Spatial Cognition, University of Bremen, P.O. Box 330 440, 28334 Bremen, Germany
peters@sfbtr8.uni-bremen.de

Abstract. Research in cognitive sciences suggests that orientation and navigation along routes can be improved if the graphical representation is aligned with the user's mental concepts of a route. In this paper, we analyze an existing 2D schematization approach called wayfinding choremes and present an implementation for virtual 3D urban models, transferring the approach to 3D. To create the virtual environment, we transform the junctions of a route defined for a given road network to comply with the eight sector model, that is, outgoing legs of a junction are slightly rotated to align with prototypical directions in 45° increments. Then, the adapted road network is decomposed into polygonal block cells, the individual polygons are extruded to blocks and their façades are textured. For the evaluation of our 3D wayfinding choreme implementation, we present an experiment framework that allows to train and test subjects by route learning tasks. The experiment framework can be parameterized flexibly, exposing parameters to the conductor. We finally give a sketch of a user study by identifying hypotheses, indicators, and, hence, experiments to be done.

M. Sester et al. (eds.), *Advances in GIScience*, Lecture Notes in Geoinformation and Cartography,
DOI 10.1007/978-3-642-00318-9_21,

1 Introduction

The spreading applications of 3D geovirtual environments show that they are evolving from expert tools for selected applications to platforms addressing everyday needs of average users. For example, virtual 3D city models are used as integration platforms for real estate search, vacation planning, and traffic information. Thus, they serve as a spatial index structure for users to access data from different domains, revealing underlying spatial relations in the data.

However, in order to compete with 2D maps for navigation and orientation tasks, applications using 3D geovirtual environments need to provide additional benefits that legitimate the higher effort of creating and maintaining the used models, providing the infrastructure, such as a PC or a mobile device, and actually interacting with them. Given the higher complexity as one reason, many people have problems with navigation and orientation in virtual environments. For example, the acquisition of survey knowledge is difficult, as the impression from egocentric perspective has to be transferred to an allocentric model of the situation (Nash et al. 2000).

1.1 Virtual Environments

Viewed from a cognition science point of view, a virtual environment (VE) "[...] offers the user a more naturalistic medium in which to acquire spatial information, and potentially allows to devote less cognitive effort to learning spatial information than by maps." (Montello et al. 2004, pp:275).

Encouraged by this argument, lots of research has been done regarding navigation in virtual VEs. As interactive, real-time, 3D graphical renderings of spatial data, VEs are controlled by the users, thus directly changing and responding to users' behavior, e.g., through joystick or mouse interaction. In a VE, users typically have a dynamic first-person perspective on the scene. They acquire information sequentially and have to integrate them over time to build up a mental representation of the environment. Many studies show that people are able to learn spatial information from a VE (Montello et al. 2004).

There is a long tradition to present spatial information in a static pictorial way, e.g., in maps, and using verbal descriptions. Several approaches aim at easing the use of navigation assistance systems to increase user's performance in navigation through an environment. Most of these approaches apply principles of information reduction or abstraction to the spatial data to emphasize the necessary information elements while hiding unnecessary or disturbing parts.

One of these approaches are the wayfinding choremes introduced by (Klippel et al. 2005). *Choreme* is coined from the greek word for space, *chor*, and the suffix *-eme*, and means elementary primitives of space in analogy to phonemes for speech, or graphemes for written language. These wayfinding choremes represent prototypes of turning actions, which can also be used to emphasize turning actions in wayfinding maps (Klippel et al. 2005).

In this paper, we transfer the schematization principles of wayfinding choremes, i.e., the concept of cognitively adequate route representations developed for maps, to 3D VEs (Figure 1). We want to investigate, if assumptions made for 2D plans work in 3D visualization in spite of the fact, that space reception is fundamentally different. After the introduction of the theoretical basis of 3D wayfinding choremes and a sketch of the experiment to be done, we will present an implementation of wayfinding choremes in a 3D VE, followed by our experiment application. Then, we will discuss the concept as well as relate our solution to other approaches. We will conclude this paper with an outlook.

This paper focuses on the concept of 3D wayfinding choremes, their technical realization and the implementation of an experiment application. We expect to get first results of our ongoing empirical studies in the next months.

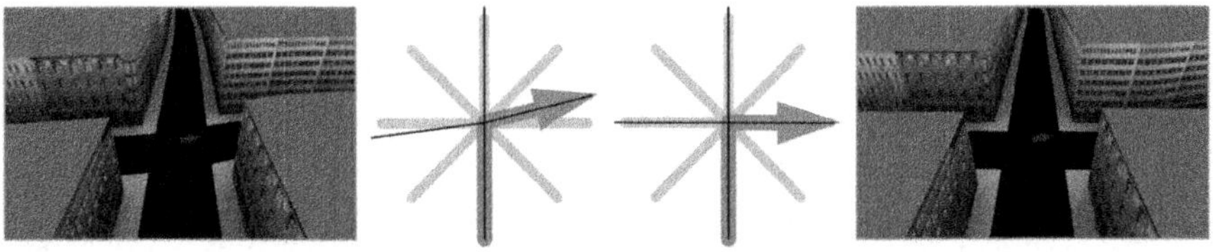

Fig. 1. Transformation of a junction into a prototypic configuration: The original roads (left) are analyzed, sorted and processed according to their angle to create a cognitively adequate representation of the junction (right).

2 3D Wayfinding Choremes

When VEs are used for tourist information, the most important purpose is communication of knowledge of the environment to users, i.e., tourists, to enable them to navigate in the VE and/or to learn a route. Unfortunately, navigation is often difficult in VEs even for trained users, especially if they have to navigate in large environments. The difficulties arise from users' problems to orientate themselves in VEs; also the acquisition of survey knowledge of the environment is poorer compared to navigation in the real world. These problems to some extent originate from the absence of vestibular and proprioceptive stimuli; though, many difficulties and spatial behavior issues within virtual environment are only fragmentary understood (Nash et al. 2000).

There are different approaches aiming at improving users' navigation performance in VEs, for example, providing a map or positioning additional landmarks within the VE, e.g., (Darken and Sibert 1993, 1996), or redesigning structural elements of VEs founded by the theory of Lynch (1960), e.g., (Omer et al. 2006). So far, these studies present a large range of different setting in terms of subjects' exposure time, their level of expertise and also in the details of the represented VE, such as size and naturalism of representation, e.g., (Goerger et al. 1998; Richardson 1999). Hence, existing approaches have largely contradicting navigation performance results.

2.1 Schematization

In our approach, we decide to apply the principle of schematization for enhancing the structural information of our urban VE. Definitions of the term schematization strongly vary between science disciplines. In cognitive sciences, especially linguistics, schematization is interpreted in the context of information processing. Herskovit (1998) postulates that three distinguishable processes are involved in schematization: abstraction, idealization, and selection. In computer science and artificial intelligence, schematic representation focuses on the identification and extraction of information that is relevant for a task (Berendt et al. 1998). Hereby three types of knowledge are defined: Knowledge that is essential and therefore needs to be represented unaltered; knowledge that can be altered, but has to be presented; and knowledge that should be omitted (Goerger et al. 1998; Palmer 1978). We understand schematization beyond the named definition as a process of intentionally simplifying a representation beyond technical needs to achieve cognitive adequacy, as defined in (Klippel et al. 2005). Cognitive adequacy, defined by Strube (1992), means on the one hand the representation of cognitive processes or on the other hand representations that have the quality to support cognitive processes. Compared to generalization, as understood in cartography, schematization is more than a simplified representation – it is a cognitively motivated representation.

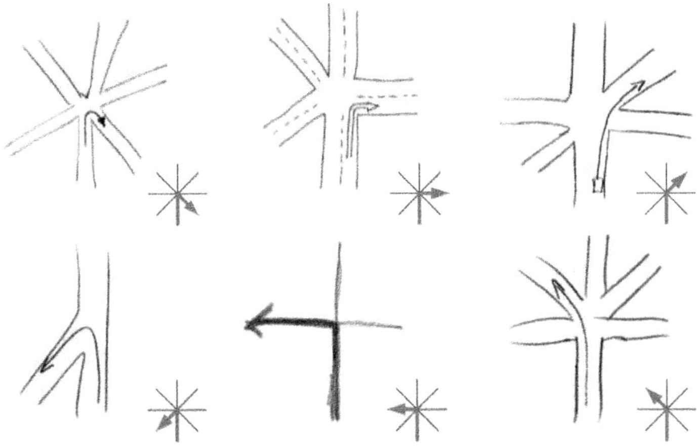

Fig. 2. Sketches of intersections with turning descriptions and their prototypes in the bottom right. (adapted from Klippel et al.(2005))

2.2 2D Wayfinding Choremes

Representing spatial information in a two-dimensional way as maps has a long tradition. Usually, cartographic maps depict selected aspects of the environment on a spatial scale much smaller than 1:1. Schematization is one way of selecting the represented aspects beyond the technical need towards a cognitively motivated representation. Schematization principles are applied to improve the legibility of maps depending on the specific tasks.

One example for cognitively adequate schematization principles for wayfinding maps are wayfinding choremes introduced by (Klippel et al. 2005). They define wayfinding choremes as mental conceptualization of functional wayfinding primitives. This schematization principle is aimed at easing the decision process at a junction by replacing the original curve by prototypes. Klippel et al. empirically identified mental conceptualizations of turning situations. Figure 2 shows sketches of resulting prototypical turning directions and the correlated wayfinding choremes. Participants tend to represent a turning action in a prototypical way.

Wayfinding choremes are prototypes of turning actions. Externalized graphically, they are useful to emphasize a turning action in a wayfinding situation. This method can be applied to route-following maps. Empirical data (Meilinger et al. 2007) show that schematic maps with prototypical junctions improve navigation performance for wayfinding compared to normal, that is, unchanged floor plans.

Fig. 3. A sketch of 3D wayfinding choreme. The original intersection (left) is compared with the prototype of this intersection (right). On the lower part, the respective 2D representation is shown. (from Peters and Richter (2008))

2.3 Transferring Wayfinding Choremes to 3D

In our approach, we apply the schematization principle of wayfinding choremes to VEs. As explained in the beginning, navigation in VEs is a difficult navigation task. For enhancing navigation performance, 3D wayfinding choremes have been suggested (Peters and Richter 2008), applying the principle of wayfinding choremes to 3D representations by locally changing the angular configuration of road junctions. Furthermore, not only angles of the intersecting roads but also angles of buildings have to be changed locally at the intersection, as sketched in Figure 3. For the implementation of 3D wayfinding choremes, it is not necessary to change the global configuration of the street network. We will explain our implementation in detail in Section 3.

"Much of the knowledge about time and space is qualitative in nature. Specifically, this is true for visual knowledge about space." (Freksa 1991, pp:365). In our approach, we aim to emphasize the qualitative nature of VEs by applying the schematization principle of wayfinding choremes. By replacing junctions with prototypes, this schematization approach eases the decision for actions to be performed at these intersections, as a displayed prototypical direction instead of the real angle (e.g., 90° instead of 79.3°)

can be understood easily. This can be argued, as the estimation of angles is a difficult task in VEs (Riecke 2003). It can be expected that orientation is more accurate using prototypical angles.

2.4 Experiment

To define requirements for software that creates a VE and supports conducting experiments to evaluate the wayfinding choreme concept in 3D, we give a sketch of an experiment setup. We hypothesize that 3D wayfinding choremes in VEs improve navigation performance. Therefore, we will compare route learning performance in an unchanged VE with the transformed VE. The subjects will learn the route by actively navigating through the VE, following the indicated route that is highlighted by arrows. Then, the participants have to navigate along the route without additional route indicators being presented. We will analyze the number of errors of the participants when reproducing the route as a measure for their performance. An error is defined as a wrong turning decision that is not recognized within a predefined distance to the route. The subjects will be divided in two groups: One group will learn the route in an unchanged world and the other one in a choremized world.

Additionally, we will perform some pre-testing of the perceptual and also qualitative nature of intersections in VEs. We will analyze the quality of estimation of angles in the VE by the subjects and let them draw sketch maps.

3 Creation of a Choremized World

Targeting the evaluation of the application of the wayfinding choreme concept in a user study, we design an experiment framework as follows (Figure 4): We need two main components, the world creation component to derive the VE using a given set of input data under certain parameters, and the interactive viewer application that helps conducting the experiment.

In the following, we will firstly describe the creation of the VE, and then present the application (Section 4).

3.1 Design Decisions

The VE has to look sufficiently similar to a real urban situation to be sufficiently convincing to the subjects. For the evaluation, three requirements exist to allow studying the impact of choreme transformation of junctions in the context of route navigation:

- The use of landmarks for orientation needs to be controlled and limited.
- The junctions and the course of the roads have to be emphasized.
- The terrain is planar as the concept does not cover implications for 3D terrain.

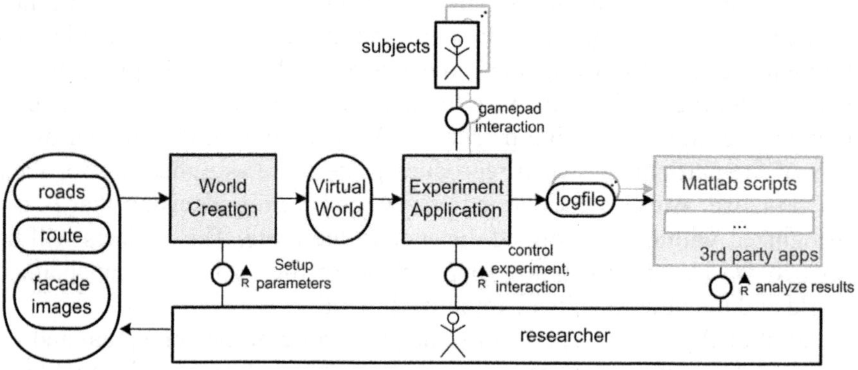

Fig. 4. Overview of the experiment framework.

Fig. 5. Façades can be visualized naturalistic, hand-drawn, or white.

Therefore, the complexity of the VE's visualization has to be reduced to limit the number of potential landmarks. To emphasize the course of roads and junction configurations, building façades are aligned with the roads. Additionally, explicit street polygons are embedded in the scene.

The creation of the choremized world can be sketched as follows. First, the input road network is processed using the input route to transform the route junctions to follow the choreme concept. The resulting transformed road network is then used to create polygonal partitions, i.e., cells. These cells are then extruded to yield 3D blocks.

3.2 Input Data

A virtual 3D city model typically integrates data from various sources and of different types, such as building models, infrastructure networks, digital terrain models, terrain textures, city furniture positions and models, and vegetation models (Kolbe et al. 2005). For our exemplary urban environment, we use just a subset of those to create an urban environment sufficient to visualize the choreme aspect along a route.

First of all we need a road network, which is in our case a part of a real world city. Data providers are TeleAtlas or OpenStreetMap, but also a manually created set of roads can be used. Our application reads them as line features from a Shapefile. The route also needs to be given as a Shapefile with a single line feature that is approximately aligned with the road network.

By creating urban VEs, we need for the building façades a set of façade images that are sufficiently generic, that is, they do not contain overly distinctive features. Using individual, remarkable façades would allow the subjects to use them as landmarks, which we want to control. In our scenario, we use a selected set of 38 different façade photos. Alternatively, completely synthetically façade images can be used, or no façade images at all (Figure 5).

Since we want to enable the controllable integration of landmark objects, we allow specifying landmark positions in a Shapefile along with a set of 3D models from typical city furniture objects. On each point in the Shapefile, one of the 3D models is placed as a remarkable landmark object on the streets.

3.3 Route Processing

3.3.1 Deriving Topology

Once the road network and the route are read in, the road network is converted into an arrangement (Wein et al. 2007; de Berg et al. 2008), which basically intersects all line features $L = \{l_i\}$ to create a half-edge data structure $G = (V,E,P)$ with vertices V, edges E, and polygons P with the usual operations (de Berg et al. 2008).

$$L = \{l_i\} \xrightarrow{arrangement} G = (V, E, P)$$

The half-edge data structure is necessary to do further processing that also regards the topology, e.g., does consider neighbor relations of junctions. In addition, it allows querying the polygonal decomposition P of the plane that is implicated by the set of linear features.

As the topological graph structure on the road network does not exist at the time of the route specification, the geometric route feature $r = \{r_1,...,r_n\}$ with $r_i = (x, y),\ x, y \in R$ has to be aligned with and mapped to the topological road network representation. This is done by searching for geometrically near and connected vertices of the computed half-edge structure.

$$(G,r) \xrightarrow{alignment} r_G = \{v_1^r,\ldots,v_m^r\} \quad v_j^r \in V$$

At the end, a topological representation of the route has been derived, e.g., an ordered list of connected nodes in the graph structure.

3.3.2 Applying Choreme Transformation

The graph together with the route information is then used to apply the geometric transformation to junctions along the route, i.e., nodes of degree greater than two. It is therefore a transformation of the half-edge structure depending on the route, $(G, r_G) \xrightarrow{chorematization} G'$.Note that the direction of the route is important, as, considering a single junction, the incoming road has to be kept unchanged, while all other outgoing roads have to be

```
applyChoremes(G, rG){
  for each vi^r in rG, i>1 do {

    if (degree(vi^r)>2){
      Bins[4] bins4
      bool isOverlap
      bins4, isOverlapp ← sortAngles(vi^r)

      if (isOverlap = true){
        Bins[8] bins8
        Bins8, isOverlap ← sortAngles(vi^r)

        if (isOverlap = false){
          transformJunction(vi^r, bins8) // use 8 sector model
        }else {}                         // leave unchanged
      }else{
        transformJunction(vi^r, bins4)   // use 4 sector model
      }
    }
  }
}
```

Fig. 6. Pseudo-code for creating a choreme representation.

changed to conform to the direction model (Klippel 2005). Hence, the algorithm iterates through all nodes along the route, starting with the first node and transforming all junctions to the destination node (Figure 6).

For one junction, the existing outgoing road segments are at first evaluated to compute the angles relative to the incoming road segment. Using the graph structure, neighboring edges can easily be queried for a certain vertex, hence we get positions of the neighboring vertices for the computation. The result is a numerical representation of the junction, containing the angles of each outgoing leg. These are sorted into bins with a defined size, first, into 90°-sized bins, i.e., a four-sector model, and then, in case of multiple legs within one bin, into 45°-sized bins. If there are still multiple items in one bin after the second iteration, the junction is not changed, as it can be argued that the junction is too specific for a generalization.

In the next step, the outgoing road segments are rotated slightly to follow the prototypic directions (Figure 7). This has a number of implications: As the transformation effect should be sufficiently local, i.e., within a radius *r*, and other junctions' positions are to be kept fix, new vertices have to be inserted along the changed edge. The prototypical direction should be remarkable, though it is not clear what this means in terms of meters. Therefore, our algorithm exposes the desired length of the prototypical direction as a parameter, being *r/2*. In case this length cannot be guaranteed, as another junction is within the desired length as shown in Figure 7, we insert the point half-way to the next junction. The connection

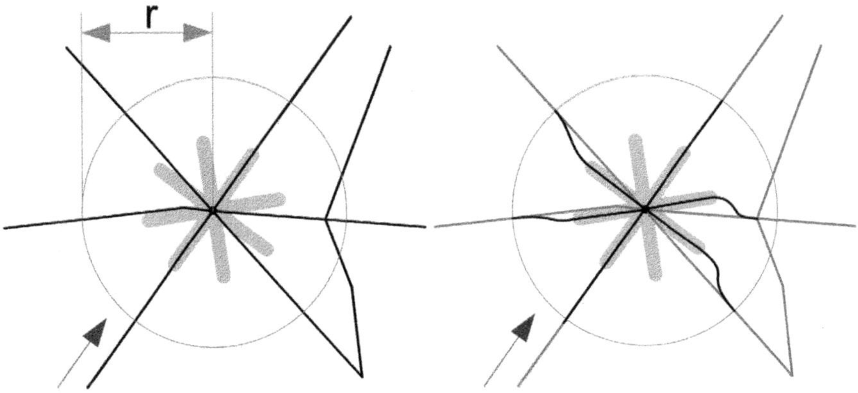

Fig. 7. The transformation applied to a junction. The prototypic directions are superimposed, the gray circle showing the local limitation of the effect. The shifted roads are smoothed using Beziér curves (right) when connected with the subsequent original edge.

of the inserted and shifted vertex with the original road network is then, as in the original paper, smoothed by applying a Bézier curve (Klippel et al. 2005). The algorithm stops, when all junctions along the route have been processed.

3.4 Creation of the 3D Geometry

After the 2D processing, 3D geometry has to be created for an explorable VE. Therefore, the half-edge data structure is queried for the polygons, which need to be further processed to be the footprints for block cells. At first, using 2D Boolean operations, the road network is subtracted from the polygons, cutting away space for the roads.

Then, the footprints are extruded to typical urban building heights, which can be given as an interval specifying minimum and maximum height. For an individual block, the height is then randomly varied within the interval, and for again randomly sized façade sections, textures from the façade textures pool are assigned (Figure 8). The texture parameterization is stored, so that we can apply approximately the same parameterization to both the choremized and the unchanged version of the world.

The resulting urban-like building blocks are finally pre-shaded and stored together with collision geometry necessary for collision free navigation, later.

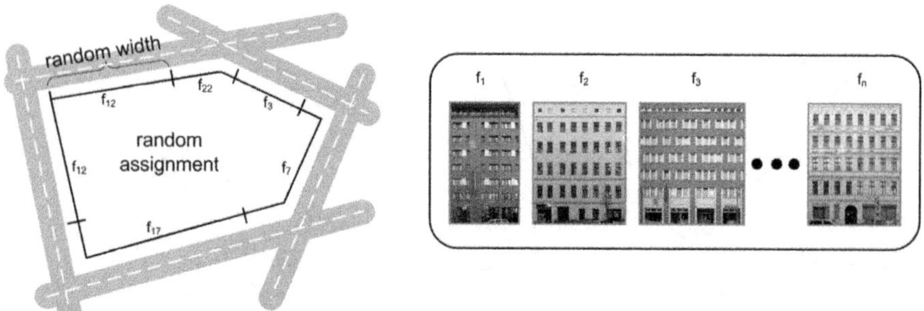

Fig. 8. A cell block's façade parameterization. A block, which is defined by its surrounding streets, is randomly textured from a pool of given façade textures.

3.5 Computational Complexity

The computation of the choreme representation starting from the road network and the route depends linearly on the size of those two data components. As the algorithm to change route junctions iteratively goes through all junctions along the route, and for each junction only the topological neighborhood is processed, the complexity is $O(n)$. The most expensive operations are done *after* the choreme transformation for the creation of the virtual environment, i.e., the Boolean operations and the lighting.

For an overview of the processing timings, we look at two data sets of differently sized road network and two routes drawn above them.

Small World The small data set covers an area of 1500x2000 m^2 and contains 722 line features with 2330 points.
Large World The larger data set of the two covers 3000x2700 m^2 and contains 1730 line features with 5450 points.
Short Route The short route has 8 junctions on a length of 1660 m.
Long Route The long route has 51 junctions on a length of 4300 m.

The Small World data set is a subset of the Large World data set. Comparing the times (Table 1) to compute these different data sets reveals, that most of the time is spent with geometry creation. During the geometry creation phase, the polygons are trimmed through the surrounding streets by performing buffering and Boolean difference operations. Then, the resulting polygons are extruded and textured and collision geometry is created. However, the Boolean operations take the most time.

Table 1. Timings for the computation of differently sized worlds.

	Small World short route [s]	Large World short route [s]	Large World long route [s]
Choreme Transformation	0.05	0.11	0.13
Geometry Creation	10.73	25.83	29.94
Pre-Lighting	6.03	9.61s	10.05

4 An Interactive Application for Experiments

For evaluating the concept of applying wayfinding choremes on a route in a VE, we have created an interactive application. It allows the user to explore the VE with and without choreme transformation, thus allowing to compare both navigation performances. Furthermore a number of parameters of the visualization for the subject can be controlled. The application works as a tool to conduct and trace experiments with different subjects and different settings. In an experiment, the subjects' behavior, i.e., movement trajectories and heading information are stored in a log file together with name, date and the current visual parameter settings. Therefore it is suitable for arbitrary user studies targeting VEs that compare subjects' performance along a route in two different configurations of a scene.

The application is written in C++ and uses the Virtual Rendering System (VRS)[1] for 3D scene visualization and interaction, and Qt for the user interface.

4.1 Exploration Mode

After loading a data set, the application is in exploration mode, allowing free navigation in the VE. It is targeted at the experiment conductor to inspect the chosen route and the choreme effect. Therefore, the two representations of the world, that is, with and without the choreme transformation along the route, can be switched instantly. The user can also jump directly to junctions along the route for quick inspection.

A separate properties panel allows the configuration of the VE and experiment parameters independently from the main window showing the 3D scene. In addition to switching between the two world representations and loading another data set, the landmark objects and guiding arrows along the route can be disabled to compare the subjects' behavior (Figure 9). Especially, the field of view of the virtual camera can be configured to permit

Fig. 9. The display of additional elements of the VE can be toggled, e.g., route indicating arrows (left) and the destination (center), as well as landmark objects (right).

the adaptation of the visualization to the experiment's setup. This is important, as it directly impacts humans' ability to estimate angles as well as the feeling of presence within the VE (Lin et al. 2002).

While the virtual environment can be explored freely using mouse and keyboard, for the experiment subjects, we chose a limited, gamepad-based interaction. As the navigation should be as simple and direct as possible, requiring the least possible learning effort, we decided to use an input de-

[1] The Virtual Rendering System, www.vrs3d.org

vice known to many from computer games. Further, the movement using the gamepad is restricted to a horizontal plane parallel to the virtual terrain at 1.80 m height. To keep the subjects from going behind the façades, a simple collision handling mechanism is applied.

4.2 Experiment Mode

During the experiment, the application guides and documents the process to support the conducting researcher. Depending on whether route arrows are displayed, we differentiate, if the subjects are to learn or to reproduce the route from memory.

When the researcher starts the experiment, the application sets the position to the beginning of the route and displays a start message. Then, the subject can move along the streets, possibly following the route until the destination is reached. The movement is limited by the route in that the subject is invisibly caged within a medium distance, e.g., 30 m around it.

The current status of the experiment, i.e., started, paused, finished, or canceled, can be controlled with the properties panel, or changes are triggered by user actions, i.e., when the destination is reached. The attempt of leaving the route is also communicated to the user by flashing red.

Most important for the experiment's analysis is the potential for automatic evaluation. Therefore, experiment relevant data is documented in a well defined ASCII file for later post-processing in tools such as Matlab. Each experiment is documented with a header containing a given subject identifier and the environment's configuration. The progress of the subject is tracked with respect to position and heading in regular, configurable time intervals, such as 1/10 s. It is supplemented by status change events and collision events when the route is left. Thus, errors of the subjects while reproducing the route can be counted as an indicator for the adequacy of the representation.

4.3 Experiment Questions

Using our experiment application, a number of hypotheses can be evaluated both quantitatively and qualitatively.

Estimating Angles The estimation of angles of outgoing legs of a junction is the basis for a determination of the feasibility of the choreme approach in 3D. The situation is fundamentally different compared to 2D maps, as the subjects have to transfer the egocentric impression from the 3D visualization to a mental allocentric concept of the junction. In our application, the subjects can be told to explore a junction and then asked to sketch a top-down view of it. We are interested, to which degree the estimation of the angles is correct.

Choreme Impact on 3D Route Comprehension While cognitively adequate representationof spatial data was shown to be effective for 2D floor plans (Meilinger et al. 2007), our application addresses the 3D case. It is possible to assess the choreme transformation of a 3D city environment along a route. The application automatically tracks attempts to leave the route, e.g., when a subject has forgotten the correct way. By counting these attempts, we have a quantitative measure for the confidence of the learned route, and, hence, can compare a subject's performance in the unchanged and the changed world.

Landmarks along the Route Landmarks undoubtedly play an essential role in the wayfinding process. Our application allows for the definition of landmark objects' positions and models to control their integration in the VE. Therefore, the subjects' wayfinding performance in different configurations can be measured and compared.

As the application only uses a small set of sufficiently generic, photo-realistic or synthetic façades or even no façades at all, we can exclude as far as possible the effect of façades used as landmarks. In addition, the flat terrain and the simple scene structure with low detail ground complexity limit memorization of the route to the sequence of junctions and their configuration.

5 Discussion

The presented framework for the creation of a virtual city environment externalizes the mental concept of a route and can be used in different settings. As our focus is on experiments evaluating the usefulness of 3D wayfinding choremes, the presented implementation creates a geometrically simple urban environment that still provides first insights into this question. In addition to using the experiment application in our own experiments, it is possible to export the created VEs for use in other applications in standard formats, e.g., CityGML (Kolbe et al. 2005) or VRML. Thus, the 3D environments can be evaluated by the community, or other experiments can be done. Another interesting opportunity is to use the transformed road network as input data for a procedural city model creation (Müller et al. 2006) as it would lead to much higher detailed buildings and ground level geometry to increase the photo-realistic impression. Still, the environment could be created in a controlled way.

5.1 Concept

Regarding the potential to transfer the wayfinding choreme concept to 3D, we are not sure, what the results of the experiments will be. The applicability of the concept in 3D is contradicted by our subjective impression that the changes of a junction's configuration are quite small. In addition, it is questionable, if the simplification of the route junctions in the sense of the choreme concept leads to a better comprehension and memorization of the route, or if it removes unique properties necessary for memorization. Finally, in our world 3D junctions are especially perceived through the closed façades. In a more photo-realistic 3D world, apart from very dense central city areas, buildings have free spaces between them, making perception and judgment of a junction more difficult. This effect does not occur on 2D maps, as roads are typically printed as lines and a possible transformation directly gets visible.

However, previous experiments have shown that route junctions are not remembered with exact angles regarding their configuration (Tversky and Lee 1999). Therefore, a simplification according to the wayfinding choremes reduces the presented information to the necessary aspects, thus easing the process of picking up the path. Therefore we are interested in conducting the experiment for 3D environments to create a more sound base for further discussion.

5.2 Comparison with Existing Work

If we compare our work with existing approaches, routes are usually visualized using additional scene elements that give directional hints. Typically, line features aligned with the course of the route or arrows at decision points are integrated in the visualization to guide the user. These elements can be integrated, for example, within live video imagery as an overlay showing route indicators in a perspectively correct way (Narzt et al. 2004). In (Coors et al. 2005), multi-modal representations of routes are analyzed, including verbal instructions, 2D and 3D route hints. Our solution uses integrated elements during learn phase.

Current commercial solutions for navigation systems typically do rely on these explicit route indicators and integrate them in 2D views as well as, increasingly, 3D views. However, 3D visualization in navigation systems is in its infancy. Today's solutions just show annotated 3D models of the environment and use only few techniques to support perception and comprehension of the route.

In addition to explicit route indicators, a route and its environment can be visualized in a deformed way to ease its communication. The classical example for 2D route visualization, (Agrawala and Stolte 2001), simplifies the route by shrinking distances of straight route segments while preserving angles. In (Böttger et al. 2008), a metro plan, i.e., a schematic, topological route representation, is used to transform a topographic map accordingly with image warping techniques, revealing the spatial relations between the metro stations. For short routes within a 3D VE, (Degener et al. 2008) present an image warping technique to create a single image describing the way. Focus + context visualization as in (Trapp et al. 2008) is another way to support wayfinding along routes in 3D VEs. Our approach focuses on the geometric transformation of junctions to conform to the cognitive concept of a route.

The automatic creation of artificial virtual 3D city models is typically done using procedural techniques. Based on a grammar, that is, a set of rules, whole city models including a road network can be created (Parish and Müller 2001). This technique can also be extended to create different, complex building types (Müller et al. 2006). A similar approach with the focus on interactive road network creation is presented in (Kelly and McCabe 2007). As we create a VE aiming at the conduction of an experiment, we currently do not need a very high degree of realism.

6 Conclusion and Outlook

We present a framework to create a cognitively adequate representation of a route in a 3D virtual environment following the concept of wayfinding choremes, transferring the concept from 2D maps. The framework includes an application to conduct experiments for the evaluation of this transferred concept and support the conducting researcher with the automatic logging of relevant events during one experiment.

Our next steps will be the accomplishment of experiments, testing the memorization of a route in the unchanged and the transformed environment. In addition to monitor and flat-screen projector based tests, we plan to do the experiment in a half cylinder. We presume, that the results will be different compared to the flat screen experiment, as the immersion is enhanced, and, standing on a junction the subjects actually can look into the outgoing roads without further interaction. This will probably influence the ability to estimate angles, and, hence, the impact of the choreme transformation.

Technically, we plan to change our implementation in some aspects. An enhancement for the choreme concept refines the eight-sector-model and suggests a nonuniform size of the different sectors (Klippel et al. 2004). Especially aiming at verbal communication of directions, for example, the sector straight ahead is smaller compared to the directions "veer left" or "veer right". Another interesting idea is the dynamic adaptation of junctions as the user gets near, possibly using image warping techniques (Böttger et al. 2008). This would be needed for dynamic rerouting in case of a lost way and thus is relevant for car navigation systems.

Finally, the results of the experiments will eventually be integrated into the experiment application and/or be used in a refined concept.

Acknowledgement

We want to thank Sebastian Pasewaldt for supporting us with the implementation. We also thank the reviewers for their comments.
This work has been funded by the German Federal Ministry of Education and Research (BMBF) as part of the InnoProfile research group "3D Geoinformation" (www.3dgi.de).

This work is also supported by the Transregional Collaborative Research Center SFB/TR 8 Spatial Cognition, which is funded by the Deutsche Forschungsgemeinschaft (DFG).

References

Agrawala, M. and Stolte, C. (2001) Rendering Effective Route Maps: Improving Usability Through Generalization, in: Proceedings of ACM SIGGRAPH, ACM press.

Berendt, B., Barkowsky, T., Freksa, C., and Kelter, S. (1998) Spatial Representation with Aspect Maps, in: LNCS: Spatial Cognition - An Interdisciplinary Approach to Representing and Processing Spatial Knowledge, Springer, pp. 313–336.

Böttger, J., Brandes, U., Deussen, O. and Ziezold, H. (2008) Map Warping for the Annotation of Metro Maps, in: IEEE Computer Graphics and Applications, 28(5), IEEE Computer Society Press, pp. 56–65.

Coors, V., Elting, C., Kray, C. and Laakso, K. (2005) Presenting Route Instructions on Mobile Devices: From Textual Directions to 3D Visualization, in Dykes, J., MacEachren, A.M. and Kraak, M.-J (eds.): Exploring Geovisualization, Elsevier, pp. 529–550.

Darken, R. P and Sibert, J. L. (1993) A Toolset for Navigation in Virtual Environments, in: UIST: Proceedings of the 6th annual ACM Symposium on User Interface Software and Technology, ACM press, pp. 158–165

Darken, R. P. and Sibert, J. L (1996) Wayfinding Strategies and Behaviors in Large Virtual Worlds, in: Proceedings of the SIGCHI conference, pp. 142–149.

De Berg, M., Cheong, O., and van Kreveld, M. (2008) Computational Geometry: Algorithms and Applications, Springer.

Degener, P., Schnabel, R., Schwartz, C. and Klein, R. (2008) Effective Visualization of Short Routes, in: IEEE Transactions on Visualization and Computer Graphics 14(6), IEEE Computer Society Press, pp. 1452–1458.

Freksa, F., (1991) Qualitative Spatial Reasoning, in Mark, D.M and Frank, A.U. (eds.): Cognitive and Linguistic Aspects of Geographic Space, Springer, pp. 361–372.

Freksa, F. (1999) Spatial Aspects of Task-Specific Wayfinding Maps, in: Gero, J.S. and Tversky, B. (eds.): Visual and Spatial Reasoning in Design, University of Sidney, Key Centre of Design Computing and Cognition, pp. 15–32

Goerger, S.R., Darken, R.P.,. Boyd, M.A, Gagnon, T.A., Liles, S.W., Sullivan, J.A., and Lawson, J.P (1998) Spatial Knowledge Acquisition from Maps and Virtual Environments in Complex Architectural Spaces, in: Colorado Springs US AirForce Academy (ed.): 16th Applied Behavioral Sciences Symposium, pp. 6–10.

Herskovits, A. (1998) Schematization, in Olivier, P. and Gapp, K.P (eds.): Representation and Processing of Spatial Expressions, Lawrence Erlbaum Assiociates, pp 149–162.

Kelly, G. and McCabe, H. (2007) Citygen: An Interactive System for Procedural City Generation, in: Proceedings of Fifth International Conference on Game Design and Technology, ACM press, pp. 8–16.

Klippel, A., Dewey, C., Knauff, M., Richter, K.-F., Montello, D.R., C. Freksa, and Loeliger, E.-A. (2004) Direction Concepts in Wayfinding Assistance Systems, in Workshop on Artificial Intelligence in Mobile Systems (AIMS'04).

Klippel A., Richter, K.-F., Barkowsky, T. and Freksa, C. (2005) The Cognitive Reality of Schematic Maps, in: Meng, L., Zipf, A. and Reichenbacher T. (eds): Mapbased Mobile Services - Theories, Methods and Implementations, Springer, pp 57–74.

Kolbe, T. H., Gröger, G. and Plümer, L. (2005) CityGML Interoperable Access to 3D City Models, in van Oosterom, P., Zlatanova, S. and Fendel, E.M. (eds.) Proc. of the 1st International Symposium on Geoinformation for Disaster Management, Springer.

Lin, J.J.W., Duh, H.B.L., Parker, D.E., Abi-Rached, H. and Furness, T.A. (2002) Effects of Field of View on Presence, Enjoyment, Memory, and Simulator Sickness in a Virtual Environment, in: Proceedings of the IEEE Virtual Reality Conference , IEEE Computer Society press.

Lynch K. (1960) The Image of the City, MIT Press.

Meilinger, T., Hölscher, C., Büchner, S.J. and Brösamle, M. (2007) How Much Information Do You Need? Schematic Maps in Wayfinding and Self Localisation, in: LNCS: Spatial Cognition, Springer.

Montello, D. R., Hegarty, M. and Richardson, A. E (2004) Human Spatial Memory: Remembering Where, Chapter Spatial Memory of Real Environments, Virtual Environments, and Maps, Lawrence Erlbaum Associates, pp. 251–285.

Müller, P., Wonka, P., Haegler, S., Ulmer, A. and van Gool, L. (2006) Procedural Modeling of Buildings, in: Proceedings of ACM SIGGRAPH 2006 / ACM Transactions on Graphics, 25(3), ACM press, pp. 614–623.

Narzt, W., Pomberger, G., Ferscha, A., Kolb, D., Müller, R., Wieghardt, J., Hörtner, H. and Lindinger, C. (2004) A New Visualization Concept for Navigation Systems, in LNCS: User-Centered Interaction Paradigms for Universal Access in the Information Society, Springer.

Nash, E. B., Edwards, G. W., Thompson, J. A. and Barfield, W. (2000) A Review of Presence and Performance in Virtual Environments, in: International Journal of Human Computer Interaction, 12(1), Taylor & Francis, pp. 1–41.

Omer, I., Golblatt, R., Talmor, K. and Roz, A. (2006) Enhancing the Legibility of Virtual Cities by Means of Residents' Urban Image: A Wayfinding Support System, in: Complex Artificial Environments Simulation, Cognition and VR in the Study and Planning of Cities, Springer.

Palmer, S.E. (1978) Fundamental Aspects of Cognitive Representation, in Rosch, E. and Lloyd, B.B.(eds.): Cognition and Categorization, Lawrence Erlbaum Assiociates, pp. 259–303.

Parish, Y.I.H. and Müller, P. (2001) Procedural Modeling of Cities, in: SIGGRAPH'01: Proceedings of the 28th Annual Conference on Computer Graphics and Interactive Techniques, ACM press, pp. 301–308.

Peters, D. and Richter, K.-F. (2008) Taking off to the Third Dimension - Schematization of Virtual Environments in: International Journal of Spatial Data Infrastructures Research, Vol 3, Joint Research Centre of the European Commission.

Richardson, A. R., Montello, D. R. and Hegarty, M. (1999) Spatial Knowledge Acquisition from Maps and From Navigation in Real and Virtual Environments, in: Memory and Cognition, 27, Psychonomic Society, pp. 741–750.

Riecke, B. E. (2003) How Far Can We Get With Just Visual Information? Path Integration and Spatial Updating Studies in Virtual Reality, PhD thesis, Universität Tübingen, Fakultät für Mathematik und Physik.

Strube, G. (1992) Contemporay Knowledge Engineering and Cognition, in: Schmalhofer, F., Strube, G., and Wetter, T.(eds): The Role of Cognitive Science in Knowledge Engineering, Springer, pp. 161–174.

Trapp, M., Glander, T., Buchholz, H. and Döllner, J. (2008) 3D Generalization Lenses for Interactive Focus + Context Visualization of Virtual City Models, in Proceedings of the 12th International Conference on IEEE Information Visualization, IEEE Computer Society Press, pp. 356–361.

Tversky, B., and Lee, P. U. (1999) Pictorial and Verbal Tools for Conveying Routes, in Freksa, C. and Mark, D.M. (eds.): LNCS: Spatial Information Theory. Cognitive and computational foundations of geographic information science.pp. 51-64.

Wein, R., Fogel, E., Zukerman, B. and Halperin, D. (2007) 2D Arrangements, in CGAL Editorial Board (ed.): CGAL User and Reference Manual. 3.3 edition.

Towards Geovisual Analysis of Crime Scenes – A 3D Crime Mapping Approach

Markus Wolff, Hartmut Asche

University of Potsdam, Research Group 3D Geoinformation,
Karl-Liebknecht-Strasse 24/25, 14476 Potsdam, Germany
Markus.Wolff@hpi.uni-potsdam.de

Abstract. This paper presents an approach towards facilitating geovisual analysis of crime scenes by introducing three-dimensional crime mapping techniques. Robbery scenes are analysed and integrated into a virtual three-dimensional environment using up-to-date methods of 3D geovisualization. For this purpose crime data as well as numerous geospatial data are integrated into a three-dimensional model of the German city of Cologne. Visualizing the results of crime data analysis in geovirtual environments eventually leads to 3D situation awareness in a civil security context. To allow for an easy transferability of methods and functions used in this approach, a GIS plug-in is developed.

1 Introduction

Digital analysis and mapping of crime events can be considered a well-established method for law enforcement agencies. In fact a wide spectrum of generating traditional crime maps has developed over time. The methodical basis of mapping and analysis comprises a wide range of disciplines from geospatial statistics, cartography, geographic information systems (GIS) and spatial epidemiology. Digital analysis and mapping of crime offers a number of benefits, particularly in the following fields of applications: operational policing purposes, crime prevention, informing and interaction with the community, change monitoring changes in the distribution of crime over time and evaluation of efficiency of crime prevention initiatives (Hirschfield and Bowers, 2001). In addition to the semantic information of a particular crime event any crime scene has a precise geo-

M. Sester et al. (eds.), *Advances in GIScience*, Lecture Notes in Geoinformation and Cartography,
DOI 10.1007/978-3-642-00318-9_22, © Springer-Verlag Berlin Heidelberg 2009

spatial location (excluding cyber-space crime possibly). Using geoinformation technology with its powerful geographic information software systems, these data can be captured and analysed. The core component, a geographic information system (GIS) is therefore widely used for spatial analysis of crime data both in academic research and in practical law enforcement (Murray et al., 2001). To document and communicate the findings of the analysis, maps are created. According to the analysis, maps vary in subject, purpose, audience and map quality. In fact, cartographic visualizations can be considered the key element to communicate the results of crime analysis. Map visualizations produced by crime analysts both in the operational and academic field are generally presented in the form of traditional two-dimensional, static maps. Most frequently these maps show feature- or pattern distributions, e.g. spatial variation of crime hotspots related to certain offences. Depending on the map topic and anticipated audience, these maps can be difficult to comprehend. A common issue deals with the question how to define adequate threshold values for choropleth maps of certain hotspots. For instance it is often rather ambiguous from which value precisely hotspots can be considered “hot” (Chainey and Ratcliffe, 2005).

In this work 3D geovisualization approaches are used to visualize crime and crime-related issues in interactive three-dimensional geovirtual environments. Consequently taking advantage of the third dimension, this contribution tries to identify benefits for a cartography-oriented design of three-dimensional landscapes of security and insecurity. As this concept links digital processing and analysis of spatially related crime data with easy-to-comprehend 3D visualizations the GIS and VIS tasks are combined in a three-step workflow specifically designed for that purpose (cf. figure 1):

- Processing and analysis of geocoded crime scene data with GIS methods. This includes grid based analysis as well as kernel density estimation (KDE) techniques. Using local indicators of spatial association the results of this KDE analysis are tested for statistically significance.
- Creation of a three-dimensional geovirtual environment. A 3D city model is used to determine building distances to the closest crime scene. This interactive environment is modelled outside the GIS. For this purpose a specialised 3D visualization system is used.
- Integration of the results of spatial analysis into the geovirtual environment.

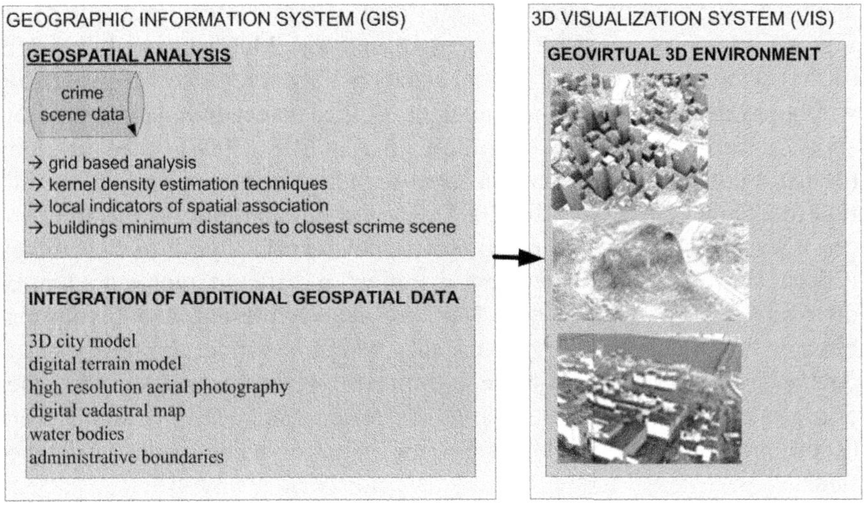

Fig. 1. Workflow applied in this study

After the introduction of related work (Section 2), we apply a scale-oriented approach: at a first level we apply 3D visualizations of crime scenes on a small, citywide scale by analysing and mapping crime scene distribution patterns (Section 3.1 and 3.2). At a second level we present some alternatives for visualization on a larger scale, focusing on the building level by using the three-dimensional city model.

The combination of analysing crime data and visualizing the results with methods from the field of 3D geovisualization can help to overcome certain shortcomings of communicating complex spatial phenomena – to both, the public and responsible decision makers.

2 Related Work

This section provides a brief overview of existing studies in the disciplines of crime mapping, 3D geovisualization and 3D crime mapping.

2.1 Crime Mapping

There is a vast amount of literature addressing crime mapping applications in theory and practical application. An introduction into theories, methods and selected software systems used to document, monitor and analyse crime data is given by Chainey and Ratcliffe (2005). Since most crime analyses are based on geocoded point data of crime sites, precise geocoding is a major prerequisite for spatial analysis. McCarthy and Ratcliffe (2005) review this topic of spatial data (in-) accuracy. However, dealing with geocoded crime data, an eminent task is to detect and map spatial hotspots of certain offences. According to Ratcliffe (2004), cited in Boba (2005), a hotspot is defined as an "area with high crime intensity". In addition to Chainey and Ratcliffe (2005), an introduction into the different approaches of detecting and mapping hotspots can be found in McCullagh (2006). With "prospective hotspot mapping" a different approach is introduced by Bowers et al. (2004). Here the authors present a bandwidth parameter for calculating hotspot surfaces which is not set arbitrarily but is determined empirically (as to be 400 meters). However, areas with higher crime rates than other areas should be tested statistically. Craglia et al. (2000) provide some insights addressing issues of statistical test and functions.

2.2 3D Geovisualization

MacEachren and Kraak (2001) present key themes and issues of geovisualization. Based on their conceptual framework Slocum et al. (2001) discuss a research agenda on cognitive and usability issues in geovisualization. Adapting techniques of information visualisation to the requirements of cartography, Kraak (2002) highlights that modern geovisualization helps to "stimulate visual thinking about geospatial patterns, relationships and trends". Furthermore he emphasizes the advantage of creating three-dimensional visualizations that allow for an "additional variable to be displayed in a single view and, as such, gives the user a direct insight into the relationship between these variables". Meng (2002) argues that the user can easily interpret spatial relationships of three-dimensional presented geo-objects without having to consult a legend. Jobst (2007) highlights the fact that three-dimensional cartographic visualizations provide "a more intuitive acquisition of space, due to an explicit use of 3D". Both, Meng and Jobst indicate also disadvantages concerning three-dimensional visualizations as for instance the absence of a single scale in perspective views, occlusion of objects, etc. However, construction of virtual three-dimensional geovirtual environments requires dedicated software systems. The LandXplorer software is an appropriate system for visualizing interactive three-dimensional maps (Döllner et al., 2006; Döllner et al., 2003).

2.3 3D Crime Mapping

In contrast to the abundant literature on crime mapping only little works can be found on the use of three-dimensional geovirtual environments in crime mapping. Though, a lot of research is carried out in the discipline of forensic three-dimensional scene reconstruction of crime sites (as for instance Se and Jasiobedzki, 2005). But there is also some work pointing in the direction of this paper. Lodha and Verma (2000) for instance present some three-dimensional visualization techniques for crime data. Based on VRML (Virtual Reality Modelling Language) the authors present an urban crime mapping application. In this regard they create predominantly 3D bar-charts, where the number of crimes is specified on the z-axis while spatial orientation is given on the x- and y-axis, describing an underlying grid. However, basic 3D visualizations of surfaces calculated on the basis of crime data, as for instance hotspot surfaces, can be found frequently – for instance in Harris (2000). Using Google Earth, even 3D city models can be used as an environment for presenting 2D or 3D crime maps (for instance http://www.geo-spatialtraining.com).

The approach presented here allows for integrating crime data into large three-dimensional city models and extensive geovirtual environments. Its key objective is to facilitate and advance geovisual analysis.

3 Mapping Crime in Three-Dimensional Geovirtual Urban Environments

This paper focuses on analysis, integration and visualization of crime data in three-dimensional geovirtual urban environments on different scales – from overall crime scene distributions to the building level. MacEachren et al. (1999), cited in Fuhrmann (2001) describe geovirtual environments (GeoVE) as particularly immersive, information intense, interactive and intelligent.

In a first step a three-dimensional geovirtual environment is created for the German city of Cologne. This provides the basis for subsequent urban crime data visualization. The Cologne GeoVE consists of a digital terrain model, a 3D city model, high resolution aerial photography (25 cm/pixel), digital cadastral map and further vector-based datasets including rivers, administrative boundaries and others (cf. Figure 2). We want to point out that the GeoVE presented here as a static screenshot image is completely interactive. A commercial GIS (ArcGIS) is used to process the datasets and to prepare them for 3D visualization. Afterwards the datasets are integrated into LandXplorer.

Fig. 2. Virtual three-dimensional environment of the city of Cologne

3.1 Identifying and Visualising Hotspots

Robbery crime scene data of 2007 are provided by the police headquarters of the city of Cologne. Each crime scene is described as a single point object, geocoded by x- and y-coordinates. In addition to these coordinates each point carries further thematic attributes describing time of the offence. Mapping those robbery scenes with the purpose of creating first overview maps on relative small scales, one has to consider, that basic positional-based point maps do not show all of the recorded crimes since several robberies can have the same coordinates (several robberies at the same registered position at different times). Visualising large spatial point datasets with point symbols is problematic as concentration of point symbols limit the perception of each separate position. Graduated symbols, instead, provide for a better comprehension of the geographical distribution of crime scenes. The use of graduated symbols requires the classification of data values (number of robbery incidents per crime scene) as a prerequisite. One class is represented by one symbol that clearly differentiates in size from other symbols depicting other class values. Therefore the position of a symbol marks the position of a robbery scene, while its size represents the number of robberies (cf. Figure 3). Relating the crime scenes to a regular grid is an alternative visualisation for point or point-related data. First, the study area is overlaid with a grid of specific cell size. Specification of the cell size is not straightforward. On the one hand this value should not be too small because this might yield too many cells with no or very few crimes. On the other hand, a large cell size might produce a map which is too coarse. Taking the spatial distribution of the robbery scenes as an experimental basis, a grid cell size of 200 meters turned out to be adequate to depict the spatial structure of robbery scenes. Subsequent to the definition of cell size the resulting grid values are classified and colour-coded according to their number of crimes calculated per grid cell (cf. Figure 4).

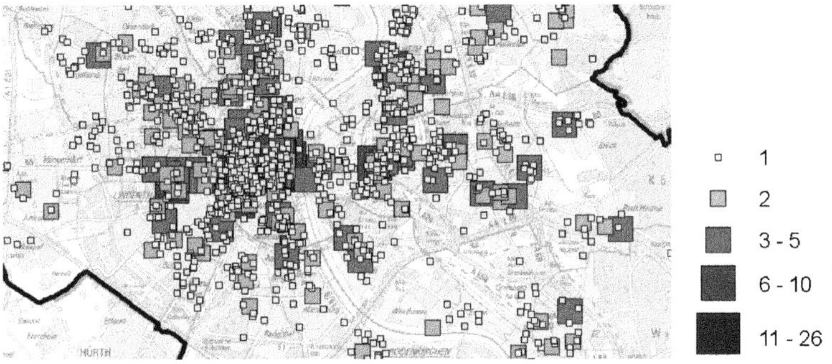

Fig. 3. Section of a thematic 2D map showing robberies in the city of Cologne

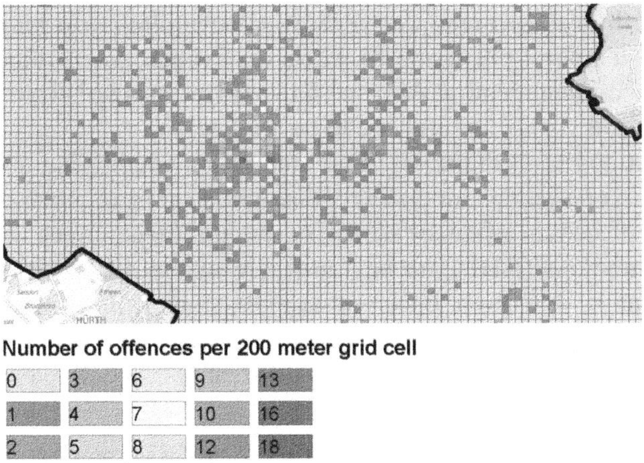

Fig. 4. Section of a 2D map showing robberies on a grid basis

In a 3D environment the number of crimes per grid cell can efficiently be visualised by assigning grid cell values to a vertical z-axis. Thus the cell height represents the number of crimes (cf. Figure 5). The 3D environment allows for intuitive, interactive exploration and analysis of both number and distribution of crime scenes.

Fig. 5. 3D visualization of total number of robberies per grid cell

The distribution of offences in space and time is one of the cardinal purposes of crime maps. Analysing hotspots is therefore of substantial interest for security agents as well as for decision makers in urban planning. Depending on the cell size and the vertical scale applied grid maps may reveal spatial-temporal clustering of particular crimes. To facilitate optimal analysis and communication of crime hotspots more sophisticated methods of analysis (and visualisation) have to be applied.

A well suited and commonly used method for identifying and visualising hotspots is based on calculating a continuous surface representing density values of certain offences[1]. Density mapping techniques using kernel density estimations (KDE) are one adequate method to identify and map crime hotspots. Although different kernel density functions exist (see Smith et al., 2006), each KDE-algorithm eventually results in a grid whose cell values represent density values (of incidents per a defined surface unit

[1] Unlike other point based data (as precipitation data for instance), crime data can not be interpolated to a continuous surface by implication.Therefore, methods such as, for instance, inverse distance weighting (IDW) or spline interpolation algorithms are inadequate to create a continuous hotspot surface.

of measure). To produce a KDE-surface map the study area is first overlaid with a grid of user-defined cell size. Subsequently density values are calculated for each cell – depending on the kernel density function applied. For analysis of robbery scenes in Cologne we use the quadratic kernel density function implemented in ArcGIS:

$$g_j = \frac{3}{4}\left(1 - t^2\right), |t| \leq 1$$

with $t = d_{ij} / h$, h as bandwidth

This function calculates for every cell of the resulting density grid g at grid location g_j, a value with distance d_{ij} from each robbery scene i. This g_j value is calculated as the sum of all applications of the kernel function over all event points in the crime scene dataset. As a result the grid generated represents density values of crimes sites related to a surface measure (for instance number of crimes sites per square kilometre). The input file for this KDE analysis is a dataset with crime scene locations. Before used as input for KDE analysis, this dataset is preprocessed: based on the robbery positions a new dataset is created that contains the total number of incidents per crime scene position as an attribute value. If there are several robberies at the same position, but at different times, these are counted and aggregated into the new dataset. The total number of robberies per identical crime scene position is written to the new datasets database as an attribute value. This dataset is used as input file for KDE analysis. Another input parameter is the cell size and the bandwidth (the search radius with distance d_{ij}). For this study a hotspot grid for the robberies 2007 is calculated using ArcGIS spatial statistics. For our study region we consider a bandwidth parameter (search radius) of 400 meters and a cell size of 50 meters as appropriate. The bandwidth value has experimentally been determined at 400 meter since it represents possible hotspots as well as the general distribution of crime scenes. For visual analysis the resulting hotspot grid is integrated into the 3D geovirtual environment. Using a colourless continuous surface instead of the 2D classified and coloured grid provides an alternative representation of the KDE grid: to allow for a visual differentiation of regions with higher crime densities from regions with lower densities, predominantly traditional two-dimensional choropleth or isopleth maps are created on the basis of the hotspot grid. This requires defining adequate thresholds for the class breaks. That, in turn, finally leads to the issue that different maps result from different threshold definitions. However, using a three-dimensional surface no thresholds have to be defined for initial visualization (cf. Figure 6).

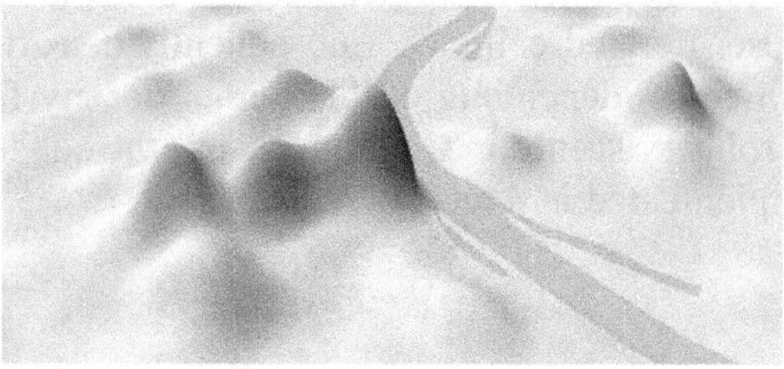

Fig. 6. 3D visualization of robbery hotspots based on a KDE surface without threshold classification

Integrated in the 3D environment this thematic relief facilitates an intuitive exploration and interactive visual analysis of crime site densities on relative small and medium scale levels. In addition the surface can be overlaid with various geocoded textures – for instance with (classified) choropleth or isopleth maps of the hotspot grid or with topographic maps (cf. Figure 7). This multiple feature coding of crime site densities can be considered as an effective visualization method to single out certain hotspot regions – for instance to brief decision makers.

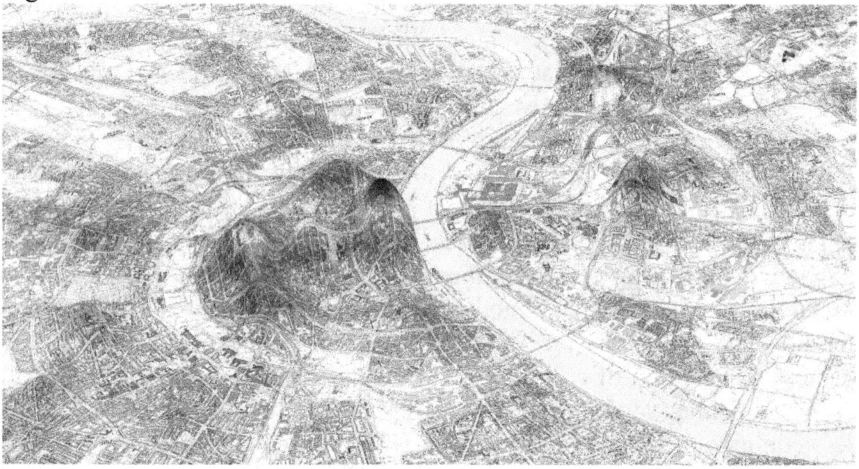

Fig. 7. 3D visualization of a classified KDE surface

One disadvantage of kernel density smoothing techniques is that areas might be covered by the surface as well where no robbery can take place. For instance no robberies will be encountered in the waterbodies of Rhine river. Using 3D visualization techniques this mistake becomes obvious. GIS functions based on map algebra allow for subtraction these areas from the KDE grid before the surface is integrated into the geovirtual environment (cf. Figure 8).

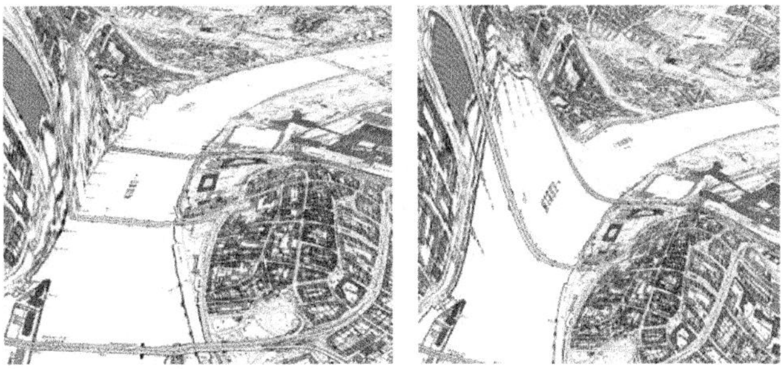

Fig. 8. Comparison of the KDE surface with (left Figure) and without (right Figure) surface correction

The geovisualization techniques applied so far facilitate an initial visual exploration of crime scene distributions. This high potential for an intuitive presentation of geographic information paves the way for an instant grasp of complex spatial situations. However, to prove the existence of a certain hotspot region from a statistical point of view, further analysis should be conducted. Therefore, the crime scene distributions are tested for statistical significance using local indicators of spatial association. Eventually, those hotspots showing high levels of statistical significance can be analysed more detailed on a larger scale level.

3.2 Validating Hotspots Using Local Indicators of Spatial Association (LISA)

Applying KDE techniques a smooth surface of crime densities is calculated. However, possible hotspot regions need to be verified statistically as well. This allows for differentiating hotspots relative to their significance level. Furthermore, visualising these results in a three-dimensional environment facilitates an interactive exploration of the hotspot surface while allowing the user simultaneously to check corresponding statistical significances.

To identify crime scenes that indicate an overconcentration of high (or low) robbery offences we apply local indicators of spatial association, also known as LISA statistics (Anselin, 1995). These indicators allow for analysing datasets on a local instead on a global level (for instance by using Moran's I or Geary's C) and allow for identifying differences between data values on a regional scale. From the methods of LISA statistics available (Local Moran's I, Local Geary's C) we calculate Getis-Ord (G_i^*) statistics, because this method detects areas where average values on a local level differ from those on a global level. In other words, areas can be identified by Getis-Ord (G_i^*) statistics where the number of observed crime incidents exceeds the number of observed incidents in the whole study area. G_i^* is calculated as following (Craglia et al., 2000):

$$G_i^*(d) = \frac{\sum_j w_{ij}(d) x_j}{\sum_j x_j}$$

In other words, G_i^* is the sum of weighted data values within a certain distance around an observation point *i* ($w_{ij}(d)$). A positive G_i^* value marks spatial clustering of high values: crime sites with high numbers of offences are surrounded by similar points (hotspot). In turn, a negative G_i^* value indicates that a crime site with few offences is encircled by crime sites showing few offences as well (coldspot). More detail on testing for spatial autocorrelation using Getis-Ords statistics can be found in Getis and Ord (1992) and Ord and Getis (1995).

To analyse the robbery dataset we apply the Getis-Ord Gi^* statistic tool implemented in ArcGIS 9.2. Compared to (2) ArcGIS calculates Gi^* statistics in a slightly modified way (ArcGis Desktop Help):

$$G_j^* = \frac{\sum_{j=1}^{n} w_{i,j} x_j - \overline{X} \sum_{j=1}^{n} w_{i,j}}{S \sqrt{\frac{\left[n \sum_{j=1}^{n} w_{i,j}^2 - \left(\sum_{j=1}^{n} w_{i,j} \right)^2 \right]}{n-1}}}$$

with x_j as the attribute value for each feature j

and $w_{i,j}$ as spatial weight between the features i and j

and n as the total number of features

and

$$\overline{X} = \frac{\sum_{j=1}^{n} x_j}{n}$$

and

$$S = \sqrt{\frac{\sum_{j=1}^{n} x_j^2}{n} - (\overline{X})^2}$$

The result is a z score for each location. This z score describes where features with high or low values cluster in space. To determine, whether a hotspot is statistically significant the local sum for each feature and its respective neighbours is compared in proportion to the total of all global features. If the calculated local sum differs from the expected local sum, and this difference is too large to be caused by random change, the result for that specific feature location will be a statistically significant z score. The higher the z score of a feature, the more intense is the clustering of high values in the neighbourhood of that feature (hotspot). Correspondingly, lower z scores imply more intense clustering of low values (coldspot).

Using ArcGIS is of advantage since the results (z values) are written to the input features database. Thus z values are linked to the map features. To test the KDE surface (Section 3.1) for its statistical significance, identical robbery incidents are used for Gi^* statistics. As the input file for Getis-Ord Gi^* statistics the dataset containing the total number of robberies per identical crime scene location (cf. Section 3.1) is used. Using the ArcGIS implementation of Gi^* statistics, a "fixed" Euclidean distance ($w_{ij}(d)$) has experimentally been determined at 200 meters.

The resulting dataset is classified according to their z scores. Subsequent class breaks are chosen to visualize significance levels of p=0.05 and p=0.01. In a final step this dataset is integrated into the three-dimensional environment. The resulting 3D visualisation allows the user to explore the KDE-hotspot surface; he also can check the hotspots for statistical significance levels. Significance levels are visualized by spheres in different colour (cf. Figure 9). This Figure shows that the hotspots in Cologne centre west of river Rhine, are also statistically highly significant.

Fig. 9. 3D visualization of a classified KDE surface and associated symbols showing levels of statistical significance

At this stage the hotspot grid is still independent of administrative boundaries. However, to facilitate further visual analysis, we provide Gi* statistics based on certain spatial entities. Therefore we spatially aggregate offences to neighbourhood district boundaries. For this purpose the number of crimes per neighbourhood is counted. In a second step this dataset is used for Gi* statistics. As a result each district-polygon gets a z-value, that describes the clustering of high/low robbery incidents in an around a neighbourhood. This dataset is integrated into the 3D environment (cf. Figure 10).

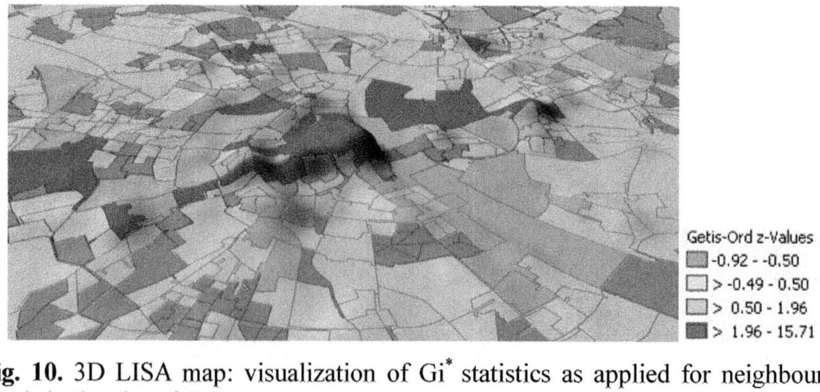

Fig. 10. 3D LISA map: visualization of Gi* statistics as applied for neighbourhoods in the city of Cologne

Figure 10 indicates that neighbourhood districts close to the centre tend to be surrounded by neighbourhoods showing high number of offences, while father districts are predominantly surrounded by neighbourhoods representing lower number of offences. Based on suchlike visualisations, decision makers as well as urban planers are able to explore each neighbourhood district robbery rate by navigating through the virtual 3D environment.

In a further step we increase the analysis scale through focussing on particular hotspots. On that level we include the urban landscape with its various interdependencies for visual analysis by integrating a three-dimensional city model into the virtual environment.

3.3 Integrating the Urban Landscape

To allow for in-depth- analysis the virtual environment is extended by a 3D city model. The use of a 3D city model facilitates visualizations of spatial relationships at the building level in an easy to comprehend way. This analytical and geovisual potential of 3D city models can be instrumental for decision makers working in security agencies for an instant comprehension of complex spatial phenomena related to urban security issues. However, city models vary in semantic and graphic detail from those with reduced level of detail (LOD1) to completely textured LOD3 models. The latter are often complemented by interior models of single rooms to complete buildings (Gröger et al., 2005, Döllner et al., 2006). In this study we use a city model that consists of approximately 22,000 buildings. Since this model was generated form airbone-LiDAR data, roof geometries are also included (LOD2).

To facilitate geovisual analysis in terms of comparing single buildings with the robbery hotspots, the city model is overlaid with the KDE-hotspot grid. Figure 11 shows the central Cologne hotspot area with a corresponding 3D city model. To broaden this visual approach and to facilitate further analysis, we calculate for each building the minimum distance to the closest robbery scene. Based on the crime scene dataset an Euclidean-distances-grid with a cell size of two meters is calculated. Each pixel of this grid represents the distance to the closest crime site. This grid is combined with the city model: for each of the 22,000 buildings those pixels are detected that are included by the respective building footprint. From this set of pixels we determine that one with the lowest value – which is the minimum distance of the building to the closest crime site. This value is added to the building database as a new attribute.

Fig. 11. 3D city model with additional hotspot texture and crime scene positions

Afterwards, the building dataset is classified and coloured according to these minimum distance values. The subsequent 3D visualization allows for exploring particular buildings of urban districts affected by a high number of robberies in their neighbourhood (cf. Figure 12). Since the distance values are stored in the buildings database, the creation of specific selection set of buildings for further analysis is supported.

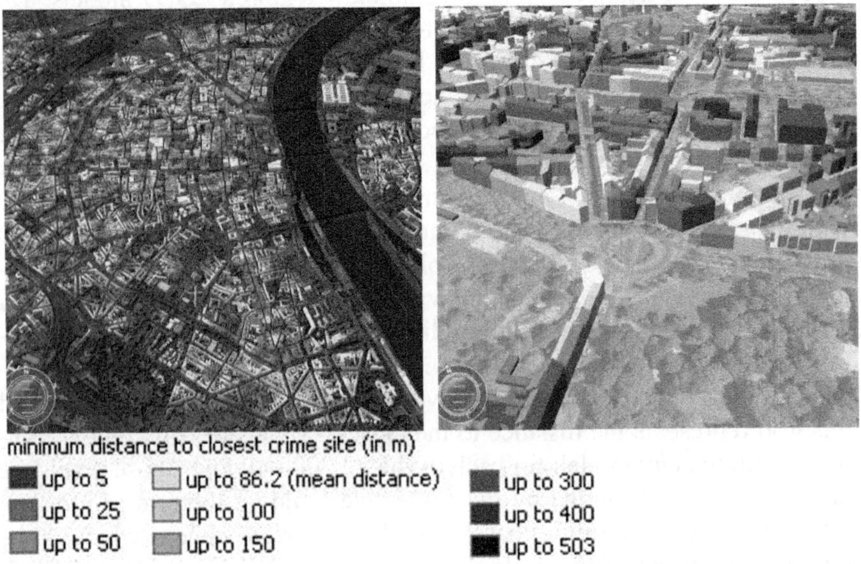

Fig 12. Minimum distances of each building to the closest robbery crime scene as overview visualization and with a smaller camera distance

This visualization facilitates an intuitive geo-communication about the distances of each building from the closest crime scene.

In addition an alternative visualization is created, visualising the "skyline of crime". For this kind of illustration the minimum distances to the closest crime scene is used as building height information (cf. Figure 13). However, this kind of visualization is suitable only for a small number of crime scenes. The higher the number of crime scenes, the more irregular the distribution of building heights and the less intuitive the graphic is.

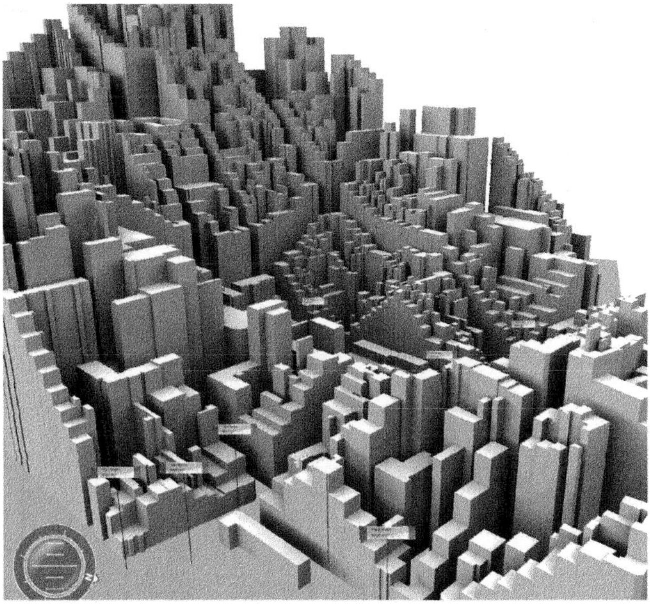

Fig. 13. The skyline of crime: the minimum distance value is used as new height information for each building. The higher the building, the greater is the distance to the closest crime scene

Finally, to facilitate maximum transferability of some of the presented city model applications a GIS plug-in is programmed. With this tool the user can easily create the proximity grid and calculate the minimum building distances to the closest crime site. Furthermore the creation of new database entries for each building regarding the membership to a distance class is supported (for instance: "distance to closest crime site >200 and <300 meters"). Colouring the buildings according to this crime site distance is straightforward by using these field values. Moreover the tool allows for creating hotspots based on KDE algorithms as well. To enhance the usability of hotspot mapping, we implement the possibility to calculate several hotspot grids with different bandwidth distances in one pass.

Therefore the user does not have to spend much time with the tedious procedure of running the same GIS function many times using different distance values. This ArcGIS plug-in is programmed as a dynamic link library (DLL) using ESRIs ArcObjects and Microsoft .Net framework with visual basic as the programming language. Figure 14 shows the dialogues of this plug-in.

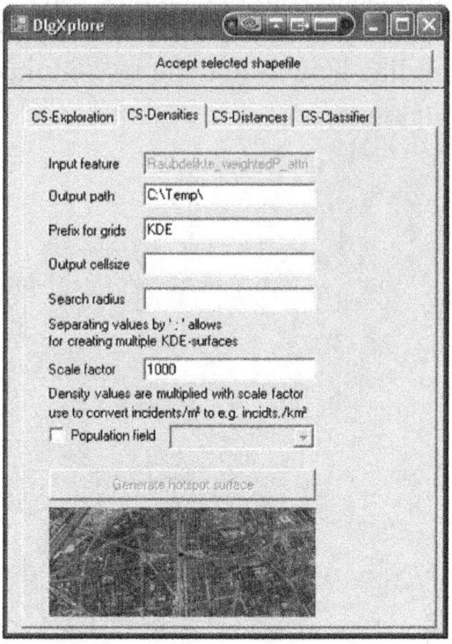

Fig. 14. Developed plug-in for ArcGIS

4 Conclusion

This paper presented several methods and functions for analysing and visualising crime data with three-dimensional geovirtual environments by the example of the city of Cologne. For this purpose we combined methods of analysing crime data with innovative 3D geovisualization techniques. Starting on an overview scale, we visualized crime scenes aggregated to a grid dataset. A second step included the calculation and 3D visualization of statistical surfaces based on kernel density estimation. Afterwards we increased the analysis scale by focussing on particular hotspots and by including the urban landscape via integrating a three-dimensional city model into the virtual environment. In the end we developed a GIS plug-in for the purpose of a user friendly execution of several of the applied techniques. However, for further studies the 4th dimension should be included in further analysis. Therefore next steps in this project will comprehend time related analysis of hotspot patterns. Further innovative three-dimensional visualization styles will have to be developed for displaying these issues with a cartographic yet appealing way.

Acknowledgements

Funding of this study by the German Federal Ministry of Education and Research (BMBF) within the framework of the InnoProfile research group '3D Geoinformation' (www.3dgi.de) is gratefully acknowledged. The authors also like to thank the police headquarters of the metropolis of Cologne for providing us extensive crime datasets. Furthermore we like to thank Virtual City Systems GmbH for providing us the 3D city model and 3D Geo GmbH (now: Autodesk, Inc.) for supplying the LandXplorer system. Boundaries of Cologne neighbourhoods areas are supplied by courtesy of INFAS GmbH.

References

Anselin, L. (1995). Local indicators of spatial autocorrelation – LISA. Geographical Analysis 27(2), 93-115.

Boba, R. L. (2005), Crime Analysis and Crime Mapping. Thousand Oaks, California, Sage Publications

Bowers, K. J., Johnson, S. D., Pease, K., (2004), Prospective Hot-Spotting The Future of Crime Mapping? The British Journal of Criminology 44(5): 641-658

Chainey, S., Ratcliffe, J. (2005), GIS and Crime Mapping. Chichester, John Wiley & Sons Inc

Craglia, M., Haining, R., Wiles, P. (2000), A Comparative Evaluation of Approaches to Urban Crime Pattern Analysis, Urban Studies 37(4): 711-729

Döllner, J., Baumann, K., Kersting, O. (2003), LandExplorer–Ein System für interaktive 3D-Karten, Kartographische Schriften 7: 67-76

Döllner, J., Baumann, K., Buchholz, H. (2006), Virtual 3D City Models as Foundation of Complex Urban Information Spaces. CORP, Vienna

Fuhrmann, S., MacEachren, A. M. (2001), Navigation in Desktop Geovirtual Environments: Usability Assessment. Proceedings of the 20th ICA/ACI International Cartographic Conference, August 06-10, Beijing, China

Getis, A., Ord, J. K. (1992), The Analysis of Spatial Association by Use of Distance Statistics, Geographical Analysis 24: 189-206

Gröger, G., Benner, J., Dörschlag, D., Drees, R., Gruber, U., Leinemann, K., Löwne M. O, (2005), Das interoperable 3D-Stadtmodell der SIG 3D, Zeitschrift für Vermessungswesen 130: 1-11

Hagedorn, B., Döllner, J. (2007), High-level web service for 3D building information visualization and analysis. 15th Annual ACM international Symposium on Advances in Geographic information Systems, Seattle, Washington, ACM

Harries, Keith (2000) Crime Mapping: Principle and Practice. Crime Mapping Research Centre. US Department of Justice, Office of Justice Programs

Hirschfield, A., Bowers, K. (2001), Mapping and analysing crime data. London and New York, Taylor & Francis

Jobst, M., Germanichs, T. (2007), The Employment of 3D in Cartography - An Overview. Multimedia Cartography. W. Cartwright, M. P. Peterson and G. Gartner. Berlin Heidelberg, Springer: 217-228

Kraak, M.J. (2002), Current trends in visualization of geospatial data with special reference to cartography. Indian Cartographer SDI-01, 319-324

Lodha, S. K., Verma, A. K. (2000), Spatio-temporal visualization of urban crimes on a GIS grid. 8th ACM international symposium on Advances in geographic information systems, Washington, D.C., ACM New York, NY, USA

MacEachren, A., M., Edsall, R., Haug, D., Baxter, R., Otto, G., Masters, R., Fuhrmann, S., Qian, L. (1999), Virtual Environments for Geographic Visualization: Potential and Challenges. ACM Workshop on New Paradigms for Information Visualization and Manipulation, Kansas City, MO., ACM

MacEachren, A. M., Kraak, M. J. (2001), Research challenges in geovisualization, Cartogra-phy and Geographic Information Science 28(1): 3-12

McCarthy, T., Ratcliffe, J. (2005), Garbage in, garbage out: geocoding accuracy and spatial analysis of crime. Geographic Information Systems and Crime Analysis, F. Wang (Ed.), IGI Global

McCullagh, M. J. (2006), Detecting Hotspots in Time and Space. ISG06

Meng, L. (2002), How can 3D geovisualization please users' eyes better, Geoinformatics - Magazine for Geo-IT Professionals 5: 34-35

Murray, A. T., McGuffog, I., Western, J. S., Mullins, P. (2001), Exploratory Spatial Data Analysis Techniques for Examining Urban Crime Implications for Evaluating Treatment, British Journal of Criminology 41(2): 309-329

Ord, J. K., Getis, A. (1995), Local Spatial Autocorrelation Statistics: Distributional Issues and an Application, Geographical Analysis 27: 286-306

Ratcliffe, J. H. (2004), The Hotspot Matrix: A Framework for the Spatio-Temporal Targeting of Crime Reduction, Routledge, 5-23

Se, S., Jasiobedzki, P. 2005, Instant Scene Modeler for Crime Scene Reconstruction Computer Vision and Pattern Recognition, IEEE Computer Society Conference on Volume, Issue 25-25 June 2005

Slocum, T. A., Blok, C., Jiang, B., Koussoulakou, A., Montello, D. R., Fuhrmann, S., Hedley, N. R. (2001), Cognitive and Usability Issues in Geovisualization, Cartography and Geo-graphic Information Science 28(1): 61-75

Smith, de M.J., Goodchild, M. F., Longley, P.A. (2006), Geospatial Analysis, Troubador Publishing

http://www.geospatialtraining.com/Newsletter/CrimeAnalysis/CrimeAnalysis.htm, last visited on February, 9th 2008